Saving the Planet with Pesticides and Plastic:

The Environmental Triumph of High-Yield Farming

"Avery says it's today's conventional farmers, the non-organic ones, who have performed an environment-saving miracle, by taking nitrogen from the air to make chemical fertilizer, and by using the often criticized pesticides and genetically engineered seeds. They feed more people on less land."

—John Stossel, ABC News

"Dennis Avery establishes himself as the most articulate exponent of high-tech farming. In easy-to-read prose, Avery documents his thesis that pesticides, fertilizers, and even plastic, properly used, are the salvation of planet Earth rather than a threat . . . . He makes a strong case."

—Darrell Smith, Farm Journal

"I strongly endorse [Dennis Avery's] call for a renewed commitment to governmental and philanthropic funding of agricultural research, including research into conventionally bred or bio-engineered new varieties of crops. A massive increase in such research is, as Avery argues, absolutely critical. Only then can the promise of high-tech breeding be combined with the social and environmental needs of the world."

—Carl Pope, Executive Director, Sierra Club

"The hungry must eat and plant life must survive. The solutions are not mutually exclusive. We desperately need an honest debate with more synergism and less conflict. Regardless of your present position, you need to read Dennis Avery's passionate plea for technology and the potential for synergy."

—Graham Kerr, International Culinary Consultant and host of syndicated TV show, "Graham Kerr's Gathering Place"

"*Saving the Planet with Pesticides and Plastic* makes a powerful case that the best chance for saving wildlife and biodiversity from the increasing demands of the growing human population is through increasing the food output from the Earth's existing farmlands . . . . All ecologists concerned about a sustainable future should consider how their own work relates to this pragmatic framework for using and conserving the Earth's resources."

—Dr. Michael Huston, Ecologist and author of *Biological Diversity* (Cambridge Press)

"Demolishing non-farm trade barriers helped unleash the longest, strongest surge of prosperity in modern U.S. history. Dennis Avery says we can unleash U.S. farm prosperity too by demolishing farm trade barriers even as we help save species-rich tropical forests in the increasingly-affluent, densely-populated countries of Asia."

—Senator Pat Roberts, (R-KS), Senate Agriculture Committee member and former chairman, House Agriculture Committee

"Dennis Avery's contention that modern high-tech food is safer than "organic" and "natural" food is, in general, accurate. Whilst early claims about the health and environmental risks of pesticides and food additives were based on plausible theories, backed up by semi-plausible scientific data, most of these have since proved unworthy of public concern. Better data and better theories indicate that the risks posed by pesticides and food additives are small compared with the benefits of their use. The more recent scares, however, are based on scientific theories that are only semi-plausible and are backed up by almost no data at all."

—Julian Morris, London, author of *Fearing Food: Risk, Health and the Environment*

Saving the Planet with Pesticides and Plastic:

The Environmental Triumph of High-Yield Farming

Dennis T. Avery
Director of Global Food Issues

SCIENTIFIC PUBLISHERS (INDIA)
P.O. Box 91 JODHPUR

Hudson Institute
Indianapolis, Indiana

Hudson Institute
Indianapolis, Indiana

First Reprinted in India, 2000
ISBN : 81-7233-240-8

The views in this book are solely the views of the author. No opinions, statements of fact, or conclusions contained in this document can be properly attributed to Hudson Institute, its staff, its members, its contracted agencies, or the other institutions with which the author is affiliated.

Reprint in India by
PAWAN KUMAR
SCIENTIFIC PUBLISHERS (INDIA)
5-A, New Pali Road, P.O. Box 91
JODHPUR - 342 001
E-mail : artrans@jp1.dot.net.in
Fax : 91-0291-512580

Printed in India

Contents

Acknowledgments

My thanks go first and foremost to my wife, Anne. She began by cheerfully supporting my efforts to produce this book—and wound up contributing huge amounts of her own time and her keen editorial eye to its completion. Equally important, she was the only one able to tell me graciously and effectively when I was wrong in my approach to communication with the reader.

Second, I must thank Alice Killian, my tireless and errorless assistant, for long hours of work—and keeping track of everything on my disorganized desk.

I want to also thank my colleagues at the Hudson Center for Global Food Issues. My son, Alex, contributed his own original research proving that nitrate fertilizer in drinking water does not cause the infamous blue baby syndrome. Dave Juday contributed his expertise on farm trade and Washington agricultural politics. Dr. Rick Halpern was my key resource on water quality issues. Additionally, all three of them have read, researched, written, networked, debated, published, and otherwise helped to advance the cause of high-yield conservation over the past several years with an enthusiasm that belied their very modest pay. Their efforts, too, are woven into the text.

I must thank the Hudson Institute for backing a manuscript that may rank as one of the most politically incorrect in history, simply because they thought it would make an important contribution to the public debate. I must also thank my Hudson editor, Sam Karnick, who contributed key ideas that went far beyond the usual editing assistance.

Major thanks also go to Bridge News Service, for giving me a weekly column, a weekly opportunity to reach the reading public with new technologies, new markets, new debates on farm policy, new media silliness—and a weekly necessity to write. They have graciously permitted me to reprint some of the columns in this new edition.

Finally, I must thank my reviewers:

- Dr. Paul Waggoner, distinguished scientist and former director at the Connecticut Experiment Station, and author of the Council for Agricultural Science and Technology's report, *How Much Land Can 10 Billion People Leave for Nature?*
- Dr. John Osmun, former chairman of the Entomology Department at Purdue University and senior official in the Environmental Protection Agency's Office of Pesticides.
- Dr. Douglas Southgate, of the Agricultural Economics faculty at Ohio State, expert on Latin American agriculture and resources, and author of *Economic Progress and the Environment: One Developing Country's Policy Crisis* (Oxford University Press, 1994).

They have done their best to ensure that the book is conceptually and factually correct; if there are errors, they are my fault.

Preface

This book will change the way you look at the world and its future.

It is not a book about food; we in the modern world take food for granted. But Dennis Avery has identified one of the most powerful new global realities: When we get higher yields from our farms, we save room for wildlands.

Avery offers "high-yield conservation" as the only massive, proven global success strategy for simultaneously restabilizing world population growth, ending hunger and malnutrition among Third World children, saving room for virtually all of the wildlife on the planet, and providing adequate food for the world's rising pet population.

High-yield conservation should also permit the world to harvest ten times as much paper and timber in 2050—while logging less than 5 percent of the world's wild forests.

Dennis says saving the world's wildlife doesn't have much to do with recycling trash, taking away private property rights, or even bringing wolves back to suburban neighborhoods.

He likes cities because they stack most of the world's people in convenient layers so they take up less than 2 percent of the land; and we can treat their sewage more efficiently. He expects the fuel cell to begin replacing the internal combustion engine in the next decade, emitting water vapor instead of exhaust fumes. He says we can find ample space for trash landfills.

He also overturns the conventional urban view of modern farming and believes the biggest threat to the world's wildlife is not global warming or overpopulation, but the plow of that low-yield farmer.

We've lost large tracts of the tropical forests, most of it to low-yield slash-and-burn farmers. And primitive farming threatens to take half of the remaining tropical forest in the decades just ahead, along with millions of wild species.

Organic farming is the First World equivalent to traditional farming and has the same fatal flaw—low yields.

Dennis Avery uses his own lifetime of experience in agriculture and his unique global overview of the world's food and farming systems to analyze the long-term trends. He identifies key technologies, and recommends vital policy changes. He urges support for agricultural research, especially for biotechnology in food production; free trade in farm products instead of subsidies to commercial farmers; regulatory frameworks designed to encourage still more safely sustaining chemical pesticides, more antibiotics, more tractors and more new farming systems than are presently employed. Technology and research will enable us to protect the global environment and rural people all over the world. It can give us more productive power and more positive lifestyle choices than mankind has ever known.

He goes beyond conventional wisdom and current politically correct thinking to lay out the potential of well-managed technology in societies where the institutions of freedom permit people to make well-informed decisions.

In short, Dennis Avery is a model Hudson scholar, following in the path of Hudson's founder, the eminent futurist Herman Kahn.

Dr. Herbert London
President
Hudson Institute

"Then a strange blight crept over the area. . . . Some evil spell had settled on the community: Everywhere there was the shadow of death. There was a strange stillness. The birds, for example—where had they gone? . . . The apple trees were coming into bloom, but no bees droned . . . so there would be no fruit. The roadsides, once so attractive, were now lined with browned and withered vegetation as though swept by fire. . . . Even the streams were now lifeless . . . all the fish had died. . . . A white, granular powder still showed a few patches; some weeks before it had fallen like snow. No witchcraft, no enemy action had silenced the rebirth of new life in this stricken world. The people had done it themselves.

"For the first time in the history of the world, every human being is now subjected to contact with dangerous chemicals, from the moment of conception until death." [Emphasis added.]

Rachel Carson, *Silent Spring*, 1962

FAO photo by G. Tortoli

A PLAGUE OF LOCUSTS—It has been many years since American farms have experienced biblical plagues which destroy everything in their path—thanks to modern-day pesticides.

Introduction

I never meant to write this book.

I had been writing another book, about the critical importance of trade liberalization for American agriculture. Suddenly I realized this one was more important. And due to the vagaries of fate, I knew I was one of the few people who could write it. I had the necessary knowledge of global farming, and I was not beholden to any of the political or bureaucratic agendas that have so often distorted our understanding of global farming in years past.

My father was a county agricultural agent. His job was to help local farmers grow more corn from each acre and produce more milk from each cow. Dad's father had been a farmer too, and we lived on a little 80-acre farm. I grew up milking cows and baling hay. When I was a child, high-yield farming was new and glittering, and man's greatest triumph. My father was part of it, spreading word of hybrid corn and artificial insemination across the country. He even volunteered to serve in India with the original "Point Four" program to help raise India's food production (but wasn't selected).

Personally, I wasn't cut out for farming. I could handle cows, but I was hopeless with machinery. I couldn't even adjust the carburetor on our two-cylinder John Deere tractor—about the simplest mechanical task this side of pull-starting a lawn mower. So, instead of farming, I went to college and studied agricultural economics and journalism.

Then I went to work for the Department of Agriculture in Washington. I had wanted to work for one of the land-grant agricultural colleges. But the job at Penn State didn't pan out, and I figured that a year in Washington would be good experience. I stayed 30 years.

In the 1960s, when I was new in Washington, the Green Revolution was being launched. Dr. Norman Borlaug and other seed breeders were creating the miracle wheat and rice varieties that literally saved Asia from massive famine. That seemed to endorse the importance of my father's goal of making two blades of grass grow where only one grew before.

In the 1970s, I didn't pay much attention to the environmental movement. USDA was caught up in the heady era of expanding farm exports. OPEC had spread a lot of money in countries with big populations and poor diets, like Nigeria, Indonesia, and the USSR. Everybody seemed to be importing food from America. The handwringers of that day worried that Africa and Latin America would become overdependent on foreign farms.

In the 1980s, the environmental movement was gaining *real* strength—and I began getting drawn into the whirlpool. The guy in the next office mentioned that the State Department was looking for an agriculture specialist. Suddenly, I was the Senior Agricultural Analyst in the State Department. It wasn't a job that my colleagues in USDA took very seriously. State didn't deal in very sophisticated analysis by their USDA/Ph.D. standards. My job was to educate a rather disinterested bunch of Foreign Service officers on what agriculture was and could do. Still, the State Department job had all the breadth any curious farm kid could want: all the countries, commodities, and farm/food problems in the world.

One of my first challenges was to evaluate a memo from a Foreign Service colleague. That fellow had been seriously impressed with Lester Brown's latest scary pronouncement—that soil erosion was rapidly destroying the world's ability to feed itself. Brown even claimed that the Corn Belt would soon erode into a veritable desert. My Foreign Service colleague was recommending that the United States drop all of its other foreign policy concerns (Communist expansion, the trade balance, and refugees, to name a few). He wanted us to deal henceforth with the other countries of the world purely on the basis of their soil erosion!

I had known Brown when we were both at USDA. I knew he was a population control activist and not a serious analyst. Still, it took a couple of weeks closeted with the experts to write up the case against the soil erosion scare. Then I got my first lesson in the emotional intensity of environmental concerns. When my State Department colleague's memo was rejected, he tried to sneak it aboard President Reagan's airplane to the Cancun Summit through a friend in the presidential party!

At State, I found myself being force-fed a diet of 20,000 pages of overseas reports per year. I became a speed reader. I eventually visited 36 countries on four continents. I began to learn how the world fed itself and used its major natural resources. My best information sources turned out to be the USDA agricultural attachés, who were stationed in 150 or so countries. Their post reports were a gold mine of up-to-the-minute facts, trends, and possibilities.

I also got acquainted with the Consultative Group on International Agricultural Research. The group had 13 international farm research centers spread around the Third World. (There are more than 20 now.) They are supported mostly by the Agency for International Development and international banking institutions. Their job was, and is, to take the latest agricultural research methods (mainly developed in the rich countries) and focus them on the farm and food problems of the poorest countries. Norman Borlaug won his Nobel Peace Prize at one of those international research centers, in Mexico. The miracle wheat and rice varieties for the Green Revolution both came from these centers. Later, so did new miracle cassava and hybrid sorghums for Africa.

Through such work, I learned that the Green Revolution was about more than a few seed varieties. It was a research process, building on research discoveries all over the world. The most oft-repeated phrases in my agricultural attaché post reports were "record yields" and "record production." From the centers' researchers, I learned *why* we were setting food production records in nearly every country. For the next several years, I produced a steady stream of internal State Department reports designed to help our Foreign Service people understand the real problems and potentials in world food and agriculture. Most of these reports were diametrically opposed to what the environmental activists of the time had to say.

Every year, Lester Brown predicted famine on the front pages of America's newspapers. It became an annual ritual that newspaper reporters would call me for the opposing viewpoint (which they would then duly bury in paragraph 12 on page 9). In 1985, a paper I had written on the global success pattern of farm research fell into the hands of Dr. Philip Abelson, then editor of *Science* magazine. He printed it under the title, "The Global Bad News Is Wrong." It was my first successful attempt to present the broader issues of global food production outside the government.

I began getting invitations to speak to farm groups about the patterns and trends in the rest of the world. I began to realize that few Americans, farmers or no, have any real understanding of what's happening in the dynamic world beyond our shores.

My federal career ended in 1988 when I jumped through an early retirement window. I wanted to take a more active role in getting a market-oriented agricultural policy for the United States, and in the civil service my hands were tied. Toward that end, I joined the Hudson Institute part-time as their agricultural analyst.

At the time I joined Hudson, I still saw the environmental activists' criticisms of high-yield farming as a secondary problem. What I really wanted to do was tell farmers about their opportunity to help feed an Asia that was getting affluent and that would soon have nine times as many people per acre of arable land as North America. Then about 1993, I changed my mind about the priorities.

I have never meant to be an "opponent of the environmental movement." Like most rural people, I grew up caring for creatures both domestic and wild. We worried about the health of our dairy cows *and* our pheasant populations. I treasure the deer that wander through my yard, the wild turkey that call from my mountain, and the bluebirds in our two dozen nesting boxes. I even try to care about the 30-pound snapping turtles that lurk in my pond and take a toll of the new mallard ducklings every spring (though I like the ducks a lot better).

Like many Americans, I am deeply grateful to the environmental movement for raising our level of concern. But I am also deeply disappointed in many of the movement's activists.

As chapters in this book will show, high-yield agriculture is the *solution*, not the problem, for wildlife and the environment. It is the only proven way to ensure success for both people and the environment. Yet the environmental leadership refuses to see this.

Similarly, I will argue that there is no upward population spiral in the world, just a onetime surge due to modern medicine lowering death rates. Too many environmentalists seem to reject such realities. They imply that we can starve the people and keep the wildlife. In fact, as I will show, with high-yield crops and forest plantations, we can have *both* people and wildlife. Without high-yield crops, we will have neither.

Consider as a personal example my little retirement farm in Virginia's Shenandoah Valley, which was part of the "nation's breadbasket" in the 1850s. Today, America's grain is being produced on the flat, fertile fields of the Corn Belt and Great Plains, where yields are three times as high and farm machinery doesn't destroy itself on the buried rocks. The Shenandoah, on the other hand, is almost entirely in grass for dairy and beef cattle, and trees for timber and pulpwood. Wildlife is more abundant than in colonial and even precolonial times, when it was hunted intensely.

Thus, the Shenandoah has lost income, gained beauty, and ended the huge soil erosion losses that cropping inflicted on its steep, rocky slopes. Indiana has gained income and lost woodlands, and its level cropland has comparatively little risk of soil erosion. (It also had historically less biodiversity than the rougher lands.) This is a win-win alteration of the food system. Indiana's grain yields are higher, its economic costs lower, and the environmental balance in both Indiana and the Shenandoah far more sustainable.

More important, we are producing our grain on *fewer acres* by focusing production in Indiana. We can produce the food we need from fewer acres by focusing production on our best and safest cropland. That puts steep and rocky acres back into grass, forest, and wildlife in places like Virginia, West Virginia, Vermont, and Montana. The higher yields achieved by using the best land and high-yield farming methods leave more wildlife habitat than any other approach. The *acres not plowed* will be a continuing focus of this book.

To understand the environmental triumph of high-yield agriculture, we must now extend this sort of analysis beyond the borders of the United States. We must, in fact, look at the global food system as a planetary whole.

We will start that journey now.

1

Saving Lives and Wildlife

MYTHMAKERS SAY:

"Most of our environmental problems are the inevitable result of the sweeping technological changes that transformed the U.S. economic system after World War II . . . (including) the substitution of fertilizers for manure and crop rotation and of toxic synthetic pesticides for ladybugs and birds."

Barry Commoner, "Why We Have Failed," Greenpeace, September/October 1989

"How can we put an end to mass starvation and suffering in this world? There is only one answer."

From a fund-raising letter by Negative Population Growth, Inc., Teaneck, New Jersey, Spring 1994

"Just now one of the significant historical roles of the primal people of the world is not simply to sustain their own traditions, but to call the entire civilized world back to a more authentic mode of being."

Thomas Berry, *The Dream of the Earth*, Sierra Club Books, 1988

REALITY SAYS:

"Environmentalists who want to preserve wildlands from the farmer's plow as we feed a burgeoning human population should reconsider an old enemy: chemical-based agriculture."

Ron Bailey, "Once and Future Farming," *Garbage, the Independent Environmental Quarterly,* Fall 1994, pp. 42–48

"We should like to announce in advance that we have received no support from any industry, government agency or university to write or produce this book. . . . Our conscious intent has been to defend the integrity of the scientific process and to bring to attention the possibility that following the advice of some of the leaders of the environmental movement might be a path to social and economic disasters even more serious than the problems of pollution and poverty . . . "

Dr. George Claus, M.D., Ph.D. (botany) and Ph.D. (microbiology) and Dr. Karen Bolander, specialist in mass psychology. Claus and Bolander, *Ecological Sanity*, New York, McKay & Co., 1977

"While global population growth gets most of the media attention, what has been much less noted is the broad-based economic growth which has been empowering millions of poor people with the purchasing power to upgrade the quality of their diet . . . the combined effects of population and income growth are expected to double global food consumption in the next 30 years. If we double food production by doubling the number of hectares [we farm], it would create massive environmental damage . . . with large-scale destruction of forests, wildlife habitat and biodiversity."

Dr. Robert L. Thompson, Presidential Address, 23rd International Conference of Agricultural Economists Sacramento, California, 10 August 1997

"Von Braun (1995) based on a comparative study of seven countries, found that commercialization of agriculture . . . benefits the poor by directly generating employment and increased agicultural labor productivity. Both the households that are commercializing their production and hired laborers receive direct income benefits."

M. Rosegrant and C. Ringler, *Why Environmentalists Are Wrong About the Global Food Situation: Methods and Myths*, International Food Policy Research Institute, August 1997

"Merely to match the 1995 mechanical power of American tractors with horses would require us to build up the stock of these animals to at least 250 million, ten times their record count in the 1920s (Smil 1994). About 300 million hectares, or twice the total of U.S. arable land, would be needed to feed these draft animals! And yet all of the U.S. agricultural field machinery consumes annually no more than 1 percent of the country's liquid fuels"

Vaclav Smil, University of Manitoba agronomist and author of the forthcoming book, *Feeding Ten Billion* (MIT Press)

The environmental movement is valid and important. Environmentalists are forcing us to recognize that we are increasingly capable of—and responsible for—saving natural resources. Nevertheless, though they recognize the problems, they do not always clearly see the solution strategies. In the case of agriculture in particular, they have come to exactly the wrong policy solutions.

Mankind is at the most critical moment in environmental history. What we do as people and societies in the next decade will determine whether we have a more crowded but sustainable world to bequeath to future generations—or whether we will bring on the very apocalypse of famine and wildlife destruction that the gloomiest environmentalists have envisioned. Our decisions on agriculture and forestry will be the most crucial of all, because they will govern how we use two-thirds of the earth's surface. They will dictate the habitat—or loss of habitat—for 95 percent of the earth's wildlife species.

So far, we are making the wrong decisions, for the wrong reasons, based on the wrong information. The environmental movement was never more correct than when it coined the slogan, "Think globally and act locally." But today's environmentalists are not thinking about farming and forestry in global terms.

John Deer

MODERN COMBINE IN WHEAT—High yields from the best land—this combine is less important than the rust-resistant semi-dwarf wheat that puts more energy into its seed heads and less into stalk. Note the lack of weeds to compete for nutrients.

Land, one may argue, is the scarcest resource of all. We need it to produce our food and timber. It is increasingly in demand for human recreation. In addition, we recognize an almost unlimited demand for land as wildlife habitat. Virtually every bit of wildlife habitat is important. Agriculture and forestry are the only sector where we can "create" more land without sacrificing the environment. High yields are the only way to do it.

The environmental movement, however, has been recommending exactly the opposite strategy—low-yield farming. That would almost certainly trigger the plow-down of huge tracts of wildlife habitat as people attempt to avoid famine. As we shall see, the yields from traditional and organic farms are too low to feed people and still protect wildlife.

In the case of forestry, most environmentalists are similarly recommending low-yield forest management with no harvesting of trees. That would leave us with fewer trees and less wood and we would have to rely on more polluting alternatives (like steel) for construction needs. High-yield plantation forestry on a few acres is a key to having lots of forest products and still having lots of wild forests and wildlife.

Much of the green movement has also opposed international trade in general,and farm product trade in particular.They fear that trade will weaken environmental initiatives and threaten small

USDA

SLIM HARVEST—This 19th-century Russian wheat field helps us understand why Rev. Malthus was pessimistic about feeding more people in the days before farm science launched the Green Revolution.

traditional farmers. However, the lack of free trade in agriculture and forestry is likely to mean losing big tracks of wildlife habitat in some parts of the world, while safe and renewable farming resources are wasted in other places. The key problem region will be Asia, which will be nine times as densely populated per acre of farmland as North America in the year 2050.

High-yield farming is not a matter of putting small farmers out of business, anywhere in the world. A planet that must roughly triple the output of its food system has no interest in putting any farmers out of business. The question is where we must invest to expand.

High Yields Are the Way to a Better Future

The old way of producing food has been around since before Chief Massasoit taught the Pilgrims how to fertilize their corn with fish. It is called low-yield farming. Organic farmers, who after all are essentially low-yield, traditional farmers, have not suddenly discovered how to produce lots of extra food. In fact, they are recommending we produce less.

Similarly, biological pest controls are making some progress in controlling a few pests in a few places, but there is little likelihood that they will replace much of our chemical pest control. Biological controls are too narrow and too uncertain.

Integrated pest management (IPM) is also useful, and more producers are using more of it. However, it is not a way to replace pesticides, but rather a way to make them more effective.

Finally, composting and organic fertilizer can only add marginally to our plant nutrient supply. Organic farmers cannot replace the huge quantities of chemical nitrogen and mined phosphate without clearing huge tracts of land for green manure crops and thus sacrificing wildlife habitat.

The environmental movement has not brought forward any breakthroughs in food production. Nor can they bring down the world's population growth trends much more quickly than they are already coming down. The Third World's birth rates already have come more than 75 percent of the way to stability, essentially in one generation. We have a fighting chance at restabilizing population at 8 billion people and as early as 2035—mainly because of economic growth and TV. But that won't preclude the need to essentially triple the output of the world's agricultures.

No one is delivering a vegetarian world. Tropical forest is already being cleared to grow low-yielding soybeans for broiler chickens. Crop residues are being stolen to feed more dairy cows to produce more milk, despite the long-term risks to soil productivity.

Obviously, the rising food needs of the world must be met this year, and next year, and the year after that. The key question—and one that the environmental activists and organic buffs refuse to answer is this: "How many million acres of wildlife habitat are you willing to clear to have chemical-free farming?"

The question of high-yield farming is no arcane or historical debate. Its answer will almost certainly determine the future of the world's wildlife, and probably the futures of billions of people as well.

It Has Been Hard to Hear the High-Yield Message

Unfortunately, it has been tough to get a public hearing for the high-yield viewpoint. It might have been easier if Paul Ehrlich had published his 1968 book, *The Population Bomb*, before Rachel Carson wrote her powerful indictment of pesticides (*Silent Spring*) in 1962. If we had truly become concerned about population before we got frightened of pesticides, we might have been more open to the benefits of high yields. But Rachel Carson got there first. One might say she poisoned the well of public opinion against fertilizers and pesticides.

We know now that man-made chemicals are no more dangerous than natural chemicals. Most chemicals, natural and man-made, seem to be dangerous to rats in high-dose testing. But high-dose rat tests, as most scientists will admit, overstate the risk to human beings from *all* chemicals.

We also know now that Ms. Carson's fears of widespread human cancer from pesticides have not been borne out. When we adjust for the increasing age of our population, nonsmoking cancer rates have decreased as the use of pesticides has spread.

Nor do we have any examples of pesticides threatening wildlife species—or even any major wildlife *populations.* Such relatively new compounds as the sulfanylureas and the glyphosates are no more toxic than aspirin, need only a few ounces per acre, and can be used around such sensitive species as trout and quail with no harm.

Fortunately, the environmental movement no longer needs to indict farm chemicals to justify its existence. The environmental movement has demonstrated its own vital validity. The fact that it got its start in the aftermath of *Silent Spring* should not dictate environmental policy recommendations today, when we know so much more about the ecology and about cancer and when the pesticides themselves are so much safer.

Rising Costs and Tighter Budgets

As the significant costs of environmental policies are becoming clearer, environmental recommendations need to be strongly focused and cost-beneficial. The environmental movement cannot afford to waste its political capital on counterproductive policy thrusts—especially if they mean destruction of the very wildlife that is at the heart of our shared environmental agenda.

Nor can the environmental movement afford to be backed into a set of policies for which continuing popular support and real implementation are almost impossible to generate. The Third World will not accept mud huts and malnutrition as its lot in life. The First World is hardly disillusioned enough to return to them.

Virtually the whole world today has a vision of material well-being, and is pursuing it. The environmental movement and the researchers of the First World must offer technologies and policies that support peoples' aspirations *along with* their environmental goals.

Saving Room for Nature with Farm Mechanization

Draft animals usually put a heavier burden on the environment than farm machinery. It takes huge amounts of land to provide the fodder for animals that must eat every day, whether pulling a plow or not. By Dr. Vaclav Smil's reckoning, America was using some

75 million hectares of its arable land in the 1920s to feed its draft animals.[1]

Every farm had to have its own fodder production, even on the rich blacklands of Illinois. (On less favored soils, it usually took more fodder acres per animal.) Animal power also required farmers to leave wide row spacings for their feet, severely limiting the number of seeds per acre for row crops such as corn and sorghum. (U.S. farmers in the 1920s planted perhaps 5,000 corn seeds per acre; they currently plant 25–30,000 and are probably headed for 50,000.)

Machinery also produces higher crop yields, mainly by lengthening the growing season. Tractor power lets farmers plant more quickly in the spring, achieving a well-known yield bonus. Diesel engines allow quick harvesting, before the crop goes past its peak quality and minimizing losses to storms, logging, and shattering.

It might not even be possible for the world to farm the First World's huge tracts of high-quality farmland without machinery because of the high cost of labor. The United States used to need about 10 million people to plant, weed, and harvest its cotton, but those workers were originally slaves and later poorly paid sharecroppers.

America no longer has slaves or large numbers of migrant workers to perform such work. The farm kids who see only hard work for low wages as their prospect in agriculture simply leave the farms and take jobs in the cities. (Many choose to do that even though tractors, combines, and milking machines have taken over so much of the labor.) The women and children of modern America are not willing to spend their summers in the hot sun, working with the old foot-powered potato planters, hand hoes, and jute bags full of lead arsenate powder as I did in the 1940s.

Many Third World agricultures remain animal-powered, notably that of India. Indian farmers have more than 200 million cattle and nearly 100 million water buffaloes. Most of them are kept, not for milk, but for pulling plows and powering primitive irrigation pumps. Most of the stalks and straw from India's crops that should be plowed back into the soil to preserve tilth and prevent soil erosion are fed to the cattle and buffalo instead and their manure is burned as fuel. In the future, as India's food demands rise with affluence, the country will have neither arable land nor water to spare for such heavy draft animal requirements.

Africa, which lacks even animal power, has a severe labor shortage. Weeds are the biggest constraint on feed production. Even though Africa's women and children spend most of the growing season hand-weeding the crops, they barely produce enough to subsist. They cannot expand their plantings because they could not weed additional crops.

In Indonesia, the lack of mechanical power cuts yields in another way: Farmers need more than a month of heavy labor with mattocks and other hand tools to prepare their rice paddies for a crop. That means they can grow only two rice crops per year; mechanization would permit three rice crops.

Critics of mechanized farming warn that the world is rapidly running out of petroleum. However, the world's current low oil prices testify that new systems for discovering and recovering oil have expanded the supply more rapidly than the demand in recent decades. Moreover, farming uses only a tiny proportion of the world's liquid fuels.

Farming has historically used the same fuels as the non-farm economy. Recent technical breakthroughs in fuel cells indicate that the machines of the 21st century are likely to be powered primarily by limitless hydrogen rather than oil. This will radically reduce the pollution problems associated with internal combustion engines, and permit petroleum to be used in a far more limited way, as a high-value chemical feedstock.

The absolute worst thing we could do for the environment would be to shift farming back to heavier use of "biofuels," which would compete directly with wildlife for the world's scarce supply of arable land. (The solution to low farm prices is freer trade, not horse-drawn equipment or a bad energy program.)

It's Not a People Problem

Hong Kong has a population density of over 14,000 people per square mile.That is far denser than China (288 per square mile) or India (658). The environmental reality is that Hong Kong's impact on the environment is fairly easy to mitigate. Sewage treatment is cheaper in densely populated cities, while energy and land requirements are minimized. (Hong Kong is already beginning to invest in environmental cleanup.) Unfortunately, too many of the environmental activists have turned environmentalism itself into an anti-people crusade—when people are not the problem.

In the environmental context, having 8–10 billion human beings on the planet certainly represents a challenge. But short of poison gas or huge induced famines, the number of humans on earth is not ours to choose. We in the West are not in a position to make the birth decisions for the people of the Third World. Fortunately, the Third World's fertility rate is dropping more rapidly than ever before. We can expect to have a declining world population after about the year 2040.

In the meantime, too many in the environmental movement are still fixated on policy recommendations that pretend we will *not* have 8–9 billion people, and that we will not allow the Third World to become affluent. Such policies cannot possibly generate majority support. Instead, such policy thrusts risk enormous environmental losses.

Slowing World Population Growth by Higher Grain Yields

The countries that have made the most progress in raising their grain yields have also made the most progress in bringing down their birth rates. Producing more food *has not* aggravated the world's population problems. In fact, the world's population growth rate began to trend down in the very year that the world gave the Nobel Peace Prize to Dr. Norman Borlaug (1970)—for his work in creating the Green Revolution.

It is entirely reasonable that higher grain yields should be a leading indicator of lower birth rates. They help produce higher standards of living. They help give parents confidence that their first two or three children will live and that they will be looked after adequately in their old age.

Also, countries generally do not develop cities and urban industries until they have ample food available to feed non-farm populations. Urban populations almost always have sharply lower birth rates than rural ones, in every culture and on every continent.

We've known for a long time, of course, that higher yields have been associated with low birth rates in the First World:

•America's corn yields have more than tripled since 1950, compared with a 73 percent rise in our population. U.S. births per woman are at 2.1, exactly the long-term replacement level.

•French wheat yields have tripled in the same period, while its population has risen 38 percent. The French fertility rate is now below replacement at 1.7.

The same association seems to hold in the Third World as well:

•India's rice yields have risen 250 percent, and its wheat yields have more than doubled, against a population increase of 151 percent. India's births per woman have fallen from 5.8 to 3.2, with virtually all of the reduction achieved since the beginning of the Green Revolution.

•Indonesia's rice yields have tripled, against population growth of 133 percent. The fertility rate of a *Moslem population* has fallen dramatically, from 5.5 to 2.1 births per woman.

•Emerging Chile has boosted its corn yields by more than fourfold, and easily accommodated a population increase of 130 percent. Births per woman have fallen *in a Catholic country* from 4.0 to 2.3.

•Zimbabwe has long had the best corn-breeding program in Africa. Corn yields among its traditional village farmers have roughly quadrupled, matching the fourfold expansion of its population. Zimbabwe's births per woman started dropping sooner and have come down more sharply than most sub-Saharan countries, from 7.7 to 3.8.

•China's rice yields have more than doubled since 1950, outgaining a population increase of 114 percent. China's births per woman are down to a very low 1.9, well below replacement. China has been famous in recent decades for a harsh policy against large families. However, China's dense population has unquestionably driven both its population policy and the establishment of one of the Third World's best agricultural research systems.

Unsuccessful Farming Leads to More Births

It may seem counterintuitive, but the countries that have had less success in raising grain yields have also kept the highest birth rates:

- Ethiopian grain yields have more than doubled (120 percent) but the population has increased 166 percent. The fertility rate *increased* from 5.8 in the 1970s to 7.3 in 1993.

- Kenya's corn yields have risen only about 25 percent, but its population has risen more than 300 percent! A crash family planning effort by the Kenyan government recently helped start the fertility rate downward to the current 4.6.

- Ghana's rice yields have risen only 24 percent since 1950, while the population has increased more than 300 percent. The fertility rate is 5.1 births per women, down from 6.7.

- Rwanda's corn yields have risen only about 25 percent, while its population has more than doubled. (Yields for important potato and bean crops *have* risen significantly. Births per women have come down from 7.8 but are still at 6.2.)

The lesson is clear. High-yield farming helps bring birth rates down. Cutting off high-yield farming research and discouraging the use of fertilizer in the world will hamper our progress toward restabilizing population.

If the world opts for low-yield farming (or simply fails to support a wholehearted push for higher-yield farming) it will take longer to stabilize the population. Low-yield farming is likely to produce more population growth, not less. But the First World hasn't learned this yet.

The Senate Hearing

In March of 1994, I debated Lester Brown, one of our most prominent environmentalists, at a Senate hearing. For years, Brown and his Worldwatch Institute have been predicting famine and environmental disasters due to population growth. They have also contended that agricultural research and higher yields could not meet the food challenge facing the world. They have argued, instead, for population "management."

The occasion for our debate was a hearing of the Senate Agricultural Appropriations Committee on the world food outlook. The chairman was Sen. Dale Bumpers (D-Ark).[3]

Brown predicted that the world was headed for massive famine and chaos. Of course, he has been predicting these calamities virtually every year for 25 years. To date, the big famines have never appeared. Bumpers called Brown a "genius" and noted that he himself didn't think the world could sustain more than 2 to 3 billion people (roughly half of the world population at that time.)

When my turn came, I testified that there is no need for famine in the world's future. The reasons: First, high-yield agriculture is raising crop yields much faster than population growth. Second, high-yield farming has already tripled the output of land and water in farming since the 1950s, during the very period when Brown has been wrongly predicting famine. Third, plant breeding, biotechnology, and other knowledge advances continue to permit higher and higher crop yields, even in the most advanced countries. Fourth, if we *did* get famine, it would only be after starving people had destroyed virtually all of the world's wildlife in last, desperate efforts to keep their children alive.

The question was not whether we would feed more people. The real question was whether we would feed the world's extra people from a few acres or a lot of acres. Wildlife would be the key beneficiaries of our success—or pay the supreme price for our failure.

I said that, if Bumpers's Agricultural Appropriations Committee kept funding high-yield agricultural and forestry research, we would be able to feed the doubled and restabilized human population of 2050 and beyond—*and* have at least as much room for wildlife as the planet has today.

I was astonished at the senator's reaction. "Mr. Avery," he said, "your testimony makes me sorry I convened this hearing." Sen. Bumpers seemed suddenly depressed—by the idea that we wouldn't have famine!

Bumpers had apparently been contemplating a big, mechanistic, guilt-free famine solution to end the world's population growth. Meanwhile, his Senate committee was proposing to cut funds for the international agricultural research that could help produce more food quickly. The funding, instead, was going to provide more condoms and pills for the Third World—though these were unlikely to have any significant impact soon enough to stave off famine or wildlife losses.

The *Washington Post* Story

A few days later the *Washington Post*'s Boyce Rensberger wrote a story on the rapid progress in high-yield farming. "Experts on Farming Remain Optimistic That Large-Scale Starvation Is Still Unlikely," declared the headline for Rensberger's special feature on agricultural research.[4] The story covered a three-day conference of international farm research experts. Rensberger quoted Dr. Donald Plucknett, who had recently retired as the senior science advisor to the international research networks as saying: "I do believe we can continue to raise world food yields if we do the right things." Another quoted top expert, Dr. Piers Pinstrup-Anderson, agreed that the long-term world food picture looked promising because of rising crop yields. "We're not running out of (farming) resources," he said.[5]

However, Rensberger correctly noted that funding was the key problem facing the international farm research centers: "One dwindling resource, however, is financial support. Twelve countries cut their support—the United States by the largest amount, dropping [by] $6.5 million to $41.6 million. Overall, [the research group's] funding for 1993 dropped by more than 7 percent. Officials say the promise of further cuts this year could force them to close some research centers."

And, again quoting Pinstrup-Anderson, "Whether food production continues to grow may depend on the willingness of donor countries to maintain their support. It's really up to us whether we want to have a happy future or an unhappy future."

There was no firestorm of front-page media coverage on the funding crisis of the Green Revolution research institutes. There was no groundswell of public opinion demanding increased agricultural research, nor any "world conference" to organize support.

Who Will Feed China?

Late in 1994, Lester Brown made global headlines with the rhetorical question, "Who will feed China?"

Brown noted that China's meat demand was already rising by 3 million tons per year, as a result of economic growth and higher per capita incomes. Projections indicated that China might need another 200 to 250 million tons of feed grains by the year 2030 to supply the desire of Chinese consumers for more meat, milk, and eggs.

Brown seemed not to realize that his Chinese question represented the utter failure of his whole population management policy thrust. In 1974, he had written in his book *By Bread Alone* that the

Frank and Ernest

FRANK & ERNEST reprinted by permission of NEA, Inc.

world should pursue both population management and higher crop yields—but that was the last time he had anything good to say publicly about higher-yield agriculture.

Since then, Brown and the Worldwatch Institute have been totally fixated on suppressing births. Now, China confronts them with the bankruptcy of birth suppression as a stand-alone policy.

Condoms cannot raise crop yields. China is already at virtual stability in population, but it is demanding a better diet—and planning to build a big dam on the Yangtse [illegible]hat will displace one million people in order to help get it.

Population management cannot sa[illegible] wildlife. Only higher-yield agriculture will do that.

The Pattern

The truth about the world's [illegible]ly expanding ability to feed itself with higher yields has been p[illegible]ned from time to time in the media, including both the *Washington Post* and the *New York Times*. But it hasn't resonated with the public

In contrast, the misinformation about impending famine has reverberated like a cannon shot. How else can we explain Lester Brown and Paul Ehrlich? Both have been publicly, radically, and consistently wrong on hunger and population issues for 25 years.

Yet Brown sells hundreds of thousands of his books each year. His Worldwatch Institute is one of the largest outside suppliers of teaching materials to American schools and colleges.

Paul Ehrlich's *The Population Bomb* predicted in 1968 that we'd have famine in America in the 1970s, with the corpses of starvation victims piled in our streets. It is the best selling environmental book in history. In 1990, Ehrlich republished the same failed predictions, under the title *The Population Explosion*—and sold millions more copies! In reality, we have had surplus grain piled in our streets instead of Ehrlich's anticipated famine victims.

Why? After all this time, it can't be that the public believes Brown, Ehrlich, and their disciples are correct about famine. It has to be that a sizable part of the American population secretly *wants* them to be right.

Are we a nation of people-phobics? Do we have an instinctive fear of psychological crowding? There is no real crowding in America, nor is there likely to be. The people who voluntarily choose to subject themselves to life in Manhattan or the commuter traffic of I-95 around Washington are volunteers. They have made choices. If crowding is a key issue, the fact is that most of us can move to less crowded situations.

Yet the fear of crowding persists. In 1992, I wrote a column for the *Wall Street Journal* that was headlined, "Mother Earth Could Feed Billions More."[6] I got reader letters whose main point was, "Yeah, you can feed them, but pretty soon there won't be room to take a Sunday drive."

The fact is, of course, that a billion more people in Asia won't affect our Sunday drives. We will have to resolve our own traffic problems. But surely that can be done without "subtracting" four billion people forcibly from the world's population.

Virtually none of the world's population growth will occur in the United States. Our fertility rate is 0.1 percent. We would require 700 years to double our population without immigration—and we can control our own immigration if we choose.

The Turn Back to "Nature"

Is the environmental movement's turn toward wilderness, wildlife, and some sort of "natural" world due to our fear of crowding? Here are a few quotes from some environmental leaders:

"The only hope for Earth (including humanity) is to withdraw huge areas as inviolate natural sanctuaries from the depredations of modern industry and technology. Keep Cleveland, Los Angeles. Contain them. Try to make them habitable. But identify big areas that can be restored to a semblance of natural conditions, reintroduce the grizzly bear and wolf and prairie grasses, and declare them off limits to modern civilization."

David Foreman, founder of Earth First!
Confessions of an Eco-Warrior, Sierra Club Books, 1988

"Yes, wilderness for its own sake, without any need to justify it for human benefit. Wilderness for wilderness. For bears and whales and titmice and rattlesnakes and stink bugs. And . . . wilderness for human beings . . . because it is home."

Foreman, op. cit.[7]

"Restoration is a backward-looking philosophy. But unlike romanticism, which is a longing for the past, or preservation, which seeks to save what already exists, restoration implies an active participation in bringing the past back to life."

Carolyn Merchant, "Restoration and Reunion with Nature," *Restoration and Management Notes*, Winter 1986

Ironically, few people in the environmental movement have any significant firsthand experience with nature. By "experience" I do not mean carefully planned backpack trips into the Rocky Mountains, nor carefully manicured mountain retreats in the Adirondacks. I am talking about living or working in the wilderness.

Hardly anyone in the modern world has real experience with wilderness. Mostly, true wilderness residents are people like rubber-tappers in the Amazon, primitive tribesmen in New Guinea, and Lapps herding reindeer above the Arctic Circle. The wilderness-yearning in our society comes mainly from urban people who have never experienced it. Few of them actually plan to, other than an occasional comfortable visit to a national park. Virtually none of them would countenance shooting game for food, or cutting trees for cabins and firewood.

It's true that urban America today is a sobering reality. In 1997, William Bennett published a new Index of Leading Cultural Indicators for America. The Index notes that compared with 1960: The rate of violent crime in America has risen more than fourfold; birth rates among unmarried teenaged girls are up threefold; scholastic aptitude scores are down 20 points; teen suicide is up threefold; and violent crime among juveniles is up nearly threefold. And all of this despite the most prosperous economy the nation has ever seen. Given the problems in urban America, it should not be too surprising that many people would like changes in our society.

The surprise to me, however, is that environmental critics do not seem to be suggesting any societal changes—at least not in the way we organize the cities where the vast majority of the people and problems reside. Instead, the critics want to retreat—to running a nice, clean wilderness instead:

". . .[T]he wilderness world (we have) recently rediscovered with heightened emotional sensitivity, is . . . the experience of the entire human community at the moment of reconciliation with the divine after the long period of alienation and human wandering away from the true center."

Thomas Berry, *The Dream of the Earth*, Sierra Club Books, 1988[8]

"We cannot survive by planning to treat the symptoms such as air pollution, water pollution, soil erosion, etc. . . . First we must reverse the population growth. . . . The other thing we must do is to pare down to our Indian equivalents. At one end would be the starving blacks of Mississippi; they would approach unity in Indian equivalents, and would have the least destructive effect on the land. At the other end of the graph would be the politicians slicing pork for the barrel, the highway contractors, strip-mine operators, real estate developers. . . . Blessed be the starving blacks of the Mississippi with their outdoor privies, for they are ecologically sound and they shall inherit a nation."

Wayne H. Davis, "Overpopulated America,"
The New Republic, January 1970[9]

We cannot follow such advice. In the first place, it is physically impossible to yield big stretches of the country back to the grizzly bear, go back to low-yield organic farming—and still feed Cleveland and Los Angeles. Second, there is no point to it. Wilderness as such will not help us deal with our problems.

The critics observe that affluence has not brought us happiness. Nor will it. It will simply give us long lives (if we choose to live them) and comfortable circumstances in which to work out our personal and social dreams and destinies as best we can.

Few have ever found perfection in the short, mean, dangerous life of the wilderness. The American Indian found no mystic perfection in the life of the hunter-gatherer. (Our current Indians have little more understanding of life in the true wilderness than we do.)

When the Indians were running the country, myths about harmony with nature took second place to the desperate need for meat.

That meant stalking a wary deer using a homemade bow with a range of 50 yards—or facing hunger. Often, it meant pushing the "old ones" out into the winter snow to die quietly (in their fifties). Even more often, it meant killing people from the neighboring tribes to ensure that one's own tribe would have enough hunting ground to survive.

The white "pioneers" lived longer, more comfortable lives than the Indians—but not by much. Their days, spent in houses made of logs or sod, were also studded with blizzards, disease, droughts, plagues of insects, hunger, wild animal attacks, and combat with others who coveted the same piece of wilderness. For today's comfortable Americans to say they long for the wilderness is almost an insult to the people who had to contend with the dangers and desperation of the true wilderness. John Dewey, one of America's most eminent philosophers, notes that humanity has tried before to crawl back into the myth of a "natural" world.

The environmental movement derives its power from the millions of solid, responsible people who care about the environment and are trying their best to produce a sustainable, livable world. But there are more than a few highly visible people whom I will call in this book "eco-zealots." Thanks to high-yield farming, we do not need to lower ourselves to levels of inhumanity never before sanctioned in human society. We do not need to induce mass famine, nor force abortions upon the unwilling.

Nor do we need to accept the loss of any significant wildlife, wildlife habitat, or key environmental resources. We can create the new resources we need from our increasing knowledge of the natural world—and use more intensively the natural resources already supporting humanity.

The loss of high-yield farming, by contrast, would mean famine for billions and destruction of more wildlife than most of us can imagine. . . . for nothing.

FAO photo by G. Ciparisse

COW CARCASS in Drought Area

Notes

1. Vaclav Smil, *Energy in World History*, Boulder, CO, Westview Press, 1994

2. *FAO Annual Production Yearbooks;* births per woman from "total fertility rates" in World Bank Annual Development Reports, World Bank, Washington, D.C.

3. Senate Agricultural Appropriations Committee, *Hearing on the World Food Outlook*, February 23, 1994, Washington, D.C.

4. Boyce Rensberger, "Despite Horn of Plenty, Some Feasts of Famine Abound Across the World," *Washington Post*, February 28, 1994, p. A3.

5. Dr. Piers Pinstrup-Anderson is director of the International Food Policy Research Institute, the policy research unit of the Consultative Group on International Agricultural Research. IFPRI is headquartered in Washington, D.C.

6. Dennis Avery, "Mother Earth Could Feed Billions More," *Wall Street Journal*, September 19, 1991, p. A14

7. David Foreman, *Confessions of an Eco-Warrior*, San Francisco, Sierra Club Books, 1988.

8. Thomas Berry, *The Dream of the Earth*, San Francisco, Sierra Club Books, 1988.

9. Wayne Davis, "Overpopulated America," *The New Republic*, January 10, 1970, reprinted in *Learning to Listen to the Land*, Washington, D.C., Island Press, 1991, pp. 177–182.

2

Wildlife and the Acres Not Plowed

MYTHMAKERS SAY:

"During 1993, no one knows how many species of plants, animals and other living organisms disappeared forever from the Earth—probably at least 75,000 according to conservative estimates."

"More Pollution, More Loss of Species, More Business as Usual," *The Earth Times*, December 31, 1993, p. 28

"Even if humanity were to depart the earth, recovery of biotic diversity by evolutionary mechanisms would require millions of years. . . ."

Michael E. Soule, "Conservation Tactics for a Constant Crisis," *Science*, vol. 253, p. 16

"Growing numbers of Americans are concerned that farmers and consumers have become cogs in a corporate machine that pollutes our environment and values efficiency over humanity."

Earthsave, International, *Down on the Factory Farm*, August 1998

"In the United Kingdom, 77 percent of people want genetically modified crops banned, while 61 percent do not want to eat genetically modified food—attitudes typical of those in many European countries. Austria and Luxembourg are locked in dispute with the EU over a genetically modified variety of maize that the EU has approved but they have banned . . . "

Nigel Williams, "Agricultural Biotech Faces Backlash in Europe," *Science,* August 7, 1998

REALITY SAYS:

"The difference between wildlife habitat loss and cropland stability in most of the world is rising crop yields. Ecuador's crop yields declined between 1982 and 1987, driving a 2 percent annual expansion in cropland—at the expense of tropical forest. Just down the coast in Chile, cropland remained stable, despite a 17.5 percent annual growth in agricultural exports combined with 1.7 percent annual population growth. In Chile, the yields were rising."

Reported by Dr. Douglas Southgate, Ohio State University, in *Tropical Deforestation and Agricultural Development in Latin America*, London Environmental Economics Centre, 1991

"Much of the recent destruction of biodiversity in the tropics and elsewhere has been the result of government subsidies . . . that allowed the deforestation or plowing of low productivity lands that would not support agriculture. . . . "

Michael Huston, "Biological Diversity, Soils and Economics," *Science*, December 10, 1993

"The evolution of cadmium resistance [in the mudworms] could have taken no more than 30 years. . . . This capacity for rapid evolutionary change in the face of a novel environmental challenge was startling. No population of worms in nature could ever have faced conditions like the ones humankind created in Foundry Cove. . . . The rapid evolution of tolerance for high concentrations of toxins seems to be common. Whenever a new pesticide is brought into use, a resistant strain of pest evolves, usually within a few years. The same thing happens to bacteria when new antibiotics are introduced."

Dr. Jeffrey Levinton, Chairman, Department of Evolution and Ecology, State University of New York/Stony Brook, "The Big Bang of Evolution," *Scientific American*, November 1992

"Food security at the global level concerns whether the world's farmers and food system can provide twice or even three times as much food as today—at no higher real cost, and do it in a manner that does not destroy the environment. There are only three ways to increase global food availability—increase the land area planted, increase yield per hectare, and reduce post harvest losses."

Dr. Robert L. Thompson, Presidential Address, 23rd
International Conference of Agricultural Economists,
Sacramento, California, August 1997

REALITY FROM THE GOVERNMENT:

"So long as human population increases, there is no way to stop further loss of habitat unless land, water and other natural resources are used more efficiently and productively. During this century, the United States has increased productivity in agriculture and forestry spectacularly: American agriculture harvests less land today than it did in 1910, even though it feeds over two-and-a-half times more people in the United States and millions more abroad. It also feeds them better . . . improvements range from producing more net food per acre by replacing animals with mechanized power, using improved seed varieties, more net effective pest control, higher fertilizer usage and increased irrigation—to reduced wastage and spoilage using refrigeration and modern processing techniques. . . . If technology had been frozen at 1910 levels then the United States would need to harvest at least 1,200 million acres to produce the same quantities of food as it does today on 300 million acres. That would require cultivating all the crop land and forest land in the United States—leaving no productive lands for parks, wildlife refuges, nature preserves, wilderness areas or other habitat for conserving biodiveristy. . . . For example, land used for crops and pasture in the nine Mid-Atlantic States declined from 10 million hectares in 1949 to less than 5 million hectares in 1987."

Quote from a 1992 publication issued by the U.S. Department
of the Interior and Agriculture titled *America's Biodiversity
Strategy: Actions to Conserve Species and Habitats*

AND A MYTHMAKER'S REALITY:

"Passing a French Broom plant (a pest plant species) he muttered, 'I wish I had some Roundup.' This from the lips of a dedicated conservationist? Most of those battling exotic plants occasionally use herbicides. With varying degrees of distaste, every single one of the dozens of biologists, land managers and activists with whom I spoke considered judicious herbicide use a lesser evil than the harm caused by exotic plants. . . . 'What the hell do they want,' one asks. 'Do they want a short-term environmental insult or a long-term ecological catastrophe?'"

"Botanic Barbarians," *Sierra*, Sierra Club, January/February 1994, p. 57

Even the most strident naturalists agree that preventing the loss of wildlife habitat is the key to saving wildlife species. Insect experts Paul Ehrlich of Stanford and E. O.Wilson of Harvard recently co-authored an article claiming we might lose 25–50 million wilds pecies by the year 2050—through habitat destruction due to population growth and rising affluence.[1]

Eco-zealots demand that we either prove that pesticides represent zero risk to wildlife—or ban them. Unfortunately, pesticides sometimes do inflict localized harm to wildlife, particularly when they are misused or when accidental spills occur. We can't say that pesticide risk to wildlife is zero. But banning pesticides would leave us with far less wildlife habitat—and thus far fewer wildlife species.

Pesticides are a vital element of the high-yield farming systems that are already saving more than 15 million square miles from being plowed for food production. By 2050, pesticides and fertilizers could be helping to save from plow-down as much as 30 million square miles of forests, prairies, and other prime wildlife habitat.

Since habitat is the key to saving wildlife, and high-yield farming saves habitat, why aren't environmental organizations leading cheers for the companies that develop effective new pesticides for our crops? Why aren't they praising high-yield farmers? Why aren't they demanding *more* research to get still-higher yields?

It might be that the eco-activists prefer to have fewer people in the world. They might like to have so few that high-yield farming would be unnecessary. But it does no good to say that the world should have fewer people. No one is prepared to murder the people, nor forcibly abort the children in the Third World.

Nor can we expect to starve the people and let the animals live. The people would not go quietly. They would not let their children die of starvation while a wildlife preserve sat unplowed next door.

A study by Texas A&M University indicates that U.S. field crop yields would decline drastically if we substituted the currently available organic pest controls for synthetic pesticides. Soybean yields would drop by 37 percent, wheat by 38 percent, cotton by 62 percent, rice by 63 percent, peanuts by 78 percent, and field corn by 53 percent.[2]

Similar cuts in yields would be suffered on farms from France and Finland to Chile and China if they followed a no-pesticide policy. Such yield cuts would require plowing down millions of additional square miles of wildlands to replace the lost production.

If we cut out chemical fertilizers as well, we could expect world crop yields to drop back by two-thirds or more, to the yield levels harvested in the 1950s before most of the world was using high-yield chemicals.

The world is currently cropping the land area equivalent of South America (5.8 million square miles).[3] We would need the land area of both South and North America (15–16 million square miles) to produce today's food supply without chemistry. If we tried to feed the projected human population for 2050 (8–9 billion) without the help of chemistry, we must expect to plow down another 30 to 40 million square miles of wildlife for food production.

Compare the 15 million square miles of wildlife habitat saved by high-yield farming with the amount of land saved by the Nature Conservancy (one of the few environmental groups that has actually set more land aside for wildlife). It says it is administering about

USDA Photo Library

THE ACRES NOT PLOWED—There is a good deal of wildlife habitat even in intensively farmed areas. But the key to saving wildlife is farming the best land and leaving the biodiversity on the poorer land undisturbed.

78,000 square miles of land worldwide, or about the land area of Idaho. This means that high-yield farming is protecting about 200 times as much wildlife habitat as the Nature Conservancy.[4]

Gain in Crop Yields, 1950–90

	1950s	1990s	Acres saved
	(metric tons per hectare)		*(millions)*
U.S. corn	2.7	8.0	143.00
S. Africa corn	1.0	2.2	11.80
Argentine corn	1.6	4.8	15.80
Italy corn	1.8	9.5	10.50*
China rice	2.5	6.3	117.57*
Indonesia rice	1.7	4.6	46.78*
France wheat	2.2	6.6	25.20
Mexico wheat	1.1	4.6	6.30
India wheat	0.7	2.7	183.60
Ghana cassava	7.8	12.0	.79
Chile tomatoes	13.4	64.0	1.70*
Canada rapeseed	0.9	1.4	6.66
U.S. soybeans	1.4	2.6	59.95
Australia cotton	1.8	6.0	2.30*
China cotton	1.9	4.3	14.36

* Irrigated land

Source: Computed from *FAO Annual Production Yearbook* series

1999 ACRES-SAVED UPDATE:

Indur Goklany, who saw low-yield crops and wildlife habitat losses in his native India, is now Assistant Director of Policy Analysis for the U.S. Department of the Interior.

He recently authored an article in *Technology* in which he estimates that if the world's food production technology had been frozen at 1961 levels, the world would have needed to convert at least another 1 billion hectares (approx 2.5 billion acres) of wildlife habitat to cropland.

Goklany says this is true even if we assume the newly converted cropland would be equally productive (unlikely) and that people would have been content with the inadequate diets of 1961 (even less likely).[5]

Leaving the Poorer Land for Nature

We cannot evaluate the wildlife benefits of high-yield farming by the number of spiders and weeds that survive in an acre of monoculture corn. There is never much wildlife in a monoculture crop field. Nor can we condemn pesticides based on the accidental deaths of 200 birds, or even 2000.

We must give high-yield agriculture the environmental credit for billions of wild organisms thriving in and on the two acres that didn't have to be plowed because we tripled the yield on the best and safest acre.

When American pioneers began clearing the virgin forests of Ohio and Indiana in the 1800s, they were pursuing one of mankind's great conservation strategies—farming the best land.

When we farm the best land, we do three good things for conservation. First, we get the highest possible yields, so we need less total land area for food production. That is important since we're already using 37 percent of the earth's land area for agriculture. Second, the best land permits the most sustainable farming, especially from the standpoint of erosion. Third, and equally important, farming the best land saves the most biodiversity. The best quality lands typically have big populations of a few species. (America, for instance, had perhaps 60 million bison and fabulous numbers of prairie dogs, but fewer bird species than the tiny country of Costa Rica.)

More than 90 percent of the world's biodiversity lives on the lands too poor to be worth modern farming: tropical rain forests, swamps, coastal wetlands, and mountain microclimates. These are the lands with resources good enough to support species diversity and poor enough to drive intense competition between the species.

In his 1994 book, *Biological Diversity*,[6] ecologist Michael Huston, of the Oak Ridge National Laboratory, points out that all over the world the medium-poor lands harbor the greatest variety of wildlife species. Huston notes that we cleared about 100,000 square miles of wild forest in Ohio and Indiana, without losing a single wildlife species. (The passenger pigeon subsequently became extinct, but that was primarily due to market hunting, not to complete loss of habitat.)

Neither Ohio nor Indiana today harbors a single unique, endemic plant species. In contrast, Florida has 385, Texas 389, and California 1517. All are warmer states with lots of poor-quality land. The Great Plains of North America are huge, and they featured huge numbers of bison, antelope, prairie dogs, and coyotes, along with a few hundred grasses and modest numbers of birds and insects. That's not much diversity by global standards.

Even the famous grasslands and bush-veldt of southern Africa offer thousands of square miles dominated by the same key species: elephant, lion, rhino, antelope, wildebeest, kudu, warthog, zebra, and a few dozen more.

By contrast, species by the millions are found in the tropics. Gary Hartshorn, of the World Wildlife Fund, says that until recently biologists thought there were three to ten million wildlife species on the planet, with about half of them in the tropical forests.[7] However, recent efforts to count the species in particular chunks of the tropics suggest that the world may have 30 million wild species—with 25 to 27 million in the tropical forests.

Hartshorn points particularly to the work of T. L. Erwin in the Peruvian Amazon. Erwin's team counts invertebrate species by fogging tree crowns with nonpersistent insecticides. They found huge numbers of species, and very little species repetition, even on trees of the same species.

Hartshorn also notes: 311 tree species were counted on one hectare in the Amazon, which is almost half the number of native tree species in the continental United States. 236 species of vascular plants were found on one plot the size of a football field at La Selva in the lowlands of Costa Rica. La Selva has 102 species of mammals (including 63 bat species), 384 species of birds, 45 amphibians, 195 butterflies, and 54 species of sphingid moths. Another national park in Costa Rica has 3124 species of moths. In other words, a few square miles of the Peruvian Amazon contain more biodiversity than the whole of North America.

This should not be surprising. Lush, well-watered lands offer easy pickings for wild organisms. The tropical forests, with their huge seasonal variations in rainfall and the absence of winter drive intense competition for resources—and the development of specialized creatures. What this means is that to save the world's biodiversity we should get all the yield we can, safely and sustainably, from the world's prime cropland. That will leave as much of the poorer, species-rich land for nature as possible.

Fragile Land Back in Grass and Trees

We must also give high-yield farming the environmental credit for the steep, erodible, and drought-prone acres of farmland put back into grass and trees in countries like the United States and Sweden. This has been possible directly because of high-yield farming on the best and safest acres. In the 19th century, my steep, rocky little farm in the Shenandoah Valley was planted to corn and

USDA

MAKING DO WITHOUT HABITAT—These scissortail chicks are growing up in an oil refinery, but if we keep raising crop yields, most of the world's wildlife can keep their traditional habitat.

wheat—with terrible erosion losses. Once Iowa and Indiana learned to grow 100-bushel corn, my Virginia land was put back into grass, where it belonged.

Sweden has increased its forest area by 23 percent since 1970, and is continuing to shift land from farms to forests—because of rising crop yields on its best land. Germany is expanding forest by taking some of its high-risk cropland out of farming.

Tough, Tenacious Wildlife Species

Our wildlife species, meanwhile, have already proven that they are generally tough and adaptable.

According to Dr. Jeffrey Levinton, chairman of the Ecology and Evolution Department at the State University of New York/ Stony Brook, the last big explosion of species in the world occurred 600 million years ago.[8] Virtually all of our major wildlife types (Levinton calls them "body types") have been around for at least that 600 million years. During this period, they have dealt with radical changes in temperature, oxygen and CO_2, new pests, and "wild cards" like concentrations of toxic cadmium in their environments. Levinton concludes that the number of wild species has been gradually increasing throughout this long period. He says there is strong reason to believe that the number of species will continue to expand—if we do not destroy the habitats.

Levinton himself tested mudworms from a cove in the Hudson River near an old battery factory. The mudworms there had successfully adapted to levels of toxic cadmium as high as 25 percent—unprecedented in world history[9] "Overall . . . the total number of species seems to have been increasing steadily during the past 60 million years. . . . All the evidence from living groups of organisms therefore suggests that contemporary evolution proceeds as fast as ever," says Levinton.

His conclusions underscore the reality that if wildlife species do not lose their habitat, their prospects for survival are bright.

No Vegetarian World

It will not suffice to mutter about vegetarian diets because the environmental movement is not delivering a vegetarian world. No country or culture in history has voluntarily accepted a diet based solely on the relatively low-quality protein found in vegetable sources. Meat and milk consumption is rising by millions of tons per year in China and India right now as their incomes rise. Billions of people, in every newly industrializing country in the world, are increasing their demands for the resource-costly protein found in cooking oil, meat, milk, and eggs.

In the affluent and environmentally conscious United States, the *New York Times* reported in 1992 that a mere 12 million Americans considered themselves vegetarians.[10] That's a bit under 5 percent of the population (though the *Times* said it was a 20 percent increase over a decade earlier). Even that 12 million figure has to be questioned.

Also in 1992, *The Vegetarian Times* commissioned a profile and attitude survey of self-described vegetarians by Yankelovich, Skelly and White/Clancy, Schulman, a respected polling organization.[11] One of the survey questions was, "In order to satisfy my appetite, a main meal must include meat." *Half* of the vegetarians agreed with the statement. Two-thirds of them said they ate chicken, almost half reported daily or weekly consumption of some animal product, and one-third even admitted to eating red meat. *Only 4 percent of the 5 percent of America's "vegetarians" say they never eat any animal products.*

There doesn't seem to be a lot of hope that humanity will voluntarily live on porridge, beans, and onions.

REALITY SAYS:

"It is usually best to avoid saying of any idea—whether socialism, Keynesianism or Christianity—that it has not been tried. Most of these ideas have been tested to destruction by fallible human beings, and there is no point in invoking perfect versions operated by archangels."

Samuel Brittan, columnist for the *Financial Times*,
August 1, 1994, p. 12

But What About Man-Made Poisons?

Even if we grant the importance of saving wildlife habitat, the environmental hard-liners seem to believe that the pesticides will poison the wildlife *in* the habitat. That is a valid concern. The demonstrated answer to it is "no."

Naturalists all over the Western world have focused on the loss of species, and none of them have fingered pesticides for any species losses. Their major concern is habitat loss, followed by things like untreated sewage threatening coral reefs, and too much nitrate (from both farming and urban areas) overfertilizing certain lakes and streams. Pesticides haven't even eliminated the boll weevil—and we were *trying* to do that.

Recently, a devoted birdwatcher named Samuel Florman noted in MIT's *Technology Review* that the birdwatching community was perhaps guilty of crying "wolf" on bird losses.[12] Since the arrival of the Pilgrims, only four North American bird species have become extinct (the Labrador duck, the great auk, the passenger pigeon, and the Carolina parakeet). All were lost before the introduction of pesticides. Three other extinctions are suspected (ivory-billed woodpecker, Bachman's warbler, and Eskimo curlew) but again, no one is laying these at the door of farm chemicals. "Considering the speed and voracity with which the United States was developed, one looks for reasons why the destruction was not far worse," says Florman.

There is also a lot of habitat managed *for* wildlife. The U.S. Fish and Wildlife Service currently manages 20 million acres, the National Park Service another 80 million, and the national forests an additional 176 million acres. This does not count extensive wildlife management on private lands.

Roger Tory Peterson says there are now one to two *billion* more songbirds in the United States than when the Pilgrims landed.[13] Human habitat—including the suburbs—have created much more "forest edge" which is excellent bird habitat. (Florman notes that old growth forest is not congenial environment for most birds.) Also, a majority of the 254 bird species tracked by the U.S. Fish and Wildlife Service's Breeding Bird Survey *increased* in population between 1966 and 1991.

Pesticides Getting Safer for Wildlife

I recently spoke to a local chamber of commerce meeting. When I asserted that pesticides were getting safer for wildlife, one of my neighbors said he used to have more cardinals in his yard than he does now. How could I be sure that the pesticides weren't killing his cardinals?

First of all, we must understand that wildlife numbers ebb and flow due to many factors, including the amount of food for that species, competition for territory, the ratio of predators, weather (both here and in migratory paths), and any number of changes in local habitat.

However, I answered my neighbor by noting that Virginia had recently banned a chemical called Furadan 15G (carbofuran) because of "hundreds" of known bird deaths over a period of years. (The antipesticide movement alleges thousands, and they may well be right.)[14]

The chemical in question was a dry, granular insecticide and nematocide. It was supposed to be incorporated in the soil during seedbed preparation, but some granules were often left lying on the surface. It was a threat to birds if they mistook them for seeds and ate them directly.

Some birds *did* eat the Furadan granules, and they died. Some were songbirds. Worse, a number of eagles apparently died from carbofuran poisoning after eating pigeons and other prey birds that had consumed carbofuran. As a result of the bird deaths in Virginia, the manufacturer withdrew the compound from sale in areas where it might threaten sensitive bird populations.

Several things strike me powerfully about the withdrawal.

First, "hundreds" of bird deaths over a period of years, in a state with 44,000 square miles and literally millions of birds, is a tiny problem. This will probably get me condemned by the Audubon Society, but we can't make huge broad-gauge decisions about the future of the world's food systems, people, and wildlife on the basis of a small number of American bird deaths—as individually regrettable as they may be. The wildlife in Asia and the people in Africa also deserve at least some consideration.

Second, America is so rich in food productivity today that we gave up an otherwise-useful chemical that posed even that small a wildlife problem.

Third, there are other cost-effective pesticides available to attack Furadan's target pests in Virginia. If the state had to ban all pesticides and accept a 50 percent cut in its crop yields, Virginia would have to plow down another 1.5 to 2 million acres to make up the lost production. How many birds (plus other wild creatures) live in 2 million acres of forest and wild meadow?

Finally, pesticides and pesticide regulation are continuing to make the world *safer*, not more dangerous, for both people and wildlife. We eliminate any chemicals where significant risks can be proved. (DDT was banned even *without* such proof.) Pesticides today are tested far more extensively than they used to be—specifically for their safety with wildlife as well as with humans. That's a major element of the 100-plus tests that new chemicals must pass.

However, we don't dare to allay my neighbor's worries about pesticides by banning them all because the environmental cost would be too high.

Safer Handling of Pesticides

Research has shown that a very high proportion of environmental contamination attributed to pesticide spraying has come from careless mixing and rinsing of application equipment, and from accidental spills.[15] Industry research has also shown that 85 percent of all applicator pesticide exposure occurs during the mixing and loading operation. With common liquid products, the exposure is often a result of splashing or spilling while pouring and measuring.[16]

Chemical handling is now being made safer by shifting to dry formulations (to prevent damage from spills) wherever that is practical. Formulations that can't be manufactured and shipped as dry powders are increasingly delivered in premeasured water-soluble pouches that can be simply dropped into the spray tank. In other cases, the compounds are shipped in large reusable plastic containers that lock onto the farmer's spray equipment, without ever exposing the farmer or permitting spillage.

Cooperative Farmer reported "futuristic" spraying technologies, including sprayers controlled by radar to precisely calculate the tractor's ground speed in relation to spray rates. The magazine noted a direct injection spray system in which the active ingredients and the water are kept separate until they meet in the spray nozzles. This eliminates unused pesticide left in the tank. And since the water tank is never contaminated, it eliminates the problem of rinse water.

A University of Kentucky experimental sprayer uses fiber-optic cables to throw a special light beam from the sensor to a receiver lens. Anything that interrupts the wave length of the beam is identified as a weed and activates the spray nozzle. UK designers expect the sprayer to cut the application rates for post-emergence sprays by 30 percent.[17]

What About Biomagnification?

"One of the most remarkable antipesticide allegations used by environmental extremists has been the claim that pesticide levels are 'magnified' at each step of the food chain," says Dr. J. Gordon Edwards.[18] Dr. Edwards is a biology professor at San Jose State who has authored both a respected book on wildlife *(Coleoptera East of the Great Plains*) and a Sierra Club book (*Climbers Guide to Glacier National Park*).

Edwards says the concept of biomagnification as put forth by antipesticide forces is scientifically untenable and has been refuted by both experimental data and field tests. Today's pesticides are basically metabolized or excreted by fish, birds, and mammals.[19]

The DDT residues that remain in the tissues of many people and creatures have been isolated and rendered harmless; they have never been linked to any health threat. "Unfortunately," Dr. Edwards writes, "the popular press and some semi-scientific journals have been crammed with 'biomagnification' articles for many years, and antipesticide activists have made profitable use of that myth. Environmental organizations welcomed the radical concept and used it to frighten the public into donating more money to 'help fight pesticides.'"[20]

Dr. Edwards says that whenever allegations of biological magnification are brought up, the activists should be challenged to produce the actual data, from analyses of each step in the actual food chain. They should have to specify which methods of analysis were used, which tissues were analyzed, how old the samples were, and whether they were wet or dry weights. They should also have to present valid evidence that the creatures really were in the food chain being studied.

He points out that activists have made such mistakes as saying that herring gulls had to have gotten DDT in their tissues from the fish they ate, even though the subject gulls were also notorious garbage feeders in the dumps of nearby cities. They have also, he argues, taken the highest concentration from the upper element of the food chain and the lowest concentration from the lower element to get the *appearance* of biomagnification. Further, they have put herring gulls in a food chain with marine amphipods (which they don't eat) and old squaw ducks (which they don't eat either).

What About Minimizing Pesticide Use?

There are techniques that can reduce pesticide use without much production loss. However, all of them involve trade offs.

For example, we could use more integrated pest management to a fuller extent. But lots of farmers already use IPM, and more have tried it. It involves management costs and, sometimes, higher risks of pest damage. Similarly, we could scout more of our crop fields to see when and whether they *need* spraying, rather than spraying every field just in case. Again, scouting involves management costs and risks.

We could band more pesticides right along the rows where the pesticide does the most good, instead of spreading it in the mid-rows too. (This permits lower volumes to do almost as much good). However, the banding has to be done carefully or it risks extra pest damage, and the added time involved in application has to be balanced against the reduced pesticide use. We could use more crop rotation to break pest cycles without pesticides (though there is sometimes a trade-off in lowered cropping intensity).

Pesticide Volumes Already Declining

American pesticide use is already declining, as shown in Figure 2.1.

Much of this has been due to the introduction of safer new low-volume pesticides. Some of it has also been due to the spread of integrated pest management, especially in cotton.

The reduction in pesticides would have been far more dramatic except for the rapid spread of conservation tillage. Herbicides have been the only category of pesticides with any recent increase in usage volume. That's because they're used in the new soil-conserving farming systems, which rely on chemical weed control instead of "bare-earth" farming systems like plowing. But conservation tillage, now used on more than 100 million U.S. acres, is drastically cutting our soil erosion. Conservation tillage also cuts tractor fuel requirements by millions of gallons compared to plowing. So, while herbicides have increased U.S. pesticide use, the increased usage is delivering major environmental benefits.

There *have* been temptations toward prophylactic spraying in the past. One of the awful truths about past weed control has been the "neighbor factor." Farmers didn't want the neighbors to

Figure 2.1

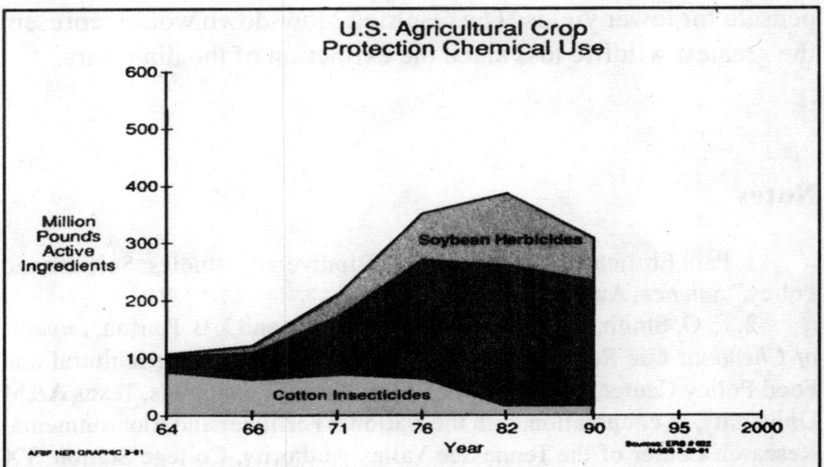

Source: American Farm Bureau Federation

see weeds in their fields. It lowered their prestige. It brought complaints that their weeds were spreading weed seeds onto neighboring fields. It is ironic now, during the low-till revolution, to see these same farmers moving en masse to farming systems like conservation tillage and no-till which feature scruffy-looking fields covered with corn stubble and dead grass.

A far stronger temptation to spray was offered for many years by the farm subsidy programs of the United States and various other rich-country governments. The subsidy programs gave farmers artificial incentives to get high yields—almost literally forcing them into heavier chemical use. Strangely, the environmental movement has been very gentle in approaching farm subsidies, even though they clearly played a huge part in boosting farm chemical use in the United States, Western Europe, and Asia.

In some parts of the world, unfortunately, pesticides *have* been misused and significant damage has been done to wildlife populations. This book is not meant to justify bad management of pesticides or to justify the over-intensification of farming produced by government farm subsidies in the First World.

However, this book confronts the antipesticide movement with an urgent reality: Shifting to a chemical-free agriculture on the basis of current knowledge would force expansion of cropland to compensate for lower yields. The resulting plow-down would represent the greatest wildlife loss since the extinction of the dinosaurs.

Notes

1. Paul Ehrlich and E. O. Wilson, "Biodiversity Studies: Science and Policy," *Science*, August 1991, pp. 758–761.

2. E. G. Smith, R. D. Knutson, C. R. Taylor, and J. B. Penson, *Impacts of Chemical Use Reduction on Crop Yields and Costs*, Agricultural and Food Policy Center, Department of Agricultural Economics, Texas A&M University, in cooperation with the National Fertilizer and Environmental Research Center of the Tennessee Valley Authority, College Station, TX (undated).

3. Cultivated land total from John F. Richards, *The Earth as Transformed by Human Action*, Cambridge University Press, 1990.

4. Land totals from Nature Conservâncy, Washington, D.C. (Regina Perkins), July 12, 1994.

5. Indur Goklany, "Meeting Global Food Needs: The Environmental Trade-Offs between Increasing Land Conversion and Land Productivity," *Technology,* vol. 6, pp. 107–130, 1999.

6. Micheal Huston, *Biological Diversity*, Cambridge Press, 1994.

7. Hartshorn, Gary S., "Possible Effects of Global Warming on the Biological Diversity in Tropical Forests," *Global Warming and Biodiversity,* Peters and Lovejoy, eds., Yale, 1992

8. Dr. Jeffrey Levinton, "The Big Bang of Animal Evolution," *Scietific American*, November 1992, pp. 84–91.

9. Ibid.

10. Columnist Marian Burros, "Eating Well," *New York Times*, July 8, 1992.

11. Kizmanic, Judy, "Here's Who," *Vegetarian Times*, October 1992, pp. 72–80.

12. Florman, S. "Progress for the Birds," *Technology Review*, July 1993, p. 63.

13. Ibid.

14. Diana West, "Taking Aim at a Deadly Chemical," *National Wildlife*, National Wildlife Federation, June/July 1992, pp. 28–41.

15. Fawcett, R.S., "Pesticides in Ground Water—Solving the Right Problems," *Ground Water Monitoring Review*, vol. 9, no. 4, Fall 1989.

16. "Futuristic Spraying Today," *Cooperative Farmer*, February 1993, pp. 20–21.

17. Ibid.

18. J. Gordon Edwards, "The Myth of Food-Chain Biomagnification," *Rational Readings on Environmental Concerns*, op. cit., pp. 125–135.

19. D. L. Gunn, *Annals of Applied Biology*, 1972, pp. 105–127; G.R. Harvey, "DDT in the Marine Environment," *National Academy of Sciences Committee Report*, August 9, 1973; and J. W. Kanwisher, "DDT in the Marine Environment," *National Academy of Sciences Committee Report*, January 30, 1973.

20. Ibid.

3

There Is No Upward Population Spiral

MYTHMAKERS SAY:

"Population, when unchecked, increases in a geometrical ratio. Subsistence increases only in an arithmetical ratio. A slight acquaintance with numbers will show the immensity of the first power in comparison of the second. . . . The power of population is indefinitely greater than the power in the earth to produce subsistence for man."

Thomas Malthus, 1766–1834,
An Essay On the Principle of Population

"In 1968, *The Population Bomb* warned of impending disaster if the population explosion was not brought under control. Then the fuse was burning; now the population bomb has detonated. . . . In the six seconds it takes you to read this sentence, eighteen more people will be added. Each hour there are 11,000 more mouths to feed; each year, more than 95 million. Yet the world has hundreds of billions fewer tons of topsoil and hundreds of trillions fewer gallons of groundwater with which to grow food crops than it had in 1968."

Paul Ehrlich, *The Population Explosion*,
New York, Simon and Schuster, 1990

"Population. I have not spoken about this before. I have been an environmentalist for 20 years, but I have never talked about population. . . . The controversy around contraception and abortion made it politically easier to speak and organize around air pollution, deforestation, toxic waste and biodiversity while ignoring the role our own burgeoning species plays in all this."

Jane Fonda, U.S. Special Goodwill Ambassador
to the United Nations Population Fund, 1994

"How long do Americans intend to watch while population growth adds both to poverty and justification for taking more from Nature? . . . With overpopulation, the clash of values grows both more bitter and more illogical. Does Kemp [former Housing and Urban Development Secretary] support open borders *and* housing the poor *and* preserving wetlands? Not possible. These are mutually exclusive goals. . . . Does Vice President Albert Gore pose as an environmentalist at the same time he plumps for immigration? . . . As an environmentalist, Gore inspires neither confidence nor trust."

Virginia D. Abernethy, *Population Politics,* New York, Plenum Press, 1993, p. 300

REALITY SAYS:

"The new African data mean we will likely get less global population than we imagined. Birth rates in sub-Saharan Africa—the last main redoubt of superhigh fertility—are plunging. . . . This is happening against a backdrop of stunning decreases in fertility over three decades in Latin America (a 50 percent drop), Northern Africa (Egypt down 42 percent) and Asia (Indonesia down 43 percent). The current UN 'medium' projection shows global population growing from 5.4 billion today to 11.1 billion in 2100. But that is based on a 2.1 Total Fertility Rate well above the current levels of modern nations . . . the UN's (more likely) 'low' scenario yields a 6.4 billion population in 2100 (and sinking) and even that is keyed to a 2.0 rate, higher than modern norms."

Ben Wattenberg, American Enterprise Institute, "Unexploding Population?" *Washington Times*, March 17, 1994

"One-quarter of world population has stopped growing. Another quarter of the population is on the verge of arriving at zero-growth fertility. . . . The rate of growth, which had been rising for the past three centuries, dropped to a little above 2 percent around 1970 and has been in decline ever since. The 20-year economic boom after World War II is clearly the reason world population growth dropped to the 1970 level. Another burst of expansion in the world economy could bring the whole population down to zero growth."

Gerard Piel, founder and longtime publisher of *ScientificAmerican*, and author of a forthcoming book on population growth, quoted in *The Earth Times*, June 15, 1994[1]

"In 1975, a typical Bangladeshi woman would have had seven children in her lifetime. Today she would probably have three. . . . In 1975, when asked how many children she wanted, a typical woman would reply four. Today she would say two. . . . Demographers agree that the fertility transition is ultimately caused by a drop in mortality. Once a couple realizes that their children are likely to survive, they can give birth to fewer infants and still be sure of being cared for in their old age."

Madhusree Mukerjee, "The Population Slide," *Scientific American*, December 1, 1998, pp. 32–33

"Has the world's fear of being overcrowded brought us to a lurking suspicion that additional food production will simply encourage more births? If so, we can set our minds at rest. The evidence shows clearly that higher-yield farming encourages fewer [Third World] births."

Dennis Avery, "Boosting Crop Yields Would Save Habitat," *Christian Science Monitor*, November 1, 1994, p. 19

The most startling thing you can tell the average American is that the world's population is *not* spiraling upward out of control. But that's the reality. What we're seeing is not a population spiral but a onetime surge. In fact, the world is in the final stages of the third—and probably last—human population surge.

The first surge began 10,000 years ago when we invented farming. Farming produced more food per acre, more dependably, than hunting and gathering. With farming, the world's human population rose fairly quickly from perhaps 10 million to about 200 million at the time of Christ.

Population stayed below 700 million until the beginning of the Industrial Revolution in the 18th century.[2] (Agriculture could have supported more people in that period, but epidemic diseases like cholera and bubonic plague kept down the urban numbers.)

The second population surge coincided with the Industrial Revolution and increased human wealth.

What we're seeing today might be called the Public Health Surge in human numbers. Death rates are being brought down quickly by safe drinking water, proper waste treatment, vaccines, and the other public health interventions. It takes longer, often several decades, for birth rates to come down to match. In the meantime, there is a surge in populations, country by country.

Figure 3.1

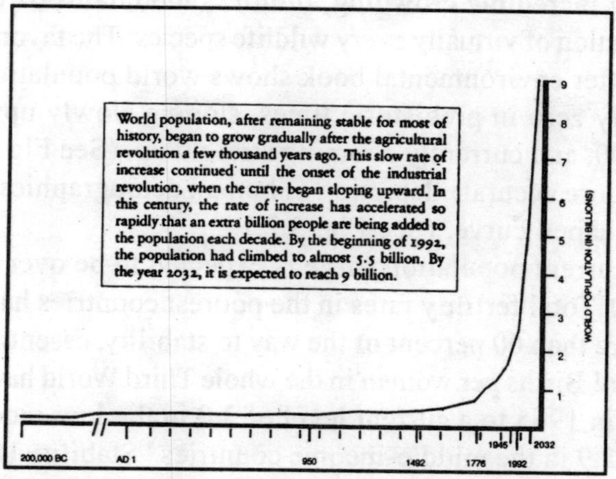

Source: *Earth in the Balance*, by Al Gore, Houghton Mifflin Company, New York, 1992, pp. 32-33.

The Dreaded L-Shaped Curve

The doomsayers claim we can't know that populations will level off again. They imply that human numbers will just go up and up,

Figure 3.2

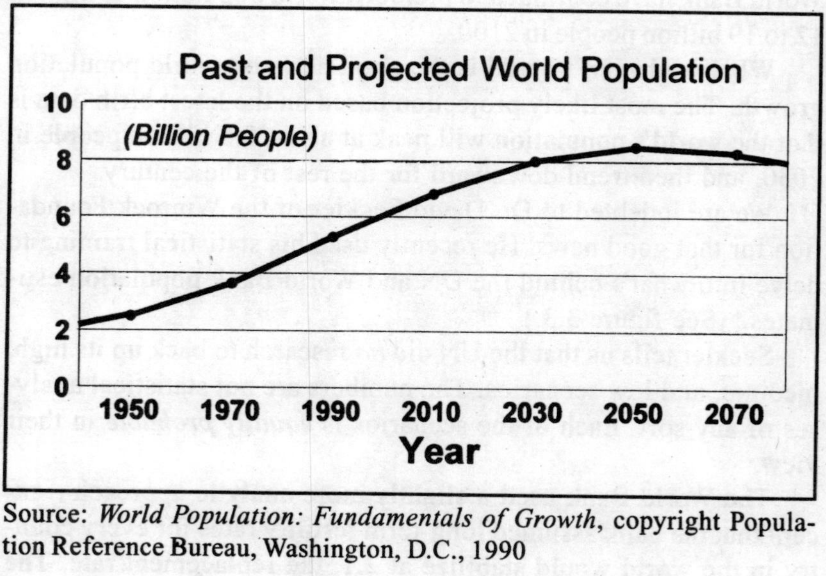

Source: *World Population: Fundamentals of Growth*, copyright Population Reference Bureau, Washington, D.C., 1990

producing incredible crowding, limitless mountains of waste, and the destruction of virtually every wildlife species. The favorite graphic in book after environmental book shows world population starting at virtually zero in prehistoric times, sloping slowly upward until about 1900, and currently zooming straight up. (See Fig. 3.1.)

The more accurate depiction of human demographics, however, is the S-shaped curve. (See Fig. 3.2.)

The current population surge is expected to be over before the year 2050. Total fertility rates in the poorest countries have already come more than 60 percent of the way to stability, essentially in one generation! Births per woman in the whole Third World have dropped from 6.1 in 1965 to a current level of 3.0 in the low-income countries and 2.9 in the middle-income countries.[3] Stability is 2.1 births per woman. The First World is at 1.7 births per woman and seems to have settled at that level.

The Good News on Global Population

People have not understood that the world's rate of population growth peaked about 1963 and is already headed rapidly toward zero. Even such "responsible" sources such as the UN and the World Bank have continued to predict a world population as high as 12 to 19 billion people in 2100.

What we haven't heard is the good news on world population growth: The most likely projection based on the latest birth data is that the world's population will peak at about 8.5 billion people in 2030, and then trend downward for the rest of the century.

We are indebted to Dr. David Seckler of the Winrock Foundation for that good news. He recently used his statistical training to delve into what's behind the UN and World Bank population estimates.[4] (See figure 3.3.)

Seckler tells us that the UN did *no* research to back up its high, medium, and low scenarios. The numbers are not statistical analyses of any sort. Each of the scenarios is *equally probable* in their view.

The World Bank used a slightly more analytic approach—except that the bank assumed long-term fertility rates for every country in the world would stabilize at 2.1, the replacement rate. The

Figure 3.3

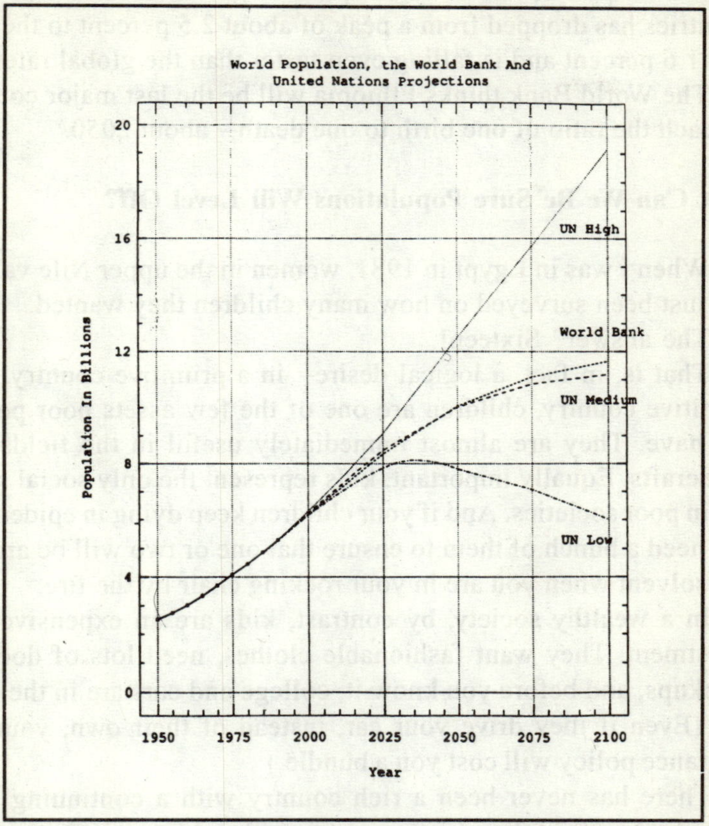

bank even assumed that the countries whose fertility rates have dipped far below 2.1 would shortly jump back up.

"There is no statistical reason to believe there is a long-term floor of 2.1," says Dr. Seckler. "On the contrary, over 38 percent of the population of the world is in countries with average fertility rates below 2.1. Many countries are at very low levels: Spain, Italy, Germany and Hong Kong are below 1.4. And many countries have been below 2.1 for a long time. We have estimated that if there is a long-term floor, it is in the neighborhood of 1.7 plus or minus 0.2. This independent estimate corresponds to the consensus estimate of a group of demographers interviewed on this subject."[5]

The world's population growth rate apparently peaked at 2.23 percent in the early 1960s. The current rate of population growth is down to 1.5 percent and falling. Population growth in the Third World countries has dropped from a peak of about 2.5 percent to the current 1.6 percent and is falling even faster than the global rates.[6]

The World Bank thinks Ethiopia will be the last major country to reach the ratio of one birth to one death—about 2050.[7]

How Can We Be Sure Populations Will Level Off?

When I was in Egypt in 1981, women in the upper Nile valleys had just been surveyed on how many children they wanted.

The answer? Sixteen!

That is, in fact, a logical desire—in a primitive country. In a primitive country, children are one of the few assets poor people can have. They are almost immediately useful in the fields and handcrafts. Equally important, kids represent the only social security in poor societies. And if your children keep dying in epidemics, you need a bunch of them to ensure that one or two will be around and solvent when you are in your rocking chair by the fire.

In a wealthy society, by contrast, kids are an expensive ego investment. They want fashionable clothes, need lots of doctor's checkups, and before you know it, college and cars are in the budget. (Even if they drive your car, instead of their own, your car insurance policy will cost you a bundle.)

There has never been a rich country with a continuing high birth rate, and rarely a poor country with a low one. But people have to *want* to limit births, which is why affluence is such a key element in lowering birth rates. Once they decide to limit births, they find ways to do it. A Turkish doctor notes that modern contraception raises the effectiveness of the family's birth decisions from perhaps 80 percent to about 90 percent.

The modern methods are very much preferred; they are used by 80 percent of the Third World women who practice contraception.[8] Condoms, pills, and female education all go with declining fertility rates. But none of them work unless people want fewer children. That comes with higher incomes. And world economic growth is surging.

Recent Income Gains in the Third World

The world stage is now set to generate more economic growth for more people in poor countries than ever before in history. More important, this is a trend, not an incident.

The General Agreement on Tariffs and Trade (now renamed the World Trade Organization) is the most successful international institution in world history. It has replaced 200 years of tariff wars with an agreement for all countries to let in the competitive products of their neighbors. This has ushered in decades of economic growth rates higher than we've ever seen before, in countries with no economic growth history. World nonfarm trade has increased more than fifteenfold since 1950, and is still rising fast. The resulting spread of economic growth is headlined by Japan and the other Asian Tigers but also include major parts of Latin America and East Europe.

Some observers—I am one—believe that the "GATT miracle" and the Asian Tiger model triggered the economic liberalization of China, helped bring down the walls of the Kremlin, and are now beginning to rescue Latin America from its traditional, stifling Spanish-style bureaucracy.

More than 30 poor countries, from India and Poland to Peru and Zimbabwe, are now trying to privatize, attract foreign capital and generate export-oriented manufacturing jobs. Most of them are succeeding, and will continue to succeed, because of the GATT/WTO free trade opportunity.

The hundred-year ideological struggle between Communism and capitalism is over, and that too is good news for economic growth. The idea that you could make people happier and better-off through government planning and control has been discredited. The idea that six people in Moscow could replace the ideas and energies of millions of on-the-spot Russian decision-makers was unlikely in 1917. It became impossible in the laptop computer age. The end of the Cold War means less money needed for defense and more cash available for civilian needs, all over the world.

Finally, the original Asian Tigers (more properly GATT tigers) are now mature enough to be engines of international economic growth themselves. The imports of Japan, South Korea, and Taiwan—along with their foreign investments—are reinforcing world economic growth.

China, with a billion people, has been boosting its industrial output at more than 12 percent annually for over a decade. "Between 1977 and 1993," the *Financial Times* said recently, "China's measured gross domestic product rose more than fourfold, the dollar value of its exports increased tenfold, and the annual inflow of direct investment rose from virtually nothing to $26 billion. Above all, hundreds of millions have experienced a transformation in how they live now, and how they hope to live tomorrow."[9]

China's per capita incomes have doubled since 1980, and in the southern part of the country they have quadrupled. Economic growth in *southern* China has been rising 20 percent per year. They are making textiles, shoes, electronics, and a host of other products. Private investors are building superhighways and expanding telecommunications. *The Economist* magazine says half of China's urban households now have refrigerators, and 70 percent have *color* TVs. A 1992 study by McKinsey & Co. estimated that there were 60 million Chinese with incomes above $1,000 (a level at which consumer buying begins to be important). By the end of the century, McKinsey expects more than 200 million Chinese to top the $1,000 income level.[10]

India's industrial output has been growing three times as fast as its population for a decade. Now India is reducing its socialist red tape and making foreign capital more welcome, so the economic growth rate could increase further. As a result of higher incomes, India's Hindus are increasing their milk consumption 2 million tons per year and are eating more eggs.

In 1970, 60 percent of Indonesians lived in absolute poverty; by 1997, less than 15 percent of them did. Meanwhile, the Indonesian poultry industry expanded at double-digit rates.

This world economic growth pattern is no OPEC boomlet like we saw in the late 1970s. This is permanent, solid growth in the world's economies. And unlike oil money which sticks to the hands of only a few, the income gains from this growth are being spread broadly among workers and active managers.

This is good news for the environment. Forget the guilt being spread by the eco-zealots who claim that the rich countries are the key environmental problem. Rich people stack themselves in cities that don't take much land, treat their sewage, buy cleaner forms of energy, and develop pollution-reducing technologies. Poor people not only have large families, they also tend to pollute more than affluent families.

Most of the Third World's population is already in the most polluting phase of economic growth by extending crops onto fragile farmland, burning trees and soft coal for fuel, and smelting lots of iron.

MYTHMAKER:

"With few exceptions, newborn humans come into a world that can offer them nothing but hunger, disease and squalor. No matter how many relief programs we set up, there is just no way to cope with mass misery that gets worse as numbers increase. . . . Our society and all others must adopt deliberate measures that will slow, halt and eventually reverse the appalling multiplication of numbers. There is no other choice . . . NPG advocates a massive increase in funding for U.S. foreign population control programs . . ."

From a fund-raising letter by Negative Population Growth, Inc., Teaneck, New Jersey, Spring 1994

New Momentum for Lower Birth Rates?

The best recent news is that lower birth rates seem to be gaining a momentum they have never had before. Until now, lower birth rates seemed to require significant gains in affluence. But now population management is becoming a reality even in countries which have not yet achieved high per capita incomes.

Kenya, long the "fertility Frankenstein" of the developing world, had a fertility rate of 8.3 children per woman in 1977–78. By 1989, however, that fertility rate had dropped to 6.7. In 1993—just four years later—it had fallen to 5.4. By 1997, it had fallen to 4.6. "The 20 percent decline in just over four years is one of the swiftest ever recorded," says the Population Reference Bureau of the latest Kenyan fertility drop.

Equally important, the demographers say, there are healthy leading indicators of change elsewhere in southern Africa, even where little economic growth is happening. Women are waiting longer to marry, getting more education, and using more contraceptives. All of this is reported in "The Fertility Decline in Developing Countries," a major article in the December 1993 issue of *Scientific American*.[11]

It is too early to assign a definitive cause for these developments. However, Dr. Seckler has broad African and Third World experience. He suggests a "modernization factor."

"This is cultural change," says Seckler. "Changes in the self-image of people—especially people in rural areas and women—due to the rapid spread of modern images and ideas through radio and television in developing countries. This factor plays a major role in lowering the 'demand for children.' We have heard, for example, that family planning messages have been inserted into the television soap-operas in Brazil with great success."[12]

What he's really saying is that with modern electronic communications, even poor women begin to see themselves in a new light. They see Western TV programs on community satellite sets that portray mothers with two children and who interact with their husbands as equals. They see women doing professional jobs *and* being women. Such exposures may now be driving social change even more rapidly than economics—or more closely in parallel with it.

The new African data mean the world is likely to get quite a lot less population growth than responsible demographers have predicted. The people-phobes will have a harder and harder time justifying their predictions of intolerable global crowding.

Zero Population Growth

I chuckle when I get the mailings touting the need to get zero population growth "right away." These mailings are not being read by the young people begetting Third World children. I laugh out loud when I read about the "rollback" of world population numbers. Are the members of these groups volunteering to subtract themselves? I know we will get zero population growth. We just won't get it fast enough to preclude the need for another tripling in world food production.

Why are we so terrified of this onetime population surge?

We've already accommodated a doubling of population between 1950 and 1990.

Certainly, if we *had* a perpetual upward spiral it would eventually defeat our efforts to supply food, forest products, and recreational space. But a onetime surge is a different matter. We will have a population peaking (hopefully at 8–9 billion) between 2035–2050, with the total of humans then gradually declining into the foreseeable future. The Third World will continue to gain affluence and join the 38 percent of the world that is already well below replacement fertility rates.

Obviously, the environmental impacts of a large human population which is also stable and affluent are far different than the zero-population lobby has portrayed.

Food, Crowding, and Violence

Much has been made recently of the idea that crowding produces violence, and that fear of hunger produces wars. Historically, it is true that tribes and nations have warred over hunting, fishing, and farmland. But we haven't seen much of that lately—or at least not since food production became a matter of technology rather than land.

Japan invaded Manchuria in the 1930s to grow soybeans—but modern Japan has become the world's largest food importer. It is buying wheat, feed grains, and meat on the world market, and exporting cars and other manufactures. Trade has proven a far more secure way for Japan to ensure its food supplies—at far less cost. Aircraft carriers and armies are hugely expensive. Combat with them is even more expensive.

The Serbs and Croatians in Yugoslavia have not been warring over farmland in the 1990s. They are warring over ancient, unresolved ethnic frictions, inflamed at least since World War II. Nor is either side gaining a better life.

Look at Ireland, which has been fighting a civil war for decades (or centuries). It was once rooted in land ownership. Now, even the diehard Irish Republican Army is pursuing peace, because it sees the rest of the world outstripping Ireland economically. Food and land are no longer even significant issues; both the Catholics and Protestants assume they'll have food. Both assume the jobs they want are urban ones. But the sons on both sides have been going elsewhere to get jobs that won't come to a war-torn land.

In Africa, recent decades have seen a good deal of hunger produced by violence—but not much violence produced by hunger. Civil wars have offered lots of examples of armies trying to keep their opponents from getting food, as in Ethiopia, Angola, Mozambique, Liberia, and Sudan.

In Rwanda, it is said that "crowding" caused the tribal genocide attempted by the Hutus against the Tutsis—with perhaps 1 million murders systematically carried out according to a plan under development for years. Why did the people of Rwanda not have faith in "population management"? Why did they not feel secure with the knowledge that contraceptive pills could cut their birth rates to more manageable levels? Any such "solution" must have seemed impossibly far distant.

But higher corn and bean yields are immediate and tangible. They provide confidence for the future here and now. Rwanda has not made big gains in corn yields, but it has doubled its potato yields in recent decades, and has recently tripled the yields of its food bean fields. (Since 1985, 500,000 Rwandan farmers have begun growing climbing beans that yield three times as much food per acre; they're the agricultural equivalent of the high-rise office building.)

If food fears *were* to blame in Rwanda, then the Western world bears a more direct responsibility than if it was unresolved ethnic hatred. We must have failed to communicate our successful experience with high-yield farming to the Rwandans. Maybe we didn't do enough to help them understand that higher-yielding corn, food beans, potatoes, and other intensive crops offered food security for their future. We didn't show them the enormous power of a little chemical fertilizer. (High-yield seeds and fertilizer sacks don't try to get revenge in the future.)

Why has there been no internationally funded agricultural research station in Rwanda, prominently displaying the potential food solutions for the future? It couldn't have cost much, if the whole international farm research system has recently gotten only $270 million per year in funding. Why were there not local farm extension agents working intensively with both Hutus and Tutsis to produce more food per acre? As an alternative to murdering people you know by their first names, it should have had a certain basic human appeal.

If we are truly worried about food shortages producing violence, then it is time for the affluent countries to put their money where their press releases have been. It is time to quit complaining about too many births, and trembling about more illegal immigration to the rich countries. If those are our concerns, it is time for us to make real investments in Third World farm research. It is time to double or triple our investments in food production systems for the developing countries. The institutional structure is already in place. The scientific tools already exist. The research thrusts are already under way. It is simply a matter of stepping up the pace and breadth.

One billion dollars per year in Third World farm research . . . as an alternative to genocide . . . ? The people arguing against high-yield farming research for the Third World may bear a heavy responsibility indeed.

Reprinted from BridgeNews Forum, January 2, 1998

Feeding the Pets of the 21st Century

We lost our Labrador retriever over the Christmas holidays. He was 12 years old and suffering from a cruel cancer, but it was hard to see him go. Fletcher was the most loyal, loving, and grateful member of the household.

Of course, that's how most people feel about their pets—and why the world will have to be prepared to feed billions of additional pets in the 21st century.

The planet's human population is now projected to peak at about 8.5 billion people, about the year 2035. This means we'll have less than a 50 percent increase from today's 6 billion. We will, however, have a much bigger increase in the population number of cats, dogs, songbirds and other companion animals.

They all eat food. Some of them eat a lot of food. Many of them eat resource-expensive foods, such as fish and meat by-products.

Americans currently keep more than 52 million dogs, almost 60 million cats, 12 million birds and 4 million riding horses. Those totals are all up from a decade ago, when the United States had 38 million dogs, 30 million cats, 6 million birds and 3 million horses.

Brazil is not yet rich, but it has 15 million dogs and 7 million cats, and Brazilian pet food manufacturers are building new factories because the country's personal incomes are rising. If Brazil reaches the current American pet density (one pet for each two people), in 2020 there will be about 115 million Brazilian cats and dogs.

Brazil's cats and dogs already eat 450,000 tons of manufactured pet food per year—in addition to untold tonnage of table scraps, milk, and other traditional pet standbys. Sales of manufactured pet foods rose 18 percent in 1996, reaching $126 million. (In 1990, they totaled only $27 million.)

China can be expected to have nearly 600 million cats and dogs within a few decades. If the Chinese government continues to have a one-child per family policy, pets will be the major outlet available for people's parental instincts. Historically, China was famous for eating cats and dogs (along with monkeys, snakes and virtually anything else that breathes). However, China's rising affluence is producing a new attitude toward pets:

- I recently met a Chinese businessman, whose sister in Beijing has a cat. Nothing is too good for her cat; even though meat is a luxury item, she buys it a fresh pork kidney every day.
- In the southern Chinese city of Guilin, which originated the Shi-tsu, I saw proud owners walking their little dogs through the city parks with as much pride and concern as any New York City dowager.
- Chinese pet stores are just beginning to get beyond the traditional caged crickets and songbirds, to the occasional kitten and small dog.

Don't make the mistake of believing that these pets can be shortchanged on either the quantity or quality of their food. Hell hath no fury like a pet-owner whose "Fluffy" has been scorned.

That's why we must be able to meet the demand for pet food as well as people food.

In poor countries, animals are treated badly. My wife lived in Ethiopia for several years, where poverty-stricken Ethiopians displayed rampant cruelty to their animals. Donkeys went about their work with festering sores—but so did the people. Tiny puppies were abandoned in the streets—but so were children. Cats were allowed to starve or killed for stealing food.

But affluent Ethiopians (there were only a few) treated their pets well. As people become more affluent, they take better care of themselves, their kids, and their pets.

The minimal approach to farming (low inputs, low yields) can't assure that we'll be able to feed the people and pets from the world's current croplands and pastures. It would be a shame if the 21st century had to sacrifice wildlands to feed its pets—or vice versa.

The best insurance for both is to continue the thrust of modern high-yield farming, with improved seeds, fertilizers, pesticides—and now biotechnology. The world's next big crop yield gains are likely to come from two surprising sources:

- Genes from wild relatives of our crop plants, which we can't use without genetic engineering techniques; and

- The "fertilizer" impact of higher CO_2 levels in the atmosphere. (Carbon dioxide is to plants what oxygen is to a long-distance runner.)

Thank Goodness.

Fletcher loved our hikes in the woods—but he also loved his dinner.

Notes

1. "Gerard Piel on Population: Assessing the Impact of Development," *The Earth Times*, June 15, 1994, pp. 28–29

2. Miller, J., Population Research Institute, Baltimore, MD, personal communication, 1994.

3. Total Fertility Rate included in the "Demography and Fertility" tables in the *World Development Report 1989* (pp. 216–217) and *World Development Report 1994*, pp. 212–213, World Bank, Washington, D.C.

4. Seckler and Cox, *Population Projections by the United Nations and the World Bank: Zero Growth in 40 Years*, Winrock International Institute for Agricultural Development, Center for Economic Policy Studies Discussion Paper No. 21, Arlington, VA, 1994.

5. Seckler and Cox, *Population Projections by the United Nations and the World Bank: Zero Growth in 40 Years*, op. cit, executive summary.

6. World Bank, Knowledge for Development, *World Development Report 1998–99, Table 3.*

7. World Infrastructure for Development, *World Development Report 1994, Table 26.*

8. Robey, Rutstein, and Morris, "The Fertility Decline in Developing Countries," *Scientific American*, December 1993, pp. 60–67.

9. "Unfinished Revolution," *Financial Times* editorial, August 23, 1994.

10. "Cracking the China Syndrome," *Financial Times*, December 31, 1992, p. 8.

11. Robey, Morris, and Rutstein, op. cit.

12. Seckler and Cox, op. cit., p. 23.

Courtesy of the Pet Food Institute

KIDS AND PETS—Let's make sure we can feed them all in the 21st century.

4

Preventing Cancer with Pesticides

MYTHMAKERS SAY:

"The problem . . . is whether any of the chemicals we are using in our attempts to control nature play a direct or indirect role as causes of cancer. In terms of evidence gained from animal experiments we shall see that five or possibly six of the pesticides must definitely be rated as carcinogens. . . . Still other pesticides will be added as we include those whose action on living tissues or cells may be considered an indirect cause of malignancy."

Rachel Carson, *Silent Spring*, p. 222

"Farmers have been using these chemicals for decades, and the chemicals have built up in the soil from repeated doses year after year. They get inside the crops and onto their outer coverings. They get into water supplies and into the fish who live in them. And they build up in the bodies of animals whose feed has been grown using these chemicals."

David Steinem, *Diet for a Poisoned Planet*,
New York, Harmony Books, 1990, p. 5

"The most potent cancer-causing agent in our food supply is a substance sprayed on apples to keep them on the trees longer and make them look better."

Opening line of CBS-TV *60 Minutes* segment on
Alar, broadcast February 26, 1989

"The United States . . . will be a desolate tangle of concrete and ticky-tacky, of strip-mined moonscape and silt-choked reservoirs. The land and water will be so contaminated with pesticides, herbicides, mercury, fungicides, lead, boron, nickel, arsenic and hundreds of other toxic substances . . . that it may be unable to sustain human life. . . . Thus, as the curtain gets ready to fall on man's civilization let it come as no surprise that it shall first fall on the United States."

Wayne H. Davis, "Overpopulated America," *The New Republic,* January 10, 1970

"The Truth Does Not Kill, Pesticides Do"

Placard at an antipesticide demonstration in Florida, December 4, 1992

REALITY SAYS:

"The risk of cancer from ingesting minute doses of pesticide residues is so small that it verges on the theoretical."

Harvard Health Letter, January 1994, p. 7–8

"This is nonsense. Every chemical is dangerous if the concentration is too high. Moreover, 99.9 percent of the chemicals humans ingest are natural. For example, 99.99 percent of the pesticides humans eat are natural pesticides produced by plants to kill off predators. About half of all natural chemicals tested at high doses, including natural pesticides, cause cancer in rodents. People determined to rid the world of synthetic chemicals refuse to face these facts."

Dr. Bruce Ames, University of California, Berkeley, who helped develop the Ames-Gold scale of cancer risk used around the world, in a statement issued September 1, 1993, in connection with the release of the Nation Research Council report "Pesticides in the Diets of Infants and Children"

". . . [V]irtually all chemicals are toxic if ingested in sufficiently high doses. Common table salt can cause stomach cancer. . . . The (high-dose) rodent test that labels plant chemicals as cancer-causing is misleading. . . The standard carcinogen tests that use rodents are an obsolescent relic of the ignorance of past decades."

Science editorial, Dr. Phillip Abelson, September 21, 1990

"Of the 2,598 (fruit and vegetable) samples taken in this program, 92 percent had no detectable residues . . . we feel these results clearly confirm what scientists have said for many years: the 'problem' of pesticide residues in fresh produce is more one of perception than reality."

James W. Wells, Director, California Environmental Protection Agency's Department of Pesticide Regulation[1]

Quotes from "Carcinogens and Anticarcinogens in the Human Diet,"

National Research Council, 1996, pp. 5–7

"Conclusions: First, the committee concluded that based upon existing exposure data, the great majority of individual naturally occurring and synthetic chemicals in the diet appears to be present at levels below which any significant adverse biologic effect is likely, and so low that they are unlikely to pose an appreciable cancer risk. . . ."

Translation: About half of all the natural and synthetic chemicals in our foods cause tumors in rats at high doses, but at the low levels in which these chemicals occur in our diets, they offer no significant cancer risk.

"Second, the committee concluded that natural components of the diet may prove to be of greater concern than synthetic components with respect to cancer risk. . . ."

Translation: There's no evidence that man-made chemicals are more likely to give us cancer than the natural chemicals in our foods. In fact, we consume about 10,000 times as much known cancer risk from the natural chemicals in our foods as from pesticide residues.

"The varied and balanced diet needed for good nutrition also provides significant protection from natural toxicants. Increasing fruit and vegetable intake may actually protect against cancer. . . ."

Translation: Eating five fruits and vegetables per day will cut your family's total cancer risk roughly in half, whether they're grown organically or with chemicals—and we wish more people would eat more produce.

"Current evidence suggests that the contribution of excess macronutrients and excess calories to cancer causation in the United States outweighs that of individual food microchemicals, both natural and synthetic."

Translation: Eating too much and getting fat is a big cancer risk—far bigger than pesticide residues.

"All things are poison and none without poison. Only the dose differentiates a poison and a remedy."

Paracelsus (1493–1541), German physician and the father of modern toxicology

"The overall age-adjusted cancer mortality rate had been increasing in the United States for as long as such statistics have been kept. This trend was reversed and a decline in cancer mortality began in 1991. . . . [A reduction in lung cancer] underlies the down turn in the all-cancer mortality rate, because *other-cancer mortality has been declining since at least the early 1970s.*" (Emphasis added)

Declining Cancer Mortality in the United States, presented to the American Cancer Society by Dr. Philip Cole and Dr. Brad Rodu, August 10, 1995. Copyright, American Cancer Society, 1996

"The Cancer Killer: It's a cell's most elegant defender, a gene called p53. It stops tumors before they grow. But if damaged, it is involved in 60 percent of cancers. . . . All told, p53 has been implicated in more kinds of cancer than most people knew existed; 52 and counting."

Sharon Begley, "The Cancer Killer," cover story, *Newsweek*, December 23, 1996

In a sense, the public can be forgiven for fearing pesticides. In the early days, when we knew less about cancer, we adopted (out of fear and ignorance) the most drastic test we could find. That was the high-dose rat test. Then the high-dose rat tests "proved" that the pesticides caused cancer.

There is mounting scientific evidence, however, that high-dose rat tests tell scientists little about the real risks in our food. The rat tests in fact mislead the rest of us into fearing the wrong foods and the wrong dangers.

Rachel Carson was wrong in her belief that five or six pesticides then in use were causing cancer, and that more would be proven carcinogenic. In fact, to date we have found one pesticide that caused cancer—lead arsenate—which was displaced by DDT and the other modern pesticides. Arsenic is carcinogenic. But we no longer use lead arsenate because of its broad, immediate precancerous toxicity.

The high-dose rat tests find "cancer" in half of the chemicals tested (natural and man-made). Few consumers realize that there are no pesticides approved for use in the United States today that are known causes of human cancer.

There are some approved chemicals on the "suspected" carcinogen list. They are mostly there, however, because *the heavy skewing of the diets* in the high-dose tests *itself* apparently causes cell division and increased cancer risk. Furthermore, the EPA overstates the dietary risk of pesticide residues to humans by 10,000-fold. The EPA assumes all foods carry the maximum pesticide residues, even though only a small proportion carries any residues at all.

More Food Through Chemistry

Today, farm chemistry means life itself for billions of people. All over the world, pesticides and fertilizers mean ample food, free from destruction by locusts, armyworms, witchweed, crabgrass, Johnsongrass, aphids, stem rot, leafrollers, planthoppers, corn borers, dangerous natural toxins, and thousands of other pestilential plagues. Most countries would not be able to sustain life for their populations without farm chemicals. What is the health trade-off for this abundance, produced with the help of chemicals? None.

In fact, there are health gains:

- Pesticides help prevent the grains and oilseeds we eat from becoming contaminated with bacteria or fungi which produce toxins and mycotoxins that *can* cause cancer. One-fourth of the world's cereals and oilseeds are contaminated with such toxins. Our crops carry much lower risks, because of pesticides.

- Even more important, pesticides permit us to provide an abundance of the low-cost, attractive fruits and vegetables, which are humanity's best defense against both cancer and heart disease.

The eco-zealots demand that we prove zero human risk in pesticides—or ban them. In fact, zero risk cannot be proven in anything. The rapidly lengthening life spans in countries using farm chemicals are a strong indicator that we thrive on the foods containing both natural and man-made "carcinogens" labeled risky in the high-dose rat tests. It makes little sense to suggest that we would be living even longer without the pesticide residues—until after the chemophobes have found ways to deal with the "rat-risky" natural compounds which are ten thousand times more abundant in our diets.

Suppressing Toxins

The United Nations Food and Agriculture Organization estimates that up to 25 percent of the world's food crops are contaminated with mycotoxins, many of which have the potential to harm people. A recent report from the Council for Agricultural Science and Technology examined the danger these toxins can represent, and the ways in which Americans avoid them.[2]

Aflatoxin, for example, found most often in moldy corn, peanuts, and cottonseed, is a potent carcinogen. In western India, about 100 people died in 1974–75 when 200 villages tried to subsist on heavily molded corn. In England, thousands of turkeys died—of acute liver failure—in the 1960s when fed aflatoxin-contaminated peanut meal from Brazil. When dairy cattle are fed contaminated feed, aflatoxin can be transmitted to milk, and even to such products as cheese and yogurt. In humans it is associated with cancer of the liver and—in severe exposures—with gastrointestinal hemorrhage.

Ergotism, caused by fungal alkaloids which most commonly infect rye, periodically reached epidemic proportions in the Middle Ages and as recently as the 18th century. Whole populations seemed affected with giddiness. Mass movements with irrational goals appeared suddenly and then faded from sight. (Some historians speculate the Salem witch trials in colonial Massachusetts may have been triggered by ergotism.) In a severe 1978 epidemic in Ethiopia, caused by infected sorgum, half of the victims died. Many others developed dry gangrene and lost arms or legs.

Avoiding and minimizing these mycotoxins starts with controlling insect and rodent damage in the field; insect and rodent damage creates openings for the fungi which produce the mycotoxins.

Pesticides are thus the frontline defense against these toxins. Even after harvest, however, the crops must be protected from storage pests. We have to build safe, dry storage—and sometimes we also have to use low-toxicity antifungal agents such as propionic acid and acetic acid, or even fumigants.[3]

Fighting Cancer and Heart Disease with Produce

Pesticides have been critically important in enabling us to produce ample supplies of attractive and reasonably priced fruits and vegetables for year-round consumption. Eating more fruits and vegetables, studies show, can cut cancer risks in half and markedly reduce the ratio of heart disease and immune dysfunction.[4,5] Thus, *anything* which boosts fruit and vegetable consumption saves lives.

Dr. Anthony Miller is an epidemiologist at the University of Toronto and a member of the committee which drafted the National Research Council's comprehensive study, *Diet and Health*. He cites examples of findings to date:

- Carotene, found in carrots, broccoli, and many other orange and dark-green leafy vegetables, is converted in our bodies to retinols. The retinols suppress a whole range of cancers—skin, breast, bladder, esophagus, colon, pancreas, lung, and prostate.

- Vitamin C from citrus fruits and vegetables protects against stomach cancer. Stomach cancer was once our most prevalent cancer but the rate of affliction has been cut by three-fourths over the last 40 years. We know that vitamin C inhibits a group of carcinogens called nitrosamines.

- People who eat cabbage once a week have only one-third as much colon cancer as people who eat it less than once a month.

- Nitrate, associated with many vegetables, cuts the risk of lung cancer.

- Oral cancer risks are lower in people who eat lots of citrus fruits and in those who eat lots of dark yellow and cruciferous vegetables.

- Pancreatic cancer is lower in people who eat lots of dietary fiber and vitamin C.[6]

We are only beginning to understand the protective benefits of fruit and vegetable consumption.

For instance, two researchers at Johns Hopkins University in Baltimore recently isolated a single chemical—sulphoraphane—that has "remarkable tumor-fighting ability." It is present in most green and yellow vegetables. Sulphoraphane encourages protective enzymes that work to prevent cancer. There is a strong likelihood that this is only one of several (or many) such compounds.[7]

The Journal of the National Cancer Institute in June, 1990, detailed work done at the Institute for Hormone Research in New York City. "We have identified the compound in cruciferous vegetables which appears to act in a way which should be protective," said doctors Leon Bradlow and Jon Michnovicz. The compound is known as indole-3-carbinol. Their study found that the compound speeded a particular process of estrogen metabolization, thus theoretically reducing the threat of breast cancer. They have since found it reduced cancer risk in laboratory mice. They have begun the lengthy process of documenting whether the chemical actually works in humans.

It is this pattern of protection from fruits and vegetable consumption which has led the National Research Council to recommend that Americans double their current average consumption of fruits and vegetables. The National Institutes of Health, dieticians, and most other health professionals support that recommendation. (Incidentally, there is no evidence that consuming chemically synthesized vitamin supplements has the same protective impact.)

Only about 9 percent of Americans currently eat the five full servings of fruits and vegetables a day recommended for optimum health. This raises an interesting point. If only 9 percent of the people eat enough fruits and vegetables when they are relatively cheap and attractive, how many of us would eat enough if they were twice as expensive and shabby-looking? What, in sum, is the public health impact of eco-activists frightening consumers about nonorganic fruits and vegetables?

REALITY:

Green and Yellow Veggies Protect Against Blindness

"Two new studies offer tantalizing evidence that chemicals in green and yellow vegetables may protect against heart disease and the most common cause of blindness in the elderly. . . . The compounds, called carotenoids, are the colorful pigments that make squash yellow and spinach green. . . . Nonsmokers with the highest blood levels of carotenoids had 70 percent fewer heart attacks than nonsmokers with the lowest levels. . . . "People who consumed the most dark green, leafy vegetables were 43 percent less likely to have developed [age-related macular degeneration] . . . . AMD causes vision loss in an estimated 13.1 Americans and accounts for up to a third of the 900,000 U.S. cases of blindness. . . ."

From an Associated Press story published in the *Washington Times*, November 9, 1994, p. A4

MORE REALITY:

"In separate studies, scientists in the Netherlands have found that apples are a great source of flavonoids—a naturally occurring chemical in many fruits, vegetables and beverages—which may reduce the risk of death from coronary heart disease and inhibit the development of certain cancers. The October 23 issue of *The Lancet* reports on a five-year study of 805 men, ages 65 to 84. The . . . men who consumed the highest amounts of flavonoids suffered about half as many heart attacks as those who consumed the lowest amounts."

The Packer, January 15, 1994, p. 4A

Bottom-Line Reality: Your mother and grandmother were right—carrots are good for your eyes and "an apple a day . . ."

How Much More Costly Without Pesticides?

A group of horticultural researchers associated with Texas A&M University did a recent study of the impact of giving up pesticides for fruit and vegetable production in the United States. They

Figure 4.1

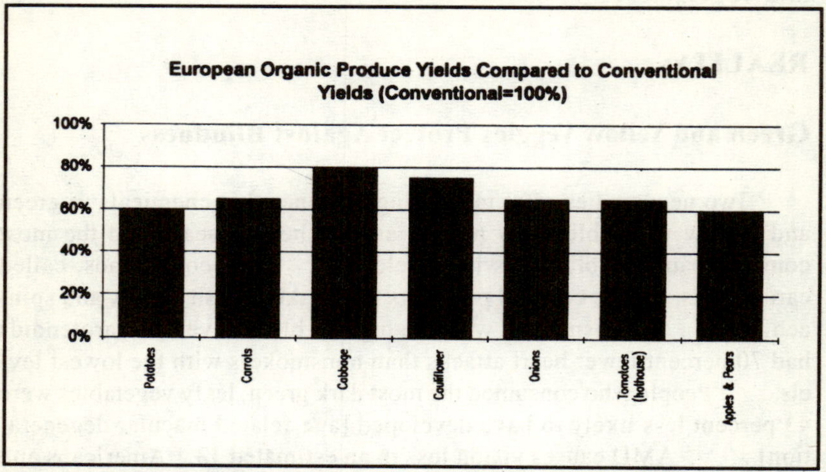

Source: *Organic Farming: Summary of Findings from a Study of Seven European Countries by the Landell Mills Research Group,* copyright European Crop Protection Association, 1992

Figure 4.2

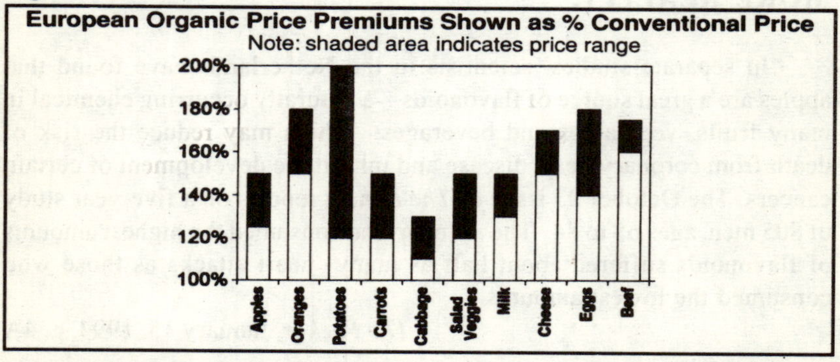

Source: *Organic farming: Summary of Findings from a Study of Seven European Countries by the Landell Mills Research Group,* copyright European Crop Protection Association, 1992

compared the organic alternative with the costs, yields and practices for commercial producers (most of whom already use various forms of integrated pest management).[8]

They concluded that yields for both fruits and vegetables would drop substantially without pesticides—forcing major increases in cost.

Organic vs. Mainstream Produce Prices in Washington, D.C., June 1994

Fruits	*Mainstream Price*	*Organic Price*	*Premium*
oranges	.45 each	.45 each	0%
apples (Red Delicious)	.50/lb	1.35/lb	170%
lemons	3/.99	1.59/lb	60%
pears	.99/lb	.99 each	110%
limes	3/.89	2.89/lb	225%
Vegetables			
carrots	1.49/lb	1.67/bunch	100%
spinach	.59/lb	1.99/bunch	400%
broccoli	.79/lb	1.59/lb	100%
white potatoes	.45/lb	.99/lb	100%
yellow onions	.45/lb	1.29/lb	185%
eggplant	1.29 each	1.95/lb	100%
celery	.99 each	.99 each	0%
mushrooms	1.99/lb	3.75/lb	88%

Source: Price comparisons by the author on the same day in June 1994 at two stores a few blocks apart in northwest Washington, D.C.

Potato yields would drop "only" 50 percent. The yields of other vegetables were estimated to fall even more sharply, with sweet corn hit the hardest—a 78 percent reduction. Estimated losses in fruit yields ranged from oranges (down 55 percent) to apples (nearly 100 percent losses in commercial orchards). Fresh-market fruits and vegetables would likely suffer even bigger yield losses than processed vegetables.

The experts say a number of famous fruit and vegetable sources would apparently be knocked out completely, including Maine potatoes, Georgia peaches, Washington and Michigan apples, and California grapes. They also believe we would lose the winter production of Florida's tomatoes and sweet corn, making it difficult for the

East Coast to keep up its cold-weather fruit and vegetable consumption. (We might have to rely on imports, though there is not as yet any organic fruit and vegetable import trade.)

The Texas A&M group's study is supported by an analysis of the impact of *banning just soil fumigants*, done by the U.S. Department of Agriculture's Economic Research Service (ERS). ERS researchers estimated that the loss of soil fumigants would sharply reduce yields of potatoes and tomatoes, at a cost of perhaps $3 billion to consumers in the short term. (Over time, new organic producers would presumably enter production, somewhat easing the price increases.)[9]

These estimates of commercial yield losses are hypothetical, but they are backed by even stronger real-world evidence. Studies in seven European countries found yields on organic fruits and vegetables averaged far less than commercially grown produce (See Fig. 4.1.)

Equally important, the prices of organic fruits and vegetables are virtually always far higher than the prices of commercially grown fruits and vegetables.(See Fig. 4.2)

Prices for fruits and vegetables are notoriously volatile, but organic produce seems to be priced significantly higher than mainstream produce week in and year out. If it were not more expensive to grow organic produce, the markets would be flooded with organic items with these high price premiums.

Pesticides Aren't Even the Problem

For 30 years, Americans have worried about being poisoned by pesticide residues. Billions of dollars have been spent on studying pesticide risks. Billions more have been spent on costly organic foods and in trying to find better "organic" pest controls. Yet we are still looking for the first human victim of pesticide residues.

The safety of farm chemicals was convincingly demonstrated by Drs. Bruce Ames and Lois Gold at the University of California–Berkeley. They are the scientists who *developed* the cancer risk scale used around the world—appropriately named the Ames-Gold Scale. When both natural and man-made compounds were rat-tested at high dosage levels, not much difference was found.

That means we consume only one ten-thousandth as much cancer risk in the form of pesticide residues as we do in the form of natural carcinogens in our food.[10]

The low risk of pesticide residues is also endorsed by Dr. Robert Scheuplein, the senior cancer expert in the Food and Drug Administration's Food Safety Center. He told the American Association for the Advancement of Science in 1989 that while natural carcinogens may cause an estimated 38,000 of the nation's 500,000 annual cancer deaths, the estimate on cancer deaths from pesticide residues is less than 40.[11] Privately, Dr. Scheuplein says he doubts that anyone has ever developed a fatal cancer from pesticide residues.[12]

More than ten years ago, Congress asked two of the world's top cancer experts (Sir Richard Doll and Dr. Robert Peto of Oxford University) to evaluate U.S. environmental cancer risks. They studied American cancer deaths from 1933 to 1978—and concluded that 98–99 percent of our cancer deaths are due to (1) smoking (2) our own genetics, and (3) bad diet choices, like too much fat and too little fiber.[13]

Who believes the Doll and Peto results? Virtually every regulatory body and medical association. Including the American Institute of Food Technologists and 14 scientific societies representing over 100,000 microbiologists, toxicologists, veterinarians, and food scientists.[14]

The Environmental Protection Agency, too, confirms the Doll and Peto results. It says so in an EPA publication called *Unfinished Business:A Comparative Assessment of Environmental Problems*, published in 1987.[15]

Doll and Peto wrote, "The occurrence of pesticides as dietary pollutants seems unimportant. There has been no increase in the incidence of liver tumors in developed countries since the long-lasting pesticides were introduced. Yet liver tumors are the most common form of cancer found in animal-based toxicological studies."[16]

If pesticide residues are highly toxic to humans, they should also attack our stomachs as well as our livers. But stomach cancer rates in the countries that use the highest rates of pesticides have been dropping sharply throughout the pesticide era. Stomach cancer rates in the United States have dropped 75 percent since the

1930s.[17] Americans currently get about 24,000 stomach cancer cases per year, or one case for each 11,000 residents.[18] A similar declining stomach cancer trend is found in the other developed countries like Europe and Japan—which use farm chemicals even more heavily than the United States.

Credit for the declining trend in stomach cancer goes both to refrigeration for food storage and to modern chemically assisted farming; both help reduce the molds, toxins, and other pests in the food we consume.

No Rise in Nonsmoking Cancer

None of the nonsmoking cancer statistics have shown any increase in any of the countries where pesticides have become much

Human Exposure/Rodent Potency Index (HERP)

Hazard Index	*Daily Human Exposure*	*Risky Compound Involved*
16.0	1 sleeping pill	Phenobarbital
4.7	wine (250 ml)	Ethyl alcohol
0.3	lettuce (1/8 head)	Caffeic acid
0.1	apple	Caffeic acid
0.07	brown mustard	Allyl isothiocyanate
0.04	orange juice	d-Limonene
0.03	peanut butter	Aflatoxin
0.005	coffee (1 cup)	Furfural
0.002	Alar in one glass of apple juice (1988)	UDMH
0.001	tap water	Chlorination
0.0006	mushroom	hydrazines
0.000001	lindane	pesticide
0.000000006	captan	pesticide

Source: Gold et al., "Rodent Carcinogens: Setting Priorities," *Science*, vol. 258, October 9, 1992.[19]

more heavily used. In fact, there has been no increase in age-adjusted cancer rates—except for tobacco-related cancers. Tobacco doesn't just cause lung cancer; cancers of the pancreas, bladder, mouth, and uterus/cervix are also associated with tobacco.

The sole exception is male prostate cancer—the highest risk for prostate cancer in the world is among black Americans "for reasons not currently known."

I got a very hostile-but-chatty letter from Indianapolis. The lady credited organic foods with saving her life twice, from both ovarian and breast cancers. I had always thought that the organic foods were supposed to *prevent* the cancers, and that once the cancers appeared, the doctors were supposed to get credit if you survived. But this woman's faith in organic food and her fear of man-made chemicals were both unbounded.

Ann Misch, author of "Assessing Environmental Health Risks" in The Worldwatch Institute's *State of the World 1994,* tries to make us agonize over cancer rates that aren't rising:

> In industrial countries, cancer causes 20 percent of all deaths. . . Explanations for the higher rates of cancer in industrial countries often invoke the vast differences in diet, smoking habits and methods of preserving food. Compared with these factors, pollution may play a smaller part. Nonetheless, the role of industrial pollutants in cancer is not negligible.[20]

But the big difference between the industrial countries and the developing world is that so many Third World residents don't *live* long enough to die from cancer. They die instead from infectious and parasitic diseases, at earlier ages. (The World Bank notes that the residents of the least-developed countries now have a life expectancy of only 55 years at birth, while those in the industrial countries have a life expectancy of 77 years.)

INTERESTING REALITY:

"I told my urban caller that we could, indeed, look at pesticides as poisons. But then we must also look at the medicines we take as poisons. We take them because they kill things, like bacteria. . . . a little medicine can do enormous good . . . too much of the medicine can be dangerous. In pesticides, fortunately, the safety factors are several thousandfold, so the dangers are tiny."

Orion Samuelson, syndicated farm broadcaster for WGN-TV and WGN radio in Chicago, before theAmerican Crop Protection Association, White Sulphur Springs, West Virginia, September 26, 1994

MYTHMAKER GOES OVER THE FALLS:

"One of the most famous toxic disasters was Love Canal . . . Hooker Chemical dumped 40,000 metric tons of toxic chemicals, including dioxin, lindane and arsenic trichloride, into Love Canal, which emptied into the Niagara River, adjacent to one of the world's greatest natural wonders—Niagara Falls. The company later filled in the canal and donated the site for the construction of an elementary school."

Helen Caldicott, *If You Love This Planet: A Plan to Heal the Earth*, W. W. Norton, New York, 1992

Reality Comment: In fact, studies have found *no* health impacts from living on the Love Canal. Nor did Hooker donate the site for a school; it was taken by the school board under condemnation proceedings over Hooker's objections. The homes near Love Canal have now been reoccupied and the previous occupants show no elevated health risks.

PROFITABLE MYTH?

". . . Research shows that when people detoxify their bodies, ridding themselves of these industrial pollutants and pesticides, their IQ usually rises." (pp. 313–14).

"For the past five years I have been documenting . . . some of the world's most revolutionary research on detoxifying toxic chemicals that have accumulated in the human body. The program is being carried out at a clinic called HealthMed. . . . Thousands of individuals have gone through the HealthMed program with successful reductions in toxic levels and great improvements in overall health and spirit. . . . And it would be wonderful if everybody who needs to be purified could afford the approximately $3,000 the program costs. . . . If you suspect that you're really chemically ill, if you feel sick and disabled, I recommend that you contact HealthMed. . ." (pp. 295–298).

Excerpts from *Diet for a Poisoned Planet*, by David Steinman, New York, Harmony Books, 1990

October 22, 1994

Dear Dennis:

One of the principal problems we face continuously is how to get the public to understand that trace amounts of pesticides are seldom, probably never, toxic. Toxicologists explain it this way; it is the dose of a material that determines the poison. We accept this fact every day with medicines and food ingredients. A low dose of a sleeping pill is useful and harmless; an overdose of the same pill is hazardous and sometimes fatal. A sprinkle of table salt is healthy; too much salt has a number of negative effects on the human system.

Toxicity and hazard are *not* the same thing. *Toxicity* is the inherent capacity of a compound to be poisonous at some level. *Hazard* (degree of danger) is linked to *exposure* and toxicity; i.e., it is the chance that the material will or will not elicit a negative response. If someone is not exposed, there is no hazard. If the person is exposed to trace amounts, the chances are that there is still no hazard.

In general, it is a dose/weight relationship. The heavier the animal (or man) the less chance of a negative reaction. Toxicity is expressed as a quantity of the substance as compared to the weight of the animal. There is an amount (a threshold) for toxic response, and any amount below that level is harmless. In general, we are dealing with insignificant quantities that are not capable of being injurious. We are talking about parts per million, or perhaps even parts per billion (ppb), or parts per trillion (ppt). There is a threshold for response and any amount below that level is harmless.

When speaking of pesticides, it is a fallacy to lump them together . . . Every chemical is different, and thus the action of each chemical is unique. We would be better off and the public reassured if, like medicine, we would always identify what particular pesticide we are talking about.

Consider this: Pesticides undergo more extensive safety testing than any other product! Some pesticides that in the past caused safety concerns (e.g., ethyl parathion) have been replaced by safer, more readily biodegradable compounds. It is an insult to the public to talk about banning "pesticides" or fearing to eat vegetables that are not organically grown.

(Letter from Dr. John Osmun, former chairman of the Entomology Department at Purdue University and senior official in the EPA's Office of Pesticides, after reviewing the draft manuscript for this book.)

Here, quoted from *Cancer Facts and Figures 1994*, by the American Cancer Society, is the Society's listing of the major cancer types and their risk factors:[21]

Lung cancer: Cigarette smoking; exposure to certain industrial substances, such as arsenic, certain organic chemicals and asbestos, particularly for persons who smoke.

Colon and Rectum Cancer: Personal or family history. High-fat and/or low-fiber diet may be associated with increased risk.

Breast Cancer: Risk increases with age; personal or family history; late age at menopause; never had children or late age at first live birth; higher education and socioeconomic status. International variability in rates correlates with variations in diet, especially fat intake.

Prostate Cancer: Risk increases with age.

Pancreas Cancer: Incidence is more than twice as high for smokers; suggested associations with diabetes, cirrhosis.

Uterus (Cervix) Cancer: Early age at first intercourse, multiple sex partners, cigarette smoking.

Uterus (Endometrial) Cancer: Early puberty, late menopause, history of infertility, failure to ovulate.

Cancer in Children: Rare. Mortality rates have declined 60 percent since 1950. No risk factors cited.

Leukemia: Cause of most cases unknown. Persons with Down's syndrome and certain other genetic abnormalities have high incidence. Also linked to certain chemicals such as benzene. Certain forms caused by a retrovirus.

Lymphoma: Risk factors largely unknown, but in part involve reduced immune function and exposure to certain infectious agents. HIV virus. Other possible risk factors include exposures to *herbicides*, industrial solvents and vinyl chloride. [emphasis added]

Skin Cancer: Fair complexion. Occupational exposure to coal tar, pitch, creosote, arsenic compounds or radium.

Ovary Cancer: Increases with age. Increased risk among women who have never borne children

Bladder Cancer: Smokers incur twice the risk of nonsmokers. People living in urban areas and workers exposed to dye, rubber or leather are also at higher risk.

Oral Cancer: Cigarettes, cigars, pipe smoking, chewing tobacco and excess use of alcohol.

Low Pesticide Residues in Food

Food surveys in the First World—where most of the pesticides are used—show consistently low levels of consumer exposure. The latest large-scale monitoring studies in Germany, Ireland, Italy, Sweden, Switzerland, the UK, and the United States examined about 50,000 samples.

Most of the samples had no residues at all.[22] Only about 2 percent of the samples had residues above the maximum residue limit. Most of these "violations" were compounds not approved for that particular food item. These unauthorized residues were found more often in imported produce, but it is also true that pest control is very site-specific. Producers in Honduras may face a different set of pests on snow peas than American growers—and if the American snow pea growers haven't asked for approval on a pesticide, it raises the red flag for the U.S. inspector.

A 1990 Australian survey concluded "The fact that intakes are so much less than the [Acceptable Daily Intake] indicates that there is little health risk from pesticide ingestion in foodstuffs." A Swiss survey concluded, "Intake of most pesticides is one hundredth part or less of the ADI. Therefore according to the present toxicological knowledge the residues found do not pose a risk to the health of the consumer."

Among the studies, the only one to raise concern was in India, where a 1980–81 survey showed DDT intake was 20 percent of the ADI. (No human health effect has been linked to DDT, however, in India or anywhere else.)

In the United States, the FDA's quarterly Total Diet studies cover more than 200 food items, table-ready, and analyzed for more than 200 compounds with methods five to ten times more sensitive than the ones used in regulatory monitoring.

- No pesticide residues were found at all on 65 percent of the domestically produced food samples. Less than 1 percent had residues over the EPA tolerances, and less than 1 percent had residues for which there was no established tolerance for that particular commodity.

- On import sampling, FDA found no pesticide residues on 66 percent of the foods, less than 1 percent over tolerance, and 3 percent with residues for which there was no tolerance on that commodity.

- Traces of some 50 compounds have been found each year—but *only in the case of dieldrin did dietary intake exceed 1 percent of the Acceptable Daily Intake*. (Dieldrin residues have ranged as high as 3 percent of the ADI.)

EPA Overstates Pesticide Cancer Risk a Hundredfold

The Food and Drug Administration's annual Total Diet Study also shows the EPA's theoretical estimates overstate *everybody's* pesticide exposure by a factor of 100! The EPA, for whatever reason, assumes that all pesticides are used to the legal limit on every crop. They aren't. Not even close.

As we have just noted, the annual Food and Drug Administration food surveys have found that more than 60 percent of our foods have *no* chemical residues when ready for eating. That's because farmers don't spend cash for pesticides except where they're needed, and because many foods are washed, peeled, cooked, or otherwise modified after they leave the field. Without the "table-ready" stipulation, any residues on an orange peel, for example, would be included in "consumption" though we rarely eat the bitter orange peel.

A recent study by Dr. Sandra Archibald of the prestigious Stanford Food Research Institute and Dr.Carl Winter of the University of California–Riverside calculated how much the EPA assumption overstates the real food risk of pesticide residues.[23]

According to the study:

- EPA risk assumptions on tomatoes overstate real risk by 2,600 times.

- The risk on apples is overstated by nearly 21,000 times.

- Lettuce risks are overstated by 300 times.

Overall, the impact of reality on EPA assumptions is to cut our cancer risks from a supposed 1 in 10,000 to as little as 1 in a million.

Dr. Kenneth Olden is director of the National Institute of Environmental Health Sciences, which directs animal toxicology studies for the federal government. Dr. Olden questions the billions of dollars spent each year regulating "carcinogenic" chemicals that may pose little health or environmental risk.

Indeed, other scientists are questioning regulations governing such chemicals as dioxin, DDT, saccharin, and cyclamates, which have produced cancers in some lab animal tests but which experts believe are not harmful to humans at low doses. NIEHS officials estimate that between one-third and two-thirds of substances classed as carcinogenic because of MTD tests in rodents would be benign in humans at normal exposures.[24]

MEAT-AND-PESTICIDE MYTH:

"Cattle spend long hours at the feed troughs. . . . The feed is saturated with herbicides. Today, 80 percent of all the herbicides used in the United States are sprayed on corn and soybeans, which are used primarily as feed for cattle and other livestock. When consumed by the animals, the pesticides accumulate in their bodies. The pesticides are then passed along to the consumer in the finished cuts of beef."

Jeremy Rifkin, *Beyond Beef*, New York,
Penguin Books, 1992, pp. 12–13

What About Meat and Pesticides?

What about the safety of our meat supplies in an era of chemical pest controls for feed crops and sophisticated medicines for livestock and poultry?

Jeremy Rifkin demonstrates how far the activists are willing to depart from the truth to frighten the public.

In 1993, a team of Colorado State University researchers presented a study at the American Society of Animal Science national meeting in Spokane, Washington. The team had tested beef tissues from seven packing plants in Colorado, Kansas, Nebraska, and Texas. The beef came from a variety of sources: conventional producers, "natural" beef production (with no medicines used to prevent or treat illness), organic producers (who use neither medicines on their cows nor pesticides on the feed crops), older cull cows, and chronically ill cattle that were included only for purposes of the study.

The samples were tested for 25 different pesticides, including DDT. The researchers found *no* pesticide residues in any of the beef.[25] "Lead and cadmium were the only residues for which we detected any actual quantities," team spokesperson Julie Sherbeck said. "These were found in such small amounts as to be nearly undetectable. . . . U.S. beef is unlikely to contain levels of pesticides or any of the compounds or elements considered to be potential hazards to the public health."[26]

Why is Rifkin so wrong? There are a number of reasons.

Herbicides aren't actually sprayed on crops. They are mostly used on the bare ground where the crops *will* be grown (preemergence sprays). Or, in the no-till farming system, the herbicides are sprayed on a grass sod before the crop is planted. Even if herbicides are needed once the crops have emerged, they are sprayed or "banded" *between* the rows of crops. (Herbicide that goes onto the crop plant is counterproductive; the farmer isn't trying to kill the crop plant.)

Second, herbicides are rarely used in a field after the ears of corn or the soybean pods have begun to develop. They are used much earlier, when competition from the weeds would still cut yields.

Third, neither the ears of corn nor the soybean pods are ever exposed to sprays. The ear of corn is protected by the leafy husks that cover it. The soybeans grow in pods like peas. Both are protected until harvest.

Fourth, herbicides that do get into the crop plants do not concentrate and lurk in the ears or pods. Virtually all are metabolized rapidly by the plants. Nor do cattle "biomagnify" farm chemicals and accumulate them in their meat.

Fifth, no herbicides used on corn or soybeans are a significant danger to humans in the tiny amounts to which consumers are exposed.

The biggest of the small dangers is the small amount of atrazine that periodically shows up in stream-fed reservoirs for city water supplies. Few consumers are exposed to atrazine levels above recommended lifetime doses on even a seasonal basis—especially since the EPA raised the safety rating of atrazine sevenfold in 1993.

Finally, the medicines used on animals must be approved for consumer safety as well as the safety of the animals, and the medicines must be discontinued long enough before slaughter for them to disappear from the tissues of the meat. Obviously, farmers and ranchers are following the recommendations.

The biggest current known risk in beef consumption is E. coli bacteria, which are natural and ubiquitous in our world and our homes. The potential for this bacteria to get into our meat is the reason we should cook hamburger thoroughly instead of eating it rare. (Rare steak is OK.)

Irradiation of our beef would kill this bacteria, and scientists say irradiation is a safe process to use. But irradiation has been opposed by a number of activists—including the same Jeremy Rifkin who claims to be concerned about E. coli in the meat! (Rifkin's solution is for everyone to stop eating hamburger.)

The biggest risk in eating poultry is the danger of salmonella bacteria, not of pesticide or drug residues. Again, irradiation of poultry meat would cost-effectively eliminate even the salmonella threat and again, Rifkin opposes irradiation.

SCARE TACTICS:

"Dr. Daly believes . . . the level of a toxic chemical needed to produce significant behavioral change may be far less than the level needed to produce physical changes."

Pesticides and You, National Coalition
Against the Misuse of Pesticides, Winter 1992–93[27]

". . . [B]ut more dangerous still is the *fact* that toxic chemicals such as DDT act as neurotoxins, impairing the affected person's mental ability. . . . Even at the very low levels that are now present in the bodies of millions of Americans, chemicals such as DDT have a profound effect on mental clarity and on the ability to think, comprehend and react to outside stimuli." [emphasis added]

David Steinman, *Diet for a Poisoned Planet,* New York, Harmon Books, 1990, pp. 313–14

". . . [T]he [National Research Council] found that virtually no testing had been done on the potential for neurobehavioral damage, birth defects, or toxic effects that might span several generations by passing from parents to offspring."

Ann Misch, "Assessing Environmental Health Risks," *State of the World 1994*[28]

This latest round of chemical fearmongering is now about noncancer risks, about sublethal "behavioral modification" and impacts a generation or even two generations removed from the exposure. These kinds of claims are wonderful for the scaremongers because they can never be disproved.

Here is another example of a pesticide myth in action:

"Last year, the agency banned pesticide spraying in its more than 40 office buildings. . . . At first, the pesticide ban was widely viewed as a temporary, possibly superfluous precaution. . . . With a little checking, one woman correlated unexplained sicknesses she and her young daughter periodically experienced with the timing of pesticide applications on her floor of Building 8. . . . Soon, 50 cases of neurological problems had been identified among building occupants. . . . Eventually a number of these sick workers joined our Multiple Chemical Sensitivity Support Group. This group is actively petitioning for "reasonable accommodation" for their disability (chemical sensitivity) under the Americans with Disabilities Act and state human rights laws."

"Stopping Pesticide Abuse in Public Facilities," *Pesticides and You: News from the National Coalition Against the Misuse of Pesticides*, Winter 1992–3, pp. 15–17.

In this instance, the "research" included people remembering, months later, the dates of "stomach viruses" and their timing with office pesticide applications. That is not research. It is legal and environmental blackmail.

". . . John Graham of the Center for Risk Analysis at the Harvard School of Public Health: 'Less than 5 percent of human cancer can be traced to causes that are within the jurisdiction of the U.S. environmental

Protection Agency.' Yet according to a study by the Center for Media and Public Affairs, news stories cite man-made chemicals as a cause of cancer more than any other single cause."

Brent Bozell III, "When the Media Looks at Risk," *Washington Times*, October 17, 1994, p. A17

Finding the Real Causes of Cancer

Researchers have been increasingly finding *genetic* causes for much of our nonsmoking cancer. Recently, for example, researchers at Johns Hopkins and the University of Helsinki discovered a human gene that—when defective—apparently accounts for a high proportion of our colon cancers. The gene is carried by one in 200 people, and virtually all of them will develop colon cancer. Sporadic mutations of the same gene apparently also cause many other types of cancer including stomach, ovarian, and uterine cancer.

Researchers had previously identified a large number of other genes, called oncogenes, that play a role in initiating cancer when they are defective. The newly discovered gene, called MSH2, goes further. It effectively creates other dangerous oncogenes by allowing random mutations to accumulate in cells. Eventually, these mutations can produce tumors.[29, 30]

Also, the very process of aging plays a major role in the onset of cancer, which is fundamentally a disease of old age.

"Oxidant by-products of normal (human) metabolism cause extensive damage to DNA, protein and lipid. We argue that this damage (the same as that produced by radiation) is a major contributor to aging and to degenerative diseases of aging such as cancer, cardiovascular disease, immune system decline, brain dysfunction, and cataracts. Antioxidant defenses against this damage include ascorbate, tocopherol, and carotenoids. Dietary fruits and vegetables are the principal source of ascorbate and carotenoids and are one source of tocopherol. Low dietary intake of fruits and vegetables doubles the risk of most types of cancer as compared to high intake and also markedly increases the risk of heart disease and cataracts. . . ."

Bruce Ames, Mark Shigenaga, and Tory Hagen, Division of Biochemistry and Molecular Biology, University of California–Berkeley,[31]

Who Is Dr. Bruce Ames?

When it comes to controversy, Dr. Bruce N. Ames, . . . is no shrinking violet. The man who invented the leading laboratory test to screen chemicals for their ability to damage genes has years of solid science and the accolades of countless colleagues behind him when he makes such provocative and socially unpopular statements as these:

"I think pesticides lower the cancer rate."

"Pollution seems to me to be mostly a red herring as a cause of cancer."

"Environmentalists are forever issuing scare reports based on very shallow science."

"Standard animal cancer tests done with high doses are practically useless for predicting a chemical's risk to humans."

"Nearly all the polluted wells in the United States seem less of a hazard than chlorinated tap water."

"Ninety-nine point nine percent of the toxic chemicals we're exposed to are completely natural—you consume about 50 toxic chemicals whenever you eat a plant."

"Elimination of cancer is not in the cards, even if we get rid of every external factor."

"Nearly half of all natural chemicals tested, like half of synthetic chemicals, are carcinogenic in rodents when given at high doses."

"We're shooting ourselves in the foot with environmental regulations that cost over 2 percent of the GNP, much of it to regulate trivia."

Coming as they do from a highly respected scientist who does not do consulting work for industry, such remarks are especially irritating to those who believe that modern industry has touched off an epidemic of cancer and birth defects by contaminating the air, water, soil, and food with toxic chemicals.

Dr. Ames, a biochemist and molecular biologist at the University of California at Berkeley, where he directs the National Institute of Environmental Health Sciences Center, is a member of the National Academy of Sciences, and is the recipient of an outstanding investigator grant from the National Cancer Institute and of many highly prestigious awards for excellence in research. His hundreds of technical publications, many in an arena rife with public, political and scientific controversy and punctuated with passionate emotion, have made him one of the two dozen most often cited scientists in the world.

(Quoted from "Strong Views on Origins of Cancer" by Jane Brod, *New York Times*, July 5, 1995.)

If this trio of top cancer experts is correct, cancer is the result of oxidation—part of the process of life. The damage caused by oxidation builds up in our cells, gradually impairing their function. It's almost as though we were exposing ourselves to continual small doses of atomic radiation and our best defense is to consume antioxidants. It's not glamorous or dramatic. There's no blame to lay. No lawsuits will be filed against big, bad corporations. We know that the public responds much more dramatically when there's blame—as in a plane crash—than when it's their own behavior at fault—as in many car accidents.

On the other hand, it's not all that much fun to spend your life as a chemophobe, terrorized by cabbage, aluminum cookware, pressure-treated lumber and the cotton ticking in your mattress.

Notes

1. *Issues in Food Safety*, Fresh Produce Council, Los Angeles, May 1992, p. 2.

2. *Mycotoxins: Economic and Health Risks*, Report No. 116, Council for Agricultural Science and Technology, Ames, Iowa, November 1989.

3. *Mycotoxins: Economic and Health Risks*, op. cit.

4. Ames, Shigenaga, and Hagen, "Oxidants, Antioxidants and the Degenerative Diseases of Aging," *Proceedings of the National Academy of Science*, vol. 90, 1993, pp. 7915–7922.

5. Block, Patterson, and Subar, "Fruit, Vegetables and Cancer Prevention," *Nutrition and Cancer* 18, 1992, pp. 1–29.

6. Miller, Dr. Anthony, "Do Pesticide Scares Raise Cancer Rates?" *Global Food Progress*, Hudson Institute, Indianapolis, 1991, pp.148–154.

7. Hooker, Lisa, "Molecules for Medicine," *Johns Hopkins Magazine*, June 1992, p. 21.

8. Knutson, R.D., et al., *Economic Impacts of Reduced Pesticide Use on Fruits and Vegetables*, American Farm Bureau Research Foundation, Chicago, IL, September 1993.

9. J.R. Barse and W. L. Ferguson, "Banning Soil Fumigants: What Cost?" *Agricultural Outlook*, U.S. Department of Agriculture, Washington, D.C., June 1989.

10. Ames, "Science and the Environment: Facts vs. Phantoms," *Priorities*, Winter 1992, American Council on Science and Health, New York.

11. Warren T. Brookes, "Pesticide Phobia a Dangerous Health Threat," *Detroit News*, April 16, 1990, p. A7.

12. Personal communication with Dr. Robert Scheuplein, 1990. See also Dr. Robert Scheuplein, "The Real Cancer Risks in Our Food," *Global Food Progress*, Hudson Institute, Indianapolis, 1991, pp. 155–163.

13. Doll and Peto, *The Causes of Cancer*, Oxford University Press, 1981.

14. *Assessing the Optimal System for Ensuring Food Safety: A Scientific Consensus,* Institute of Food Technologists, 1991.

15. *Unfinished Business: A Comparative Assessment of Environmental Problems,* Environmental Protection Agency, Washington, D.C. 1987.

16. Doll and Peto, *The Causes of Cancer, op. cit.*

17. *Cancer of the Stomach*, National Cancer Institute, U.S.Department of Health and Human Services, NIH Publications 88–2978.

18. Data from U.S. Centers for Disease Control, 1993.

19. Gold, et al, "Rodent Carcinogens: Setting Priorities," *Science*, vol. 258, October 9, 1992, pp. 261-265.

20. Ann Misch, "Assessing Environmental Health Risks," *State of the World 1994*, Worldwatch Institute, Washington, D.C., pp. 119–120.

21. *Cancer Facts & Figures—1994,* American Cancer Society, Atlanta, Georgia.

22. Review conducted for the European Crop Protection Association by Dr. Helmut Frehse, former head of the Institute of Residue Analysis, Crop Protection Division, Bayer, AG, copyright ECPA, Brussels.

23. Warren T. Brookes, "Overstating Pesticide Risks by 2,600 to 21,000 Times?" *Detroit News*, February 26, 1990, p. A9.

24. *E, The Environmental Magazine*, vol. 5, no. 1, February 1994, p. 14.

25. Smith, Sofos, Morgan, Aaronson, Clayton, Jones, Tatum, and Schmidt, "Ensuring the Safety of the Meat Supply," paper presented at the Reciprocal Meat Conference of the American Meat Science Association, University Park, PA, June 13, 1994.

26. *Farm Times*, December 1993, p. 6B.

27. "Toxic Chemicals and Behavior," *Pesticides and You*, vol. 12, no. 5, Winter 1992–93, National Coalition Against the Misuse of Pesticides, pp. 6–7.

28. Misch, op. cit.. pp.120–121.

29. Michael Waldholz, "Cancer Gene Is Pinpointed in the Healthy," *Wall Street Journal*, December 30, 1993, p. B1.

30. *Cell*, December 17, 1993.

31. Ames, et al., "Oxidants, Antioxidants and the Degenerative Diseases of Aging" op. cit.

5

Children, Farmers, and Pesticides

MYTHMAKERS SAY:

"One of the profoundest shocks we have experienced during this chemical age is the realization that the womb is not the secure little niche we had imagined it to be. . . . Doctors for some time now have warned women not to spray their rooms with pesticides during pregnancies."

Frank Graham Jr., *Since Silent Spring*,
Boston, Houghton Mifflin, 1970, p. 149

"What we don't know are the long-term chronic effects of pesticide exposure on the general population. . . . And in particular—on children.

Bill Moyers, *Frontline*. PBS-TV, "In Our Children's Food",
March 30, 1993

"Our children are inheriting a dangerous world, dying from pollution and overpopulation, wired up by transnational corporations . . . like a ticking time bomb ready to blow up at any minute from a nuclear war. . . "

Helen Caldicott, *If You Love This Planet*,
New York, W. W. Norton, 1992

REALITY SAYS:

"Insufficient fruit and vegetable consumption increases the rate of most types of cancer about twofold. . . . Unfortunately, a high percentage of the American population is eating insufficient fruit and vegetables, particularly the poor and their children. . . . Synthetic pesticides have been a major contributor to health in this century by decreasing the price of fruits and vegetables. . . ."

Dr. Bruce Ames, "Comments on the National Academy Report, *Pesticides in the Diets of Infants and Children*," prepared for the California Department of Pesticide Regulation, September 7, 1993

"The Pacific Northwest's tragic outbreak of hamburger-related food poisoning (of children) . . . underscores the silliness of a current flap over agricultural pesticides. . . . EPA Administrator Carol Browner recently requested public comment on the possibility that 35 pesticides could be banned. . . . The pesticides may be carcinogenic to some animals, but EPA does not consider them carcinogenic to humans. Furthermore . . . the concentrations in which they've been found are so infinitesimal that the issue isn't health. . . ."

Editorial, "Undue Fear of Pesticides Detracts from True Threat," *Spokesman Review*, Spokane, Washington, February 8, 1993

Early in 1993, Bill Moyers hosted a *Frontline* program on PBS-TV titled "In Our Children's Food." The gist of the program was that farmers and the regulatory authorities had somehow failed to realize that *children* will eat the food produced with the help of pesticides. The program implied that pesticide residues are a ticking time bomb for our kids.

Are children really at greater risk than adults from pesticides?

They may be more at risk because of their smaller body weights. If they are, the added risk is not more than a factor of three. As we shall see, the protection factors in our regulatory system add up well into the thousands.

Let's start our inquiry into children's health by observing that cancer rates in children have fallen by more than half during the pesticide era. Childhood cancer was always rare, but it has declined by 60 percent since 1950. The American Cancer Society estimates 1,600 childhood cancer deaths in 1994.[1] Even more important, factor in the reality that children born today in the United States have a 20-year longer life expectancy than previous generations of people—who were *not* exposed to pesticide residues.

Move on to the reality that cancer is basically a degenerative disease of old age. Only a few rare types of cancers (like leukemia) attack any significant number of children. Moreover, virtually no carcinogens can trigger cancer immediately. Virtually none of the risk estimates would show *any* mortality risk without building on 70 years of lifelong exposure.

Moral: If you want your kids to live long lives, teach them not to smoke—and to eat lots more fruits and vegetables, no matter what approved pesticides have been used to grow the produce.

The National Research Council Study

In 1993, the National Research Council issued a long-awaited report: *Pesticides in the Diets of Infants and Children*. The report, which had been mandated by Congress, disappointed some observers because the authoring committee did *not* cite pesticides as a threat to children's health. In fact, the report stressed that children should eat more fruits and vegetables, even if they were grown with farm chemicals.

The NRC committee *did* attempt to raise a modest scientific point: they felt that we didn't have as much up-to-date information about exactly what American children eat today as researchers would like to have. (They were working with some data that was 10 years old.) Nor did they think we had up-to-date data on exactly how children's pesticide exposure pattern might differ from that of adults.

Whatever the merits of the committee's point about the data base, it never had a chance to be heard. Activists ambushed the NRC report. *Several days before* the report was released, activist groups and two government agencies released statements saying that the NRC report "was *expected* to say that children are at particular risk from pesticides." They used this "expectation" as justification to present their own scare report and press releases *ahead of the NRC report*.

Most of the media swallowed the bait, reporting the scaremongerings of such activist groups as The Environmental Working Group and the Natural Resources Defense Council as if they were part of the NRC. (The Environmental Working Group was a new organization. It got widely quoted—even though it refused to give out the names of its staff or their qualifications!)

Two cabinet officers also abused the NRC report to push the Clinton policy agenda: EPA administrator Carol Browner

Reprinted from the *Wall Street Journal*, April 1, 1993

Frontline Perpetuates Pesticide Myths

by Dennis T. Avery

Frontline, the Public Broadcasting System's investigative journalism show, is famous for its controversial points of view. But it's now outdone itself. In an episode titled "In Our Children's Food," which aired in most markets earlier this week, a well-meaning Bill Moyers and his PBS colleagues made recommendations that would increase our cancer and heart disease rates, increase the risk of world hunger, and plow down millions of square miles of wildlife habitat. Apparently the *Frontline* staff didn't realize that those calamities would be the result of giving up the farm chemicals it warned us against.

The show was prepared to celebrate the 30th anniversary of Rachel Carson's book, *Silent Spring*. Miss Carson blamed farm chemicals for wildlife losses that we now know were due to lost habitat and to industrial pollutants like mercury and PCBs. In her ignorance, she also feared that pesticides caused human cancer.

We now know that farm pesticide residues contain less food cancer risk than mustard and pickles or even than the environmentalists' beloved mushrooms. We now know that 99.9% of the cancer risk in our food supply come in the foods themselves. So much for the cancer risks in pesticides.

But the indictment against *Frontline* is worse than an omission of these facts. Medical practitioners across the country are telling us today that the best way to reduce both cancer and heart disease is to eat twice as many fruits and vegetables. Fruits and vegetables contain powerful chemicals that inhibit cancer. They are low in fat and high in fiber; their consumption works against heart disease.

But organic farming can't produce low-cost, attractive fruits and vegetables. It produces expensive fruits and vegetables because the insects and diseases eat most of them before they can be harvested from their organically grown gardens. The few that survive are shabby-looking, and it's hard to get kids to eat shabby-looking produce. On that basis, organic farming would produce *more* cancer, not less. . . .

Continued on next page

Biotechnology may eventually help us engineer the pest protection into plants and creatures so we won't have to spray anything anymore. But most of the ardent environmentalists say they are against biotechnology too. . . .

As evidence of farm chemical dangers, *Frontline* offers one farming town in California that for years has had an unexplained high rate of cancers. But this town is famous in medical circles because its cancer pattern is unlike any other town's. Medical studies have tried to tie the famous "McFarland Cancer Cluster" to pesticides. All have failed.

Next, Moyers cuts to a guilt-ridden California farmer whose son came down with leukemia ten years ago. He's afraid that his use of pesticides might have caused the leukemia. But farmers and farm kids have lower rates of leukemia and cancer than nonfarm kids. Where is the medical evidence to tie the California farm boy's disease to farm chemicals? The *Frontline* hosts don't tell us anything except how worried they are.

The program also ridicules a Public Health Service toxicology study which reported "There is no evidence that the small doses of pesticides that we do get are causing any harm. The only effect that can be measured . . . is the storage of one of them—DDT—in the tissues of most people. This storage has not caused any injury which we can detect."

Then Mr. Moyers crows: "DDT would be banned ten years later, just as Rachel Carson had predicted." (In the early 1970s.)

But Moyers fails to tell us that DDT was banned against the recommendation of scientists and the Environmental Protection Agency's own hearing examiner. The dozens of experts who testified at the EPA hearing overwhelmingly said DDT should keep its EPA approval because it wasn't dangerous to people or birds. The political appointee who headed EPA feared a public outcry if he concurred with the hearing examiner because so many people had read Miss Carson's book.

Is the rest of PBS' widely noted environmental reporting based on evidence this shaky?

and then Agriculture Secretary Mike Espy collaborated on an announcement that they would reduce America's future use of farm chemicals. Their statement was also issued *before* the NRC report was released. When the NRC report failed to indict pesticides as an increased risk to children, Browner and Espy were left with no rationale for their new policy—but regrettably they left the policy goal standing anyway.

MYTHMAKER:

" . . . a 1993 National Research Council report titled 'Pesticides in the Diets of Infants and Children.' The nearly 400-page book advocates more stringent pesticide laws to protect children."

"Ecohealth," *Buzzword's Earth Journal*, January/February 1994

Reality Comment: The report did *not* advocate more stringent pesticide laws to protect children. It did advocate gathering more data on children's diets and their pesticide exposures. The report also said in its opening statement:

> [Pesticide] application has improved crop yields and has increased the quantity of fresh fruits and vegetables in the diet, thereby contributing to the improvements in public health.

Apparently, when the eco-activists don't win, they simply declare victory.

MYTHBELIEVER:

A week after the National Reseach Council study was released, I got an anonymous letter.

"I was shocked when I read your recent article, 'Frontline Perpetuates Pesticide Myths' in the *Wall Street Journal*. Thank God there are people in this world who are not 'paid off' . . . by the corporations promoting the continued use of these 'killers.' It was certainly reassuring to hear this week about the study of the National Research Council, which concludes exactly the opposite of what your article stated. It's hard to realize how people like you are able to sleep at night. It certainly doesn't take much of a brain to figure out that chemicals are not good for the body. Rachel Carson ignorant? NOT!"

Overstating Pesticide Exposure—for Kids *and* Adults

Dr. Lois S. Gold, one of the country's top experts on cancer and pesticides, said even the NAS committee's call for more data was poorly justified. She said the committee should have used the data from the Food and Drug Administration's annual *Total Diet Study*.

This annual study is the only source of data on the average intake of synthetic pesticide residues from foods as they are consumed. The *Total Diet Study* is thus the most accurate assessment—by far—of real consumer pesticide exposure. It is done each year, and it already includes exactly the information the NRC panel said they needed—estimates broken out for infants, children, and adults.

The *Total Diet Study* indicates that *infants*' exposure to pesticides is roughly three times greater than that of adults. (Of course, infancy doesn't last long, and infants' and children's exposure to pesticides is already factored into the lifetime exposure estimates.)

The risk of pesticide-related cancer in children is trivial. What is *not* trivial is the pesticide scare campaign being mounted *in the name of children* that tells mothers nonorganic fruits and vegetables are dangerous. This scare campaign works directly against what should be the number one child dietary health priority—getting kids to eat lots more fruits and vegetables.

Lots of parents can't really afford the high cost of organic produce. Many children object to the appearance of organic fruits and vegetables. Without a second thought, they should be buying the attractive, reasonably priced chemically assisted fruits and vegetables that are in our markets. It's the best thing they can do for their families' health.

Cancer and Farmers

There are no known consumer health impacts from the pesticide residues in food. But what about the most-exposed populations? What about farmers, pesticide plant workers, and pesticide applicators?

We know that the pesticide risks to farmers and farm workers in the First World are low—and declining. We also know that the rates of cancer in farmers, farm workers, and their families are lower than the cancer risks of the general public.

The European Crop Protection Association recently sponsored a review of the medical research in 17 countries, covering 53 mortality studies of the "pesticide-endangered professionals."[2] They had a lower mortality rate for all causes than the general population, and their cancer mortality was not consistently different from that of the general population. Several studies of farmers showed a lower overall cancer risk.

In the first edition of this book, we reported on several questions about the farmer-related safety of pesticides, especially a suggested relationship between farmers who used phenoxyacid herbicides (such as atrazine) and soft-tissue sarcoma. For the second edition, we revisited the science on those questions under the guidance of Dr. Michael Gough, former director of the Center for Risk Management at Resources for the Future, a think tank in Washington, D.C.

***First edition quote**: "Swedish studies have firmly concluded a positive association between phenoxyacid herbicides and soft tissue sarcoma, but studies in other countries do not bear this out."*

In the late 1970s and early 1980s, a group of Swedish investigators led by Lennart Hardell reported large increases in the risks of three tumors, Hodgkin's disease, soft-tissue sarcomas, and non-Hodgkin's lymphoma, in men who were thought to have been exposed to phenoxyacetic acid herbicides. None of the associations reported by Hardell has been substantiated in the two decades since his group's original reports.

In 1994, Hardell and his colleagues[3] reported a reanalysis of the data that they had interpreted in 1981 as showing an association between herbicides and Hodgkin's disease.[4] Hodgkin's disease is not mentioned, and it is likely that the complete absence of any supporting evidence from other scientists who investigated the possible relationship influenced their decision to drop any reference to Hodgkin's disease.

First-edition quote: *"More reports are beginning to show an increased risk of lymphatic cancers, again associated mainly with phenoxyacid herbicides."*

In 1999, Hardell reported on cancer rates in herbicide-exposed workers also. He does not mention Hodgkin's disease or soft-tissue sarcomas, probably because there is no support for his original claims; however, he does report that he finds no statistically significant increase in non-Hodgkin's lymphoma associated with herbicides.[5]

First-edition quote: *"In the United States at least one research team has raised questions about a link between non-Hodgkin's lymphoma and the use of both 2,4-D and organophosphates."*

Scientists at the National Institute for Occupational Safety and Health (NIOSH) have extensively studied cancer rates in U.S. chemical industry workers who manufactured 2,4-D (2,4-dichlorophenoxyacetic acid) and 2,4,5-T (2,4,5-trichlorophenoxyacetic acid), two of the most widely used phenoxyacetic acid herbicides. The NIOSH study found no increase in Hodgkin's Disease, soft-tissue sarcomas or non-Hodgkin's lymphoma in the exposed chemical workers.[6]

The U.S. Environmental Protection Agency (EPA) has made four examinations of the animal and human data that bear on the possible carcinogenic effects of 2,4-D. The most recent one, completed in early 1997, concluded that there was no evidence to support the classification of 2,4-D as even a "possible human carcinogen."[7] 2,4,5-T is no longer manufactured in the United States; or, with the possible exception of New Zealand, in any other country. Its use was suspended in the United States because of now-discredited reports that it caused miscarriages,[8] not because of any association with any of the cancers reported by Hardell and his colleagues.

The upshot is that during two decades of intense study of possible herbicide-related cancers, several scientific papers have rung danger bells; but all turned out to be false alarms.

OFFICIAL MYTHMAKING

"EPA Restricts Pesticide Usage: Agency Targets Chemicals That May Pose Particular Health Risk to Children"

"The Environmental Protection Agency on Monday banned most uses of a pesticide applied widely for years on fruits and vegetables and tightened restrictions on another, in the first regulations intended specifically to protect children. . . . Officials said . . . [farmers] have available a variety of new, less hazardous chemicals, or they can use natural predators, such as bugs. The two pesticides banned Monday are organophosphates, used to kill insects by disrupting nerve impulses; they have the same effect on humans. . . . The chemicals have been shown to cause illness in agricultural workers, and, on the basis of animal tests, are presumed to hurt children. . . . EPA and the Agriculture Department said Monday, that though foods meeting the current standard are still safe, foods meeting the new standards would be safer. . . ."

Matthew Wald, *New York Times*, August 3, 1999

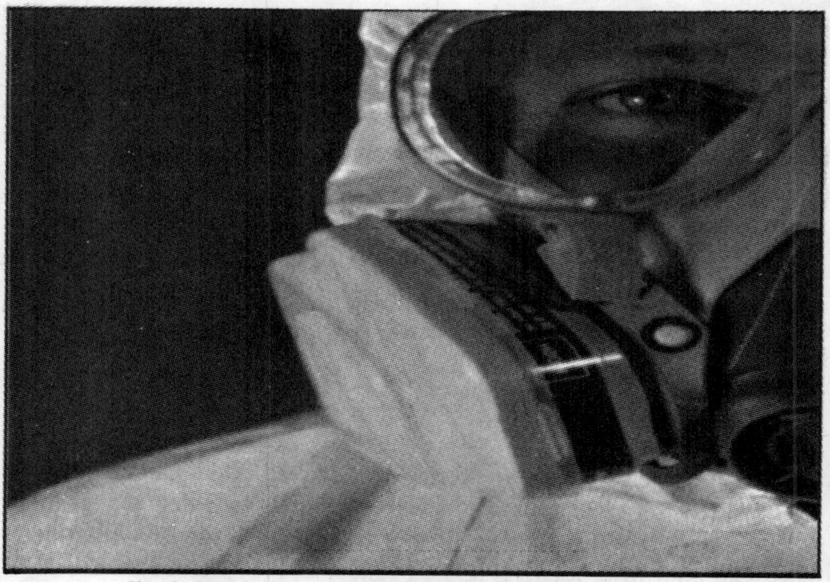

Photo by James Fashing, reprinted by permission by *Today's Farmer*.

PESTICIDE FASHION PLATE—Even as pesticides have gotten less toxic, modern equipment is expanding the applicators' safety margin.

UNOFFICIAL REALITY

Why did the EPA ban the organophosphates after the major charges against that chemical family have proven to be scientific false alarms? The giveaway is in the phrase, "The first regulations intended specifically to protect children." Can you spell "politics?"

Another study team was looking at the possibility of a link between leukemia and the use of certain insecticides on farm animals—crotocyphos, dichlorvos, famphur, the chlorinated hydrocarbon methoxychlor, and the natural pesticide pyrethrin,[9] one of the mainstays of *organic* farm production. Fortunately, no linkage was found. It would have been particually difficult for organic farmers to find a replacement for pyrethrin.

There are some farm chemicals in use that are potentially deadly, such as parathion. These are not the insidious long-term killers the scare books warn about, however; they are dangerous while they are being mixed and applied, and in the hours right after application. Farmers and farm workers must be careful in using them—and must stay out of the field for the specified length of time after spraying.

What are the real risks of pesticides for farmers? The state of Texas in 1986 had only three pesticide-related deaths. One was a suicide who deliberately consumed a potent pesticide. The second was a spray-plane pilot who crashed. The third was a farm worker who went back into a field too soon after spraying.[10] (And this was in a huge state with large acreages of fruits, vegetables, and cotton that often need high rates of chemical application per acre.)

RISK REALITY

"From 1982 to 1990, the 48 pesticide-related fatalities in California all involved the mishandling of these chemicals (accidental ingestion, suicides, termite-fumigation accidents) . . . remember the 160,000 Americans who died from tobacco related cancer in (1993 alone)."

Dr. Gordon Gribble, professor of chemistry, Dartmouth College, Hanover, N.H., letter to the editor of the *Valley News*, May 18, 1994

PESTICIDE APPLICATOR SAFETY—This First World spray rig forces the pesticide downward, behind the farmer's enclosed tractor.

Fortunately, we continue to develop still-safer pesticides that can be used in ever-smaller amounts. And we deliver them in safer packaging to reduce exposure and the risk of spills.

But even the most dangerous-to-apply pesticides are being used to protect the public health and sustain wildlife habitat, not just to line the pockets of big farmers and chemical companies. They are often needed to attack big pest infestations, and to help control the buildup of resistance to any one pesticide.

I recall as a kid dusting our half-acre of garden with a mixture of lead and arsenic powders, shaken from a burlap bag. Given the toxicity of lead and arsenic, I apparently should have been wearing one of those new EPA "moon suits" when I treated the cabbages and kohlrabi. And talk about needing to wash the produce carefully before it gets to the table!

Today's pesticide risks are not zero, but they are low and declining. The benefits are huge and increasing. When properly used, pesticides promote public health. Most are vital to wildlife conservation.

Protecting Third World Farmers

We know that the risks to farmers in the Third World are not high—but they should be a good deal lower than they are. They can be reduced, probably without food production losses, through improved training on which pesticides to use, when and how to apply

BACKPACK SPRAYER—An Asian farmer must protect his crop, but lack of protective gear and the need to walk behind his spray pattern produce health risks.

them more safely, and through better, cheaper, personal protective gear.

Third World farmers have been using more pesticides that are more dangerous to apply and some of which are not approved in the First World. The Third World farmers have tended to use pesticides more frequently than First World farmers. Third World farmers have also exposed themselves more openly to the pesticides—because they take fewer safety precautions while mixing, fail to use safety equipment where they should, and because of the backpack sprayers they typically use.

The greatest health hazard to the farm workers is during mixing and transfer of chemical concentrates, not so much the spraying. A First World farmer today typically uses dry mixtures, premeasured tablets, or water-soluble packets. If he's using a liquid, it increasingly comes in a bulk container that locks onto his spray rig in a way that makes spillage nearly impossible. He can read, and the label clearly tells him what he needs to do to protect himself. Some classes of pesticide can only be applied by a certified operator who has been trained and passed an exam on his work.

A Third World farmer, in contrast, usually mixes his pesticides in an old oil drum, and then pours them into his backpack sprayer. He is often illiterate, and he may have bought his chemical from a neighbor with no label or safety warnings on the container.

Once the pesticide is mixed, it must then be applied.

Application, too, differs enormously in the two worlds. A First World farmer is usually sitting on a tractor, with the spray aimed downward in a wide swath *behind him*. He usually encounters little of the spray himself, and is usually wearing protective gear if the pesticide calls for it.

A Third World farmer walks through the crop, spraying out in front of his own path. As the crop plants get bigger, the farmer gets more and more pesticide brushed from the plants onto his skin or clothing just from following his own spray pattern. If the backpack sprayer leaks, it leaks onto the farmer's back and legs. (I know what I'm talking about here; I control the thistles in my 70 acres of pasture by spraying 2,4-D from a backpack sprayer.)

Too few Third World farmers use protective gear while spraying, especially considering the more dangerous pesticides that are being used in the developing countries. In fact, some Third World farmers often tie cloths over their mouths and noses—which apparently absorb the chemical and *aggravate* the chemical exposure.

Many insecticides that Third World farmers use are rated in the two most dangerous categories, I and II. First World farmers use fewer of these more-dangerous chemicals, and tend to have them custom-applied by high-tech equipment when they do. (Spray planes and special spray rigs with enclosed cabs are two of the high-tech answers to applicator dangers.)

Third World farmers, as a result, have encountered far more health problems from pesticides than First World farmers. These are rarely life-threatening, but they are not trivial either: chronic eye irritation that can cloud vision over long exposure; chronic eczema on exposed skin areas; bronchial asthma; chronic stomach irritation in farmers who smoke or wipe sweat off their faces near their mouths and thus ingest pesticide orally; muscle weakness and lost nerve sensitivity (polyneuropathy), typically in the hands and feet due to absorption of some chemicals by gloves and socks; a somewhat higher rate of abnormal electrocardiogram results.

A recent study by the International Rice Research Institute found that half of the farmers putting on three sprays per year from the I and II danger categories would get chronic eye problems. (The study sample farmers had been using pesticides for 15–25 years, with heavy use of organophosphates.) Fifteen percent of the farmers in one pesticide-using test group had skin problems, and 45 percent in another. Respiratory problems also increased, especially for smokers. Nonsmoking farmers who used pesticides had a 0.3 percent probability of abnormal respiration. Those who smoked and applied two doses of insecticides per year had a 0.45 probability.

Farmers who don't use pesticides have a polyneuropathy probability of only 0.02. Farmers who applied three pesticide sprays per year had a probability of 0.24. If they drank alcohol, their probability was 0.7.

There was also a small increase in heart problems among pesticide-using farmers. Holding age and drinking constant (both produce increased heart problems) the farmers who used pesticides had 7 percent and 11 percent higher rates of functional electrocardiogram impairment.

The study found that health problems were significantly reduced when spinning disc applicators or electrodyn sprayers were used instead of the standard backpack sprayers.[11]

Another study of Ecuadorian potato farmers found 22 percent of the sample farmers had suffered a "pesticide poisoning" incident in their lives, though few of them had gone to a clinic for treatment. ("Poisoning" was defined to include dizziness and nausea.) Chronic dermatitis was almost twice as common among pesticide-using farmers, and neuropsychological tests showed reduced performance on several tests among the farmers who use the pesticides.[12]

There are no validated studies that indicate higher cancer risks for the farmers using pesticides, nor were the chemicals associated with severe neurological problems.

The eco-activists' solution, of course, would be to give up pesticide use altogether. However, this is rarely practical for either the people or the wildlife, due to reduced yields.

In Ecuador, without fungicides, late blight would destroy virtually 100 percent of the potato crop. The Andean weevil can take up to 80 percent of the potatoes in a field. Potatoes are the only crop that can produce enough food for Andean populations, given the high altitudes, limited land, and short growing seasons. As a result, Ecuadorian farmers accept the use of pesticides and the health risks involved. Only a few of them use rubber gloves, pants, jackets, or other personal protective gear.

Fortunately, there are a variety of specific things that can be done to sharply reduce Third World farmers' risks of pesticide health impairment—far short of banning the chemicals and having to select which people will starve:

- In some cases, safer chemicals are available. As an example, liquid carbofuran in Ecuador's potato fields could be replaced with the granular form which is safer for the farmer (but a greater risk to birds who might mistake the granules for seeds). In some other cases, the safer chemicals are more expensive; farmers would have to choose.

- Better protective gear (gloves, pants, jackets) can be made available, and farmers can be better-educated in the health benefits they will provide.

- Better education of the farmers can help them kill more pests with less chemical. IRRI says that the first three to five sprays that Philippine rice farmers make each season are largely wasted, because they are trying to kill sucking insects with surface sprays. IRRI is trying to help the farmers understand this. Ecuadorian farmers do not understand the life-cycle of the Andean weevil very well, and often spray the wrong pesticide at the wrong time.

- Over time, research will produce safer pesticides. More governments will ban more of the harsh pesticides when they are no longer needed, or will restrict their use. Pesticide research in the First World will lead the way to greater safety—if it is not cut off by regulation or prohibition.

- We can expect low-cost spray equipment that does not require the farmer to walk through the newly applied chemicals. Electrostatic and spinning-disc sprayers are already available that carry lower exposure risks.

To summarize, the health problems definitely linked with properly used pesticides on American farms are relatively minor. The linkages with cancer are weak, and the cancers relatively rare. The non-cancer impacts are very minor, such as temporary nausea and skin irritation.

Even in the Third World, where harsher chemicals have been used under more dangerous conditions, the pesticide health risks are rarely severe—and the pesticides are vital to human survival and protection of wildlife habitat.

Safe-Use Programs Cut the Real Health Risks from Pesticides

As our group entered the little Brazilian village, about 50 grade-school pupils and their teachers were lined up to greet us. They sang songs about preserving nature, as two little girls dressed in rabbit costumes danced. In a skit, a pupil with a backpack sprayer was musically warned to use the pesticide carefully.

The kids had comic books that warned them not to use pesticide containers to carry drinking water, or for storing food. (The comics show a farm worker puncturing the plastic pesticide container so it can't be misused as a water jug.)

At a special science-oriented grade school a few miles down the road, 4th-grade students showed us their model landscape. At one end, dead trees and dead fish in a withered, brown landscape illustrated the dangers of using too many pesticides. The other end of the diorama was green, with lush crops and many trees, graphically showing that farm chemicals, properly used, not only help feed people, but also allow them to be fed without clearing the tropical forests in Brazil's Amazon region to the north.

We were guests at the Parana State Farm Training Center. In a project jointly funded by Brazil and the global agro-chemical industry, farm workers are being trained in safe pesticide appplicator use, basic reading skills (to enable workers to read label instructions) and other agricultural skills. Lately, the program has been expanded to include schoolchildren in its "little farmers" program.

Without herbicides and conservation tillage, the region's hilly, volcanic soils would erode terribly. Without pesticides, the local coffee trees would have their yields stolen by leaf miners and nematodes. Without insecticides, the cornfields would be heavily infested with corn borers. With safely used chemicals, Parana State not only feeds its own families, but also is able to export substanial quantities of coffee and soybeans to the rest of the world.

In this part of Brazil, agriculture provides more than half of the local jobs. Most of the kids we met will find themselves eventually working on the region's mostly small farms. Virtually all of them will spend their lives around pesticides, sprayers and pesticide containers. Many of them will leave school after only two or three years—and then use pesticides on the job.

At the training center, we saw farm workers learning why they should use the protective "poncho" that will keep any liquid spilled from their back-pack sprayers from coming in contact with their skin. Many of the trainees were wearing special "moonsuits" made of chemical-repellent fabric, with plastic panels to protect their legs and backs from the hash liquids, and plastic face-masks to protect eyes and mouths.

Other workers were learning how to keep chemicals sprayed from a tractor-mounted spray boom from drifting onto adjoining fields or wildlands. Those who worked with granular pesticides were taught to wear rubber gloves, and how to make sure that the granules were properly covered by soil to protect bird wildlife.

A German scientist with our party protested that the training program was burdening the trainees with too much heavy protective gear. In the heat and humidity of the fields, they would simply take it off and leave it by the fence, he warned. At that, the instructors let us try on the new protective gear they had helped to develop; it is lighter and cooler to wear than cotton clothing, even though it repels water and chemicals. And it can be washed up to 30 times for reuse. The German asked if he could get some of the suits to show chemical companies in other countries.

The world's long-time fear that pesticides will cause cancer has been virtually laid to rest; the new replacement claim that pesticides might disrupt our endocrine systems has been sharply undercut as two prominent studies have been discredited. However, pesticides do unquestionably pose risks to farmers in the fields who must handle them in concentrated form—and possibly to their neighbors through careless handling, accidents or discarded containers.

That's why the safe-use programs are spreading rapidly. Brazil's program is huge, well organized and well funded. In the last five years, it has trained 34,000 workers and 300,000 school children in Parana alone. Along the way, 5,000 adult workers learned to read.

Safe-use programs, funded primarily by the agro-chemical industry, are now up and running in 17 Latin American countries. They are spreading in such other countries as Kenya, Tanzania, Indonesia, and Thailand. The governments of these rapidly emerging countries have concluded that pesticides, when properly used, are necessary to fulfill the future increased food demands of the country. They are also vital to protecting the environment from the habitat destruction that results with low-yield farming.

Both government and industry are quite properly willing to spend money to make sure the chemicals are used safely, and that local people don't incur health risks through carelessness or ignorance.

Notes

1. *Cancer Facts & Figures*—1994, American Cancer Society, Atlanta.

2. *Health Effects in Man from Long-Term Exposure to Pesticides: A Review of the 1975–1991 Literature*, International Centre for Pesticide Safety, copyright European Crop Protection Association, Brussels, 1992.

3. L. Hardell, M. Eriksson, and A. Degerman, 1994, *Cancer Research 54:* 2386–2389.

4. L. Hardell, M. Eriksson, P. Lenner, and E. Lundgren, 1981. *British Journal of Cancer 43:*169–176.

5. L. Hardell, and M. Eriksson, 1999, *Cancer 85:*1353–1360.

6. K. Steenland, L. Piacitelli, J. Deddens, M. Fingerhut, and L. I. Chang 1999, *Journal of the National Cancer Institute* 91, 779–786.

7. Environmental Protection Agency, Health Effects Division. 1997. Memorandum: Carcinogenicity peer review (4th) of 2,4-D Dichlorophenoxyacetic acid (2, 4-D). From J. Rowland, and E. Rinde, Health Effects Division, To J. Miller, and W. Waldrop, Special Review and Reregistration Division, January 29, 1999. [photocopy, 55 pp.].

8. M. Gough, *Dioxin, Agent Orange: The Facts,* New York, NY, 19 Plenum Press, 1986, pp. 137–148

9. National Cancer Institute, Occupational Risk of Cancer from Pesticides: Farmer Studies, National Institute of Health, Washington, D.C. , January 1991.

10. Curt Lancaster, farm director, USA Radio Network, San Angelo, TX, personal communication.

11. Pingali, Marquez, Palis, and Rola, "Impact of Pesticides on Farmer Health: A Medical and Economic Analysis," *Impacts of Pesticides on Human Health and Rice Field Biology in Rice-Growing Areas*, Kluwer, Netherlands.

12. C. Crissman and D. Cole, "Pesticide Use and Farm Worker Health in Ecuadorian Potato Production," invited paper for the Allied Social Science Association meetings, Boston, January 3–5, 1994.

6

Intensive Meat Production for the 21st Century

MYTHMAKERS SAY:

"Could it be that the whole concern about getting enough protein is actually just a relic from a less-enlightened past, with nothing to support it except the propaganda of the meat, dairy and egg industries?"

John Robbins, *May All Be Fed*, 1992

"It becomes a big issue in urban America when tons of liquid manure get into a water system serving an urban area. Nothing focuses the urban mind like a bunch of farmers leaking animal poop into the water the city folk drink."

David Yepsen, political editor, *Des Moines Register*, column of November 23, 1997

"The nearest hogs to me should be 2,000 miles away."

Litigant in a lawsuit against a Georgia hog producer

"While most Americans watch their waistlines, people in North Carolina watch their wastes. This focus on feces has never been easier . . . since the Environmental Defense Fund launched its popular Hog Watch site on the Web. A major element is the Poop Counter—a continually updated tally of the excrement that North Carolina's hogs and humans have generated so far this year. . . . Especially harmful [to water quality] are the nitrogen and phosphorous in hog waste, both tallied and displayed at the Poop Counter site."

David George Gordon, special to ABCNews.com, January 28, 1999

REALITY SAYS:

"Numerous anthropogenic studies have found that animal foods—be they scavenged (from kills by large carnivores), collected (shellfish, birds' eggs), hunted (from tropical birds to aquatic mammals), or produced eventually by domesticated species—have been universal ingredients of human behaviour. . . . An obvious conclusion is that our normal diets should contain a variety of animal foods, including meat. Indeed, human stomachs are about 40 percent smaller than they should be in a similarly sized primate. The primary reason for this difference is that—unlike the herbivores consuming large quantities of roots, leaves and fruits with low energy densities—our diets evolved to include smaller quantities of energy-dense and easily-digestible food, including seeds, nuts and meats."

Vaclav Smil, University of Manitoba agronomist and author of the forthcoming book, *Feeding Ten Billion* (MIT Press)

"Hog Farms Don't Hurt Water Quality! Data collected by the North Carolina Department of Environment and Natural Resources indicate that the water quality in the Black River, the drainage basin for North Carolina's most intensive hog producing region, has remained excellent, despite a 500 percent increase in the area's hog population. In fact, the water quality of the rivers and streams in this region may even have improved slightly. . . . The amount of hog manure applied to crop fields within the Black River watershed has increased dramatically over the last 15 years . . . [but] nutrient concentrations in the Black River have remained low and essentially unchanged. . . . This should come as no surprise, since confinement hog farms collect and use their hog manure as organic fertilizer on crop fields. This is clearly better for water quality than the traditional method of raising hogs outdoors, where the hog wastes may be washed directly into waterways."

From a Hudson Institute press advisory, September, 15, 1998

"Stunted growth has been a common human condition: indeed, even throughout today's affluent world most people born before 1950 grew up without expressing their full growth potential and have been noticeably outgrown by their children."

Vaclav Smil, University of Manitoba, op. cit.

"Pork producers operate under a zero discharge rule. Section 319 of the Clean Water Act says it is against the law for any discharge from a hog operation to find its way into groundwater or a stream, river or lake. Those who do so are subject to heavy fines. Just to put things in perspective, for every pound of nitrogen that is collected on a hog farm and applied on land as a crop fertilizer, two pounds is discharged legally by municipal sewage plants into U.S. waters."

Donna Reifschneider, President, National Pork Producers Council, letter to the *New York Times,* July 13, 1998

"Until a few years ago it was generally agreed that adult requirements for specific amino acids . . . can be met by all normal mixed diets, including the largely vegetarian ones in poor countries. . . . But . . . Milward and Rivers (1988) concluded that the estimates of adult amino acid requirements are too low. . . . A group of Massachusetts researchers . . . suggested the minimum values for amino acid requirements in adults should be set two or three times higher than the current, internationally accepted standards (Young et al., 1989; McLarney et al. 1996). . . . They also noted that the currently recommended total protein requirement for adults . . . are considerably lower than for any other mammal whose protein requirements have been studied in any detail, while in infancy they are among the highest."

Vaclav Smil, University of Manitoba, op. cit.

It is now clear that the world is in the midst of an enormous expansion of the demand for resource-costly foods that provide high-quality protein. Between 1990 and 1997, the world's meat consumption rose from 178 million tons to 215 million tons. Egg consumption increased from 35 million tons to 43 million. Demand for dairy products continues to rise in emerging economies.

Between 1987 and 1996, the Third World's economic growth averaged 4.6 percent per year (compared with a recent population growth rate of only 1.6 percent). This resulted in rapid per capita income gains in what we used to call "the poor countries."

In Asia, the most populous region in the world, economic growth for the last decade averaged 7.5 percent annually. In China, the most populous country in the world, economic growth rate has been a phenomenal 9.8 percent per year (against a recent population growth rate of 1.1 percent). Chinese per capita purchasing power increased roughly twentyfold between 1978 and 1997. Increased wealth raised Chinese meat consumption from about 23 million tons in 1980 to more than 60 million tons in 1996, according to the UN Food and Agriculture Organization. Indeed, during the 1990s, Chinese meat consumption has been rising at the unheard of pace of ten percent per year.

By the late 1990s, U.S. Department of Agriculture statistics showed that China was adding about five million tons of meat demand to the world total *each year.* Since most of the Chinese meat consumption is pork, and Chinese pork has a relatively low feed conversion efficiency, China was also adding 25 to 30 million tons worth of grain-equivalent feed demand to the world's requirements.

USDA data show the trends have been similar in most other emerging countries:

- Indonesia's poultry demand increased by 20 percent annually during the 1990s.

- India doubled its milk demand since 1980, to more than 70 million tons per year. Two-thirds of India's Hindus indicate they will eat meat (other than beef) when they can afford it. Already, India has seen a major increase in its consumption of meat and eggs. McDonald's recently began opening restaurants featuring "muttonburgers with special sauce."

- South Korean cheese consumption rose from less than 5,000 tons in 1990 to more than 15,000 tons in 1996. A pizza craze built more than 500 pizza restaurants in South Korea. In addition, Korean educators have told mothers to give their children cheese as a nutritious after-school snack high in protein and calcium.

- Japan used to consume only about 15 grams of animal protein per capita per day, compared to 65–75 grams for First World countries. People said Japan was a rice-and-fish culture. Today, however, Japan is consuming nearly 60 grams per day, and the total is still rising.

- Poorer countries such as Algeria, Pakistan, Thailand, and Malaysia are proving that they are also likely to follow the traditional European pattern of protein consumption. It looks as though virtually every other country in the world will do the same as incomes permit.

Eco-zealots and vegetarians have pretended for at least 40 years that they could make the world vegetarian. This has supposedly justified their opposition to high-yield agriculture in most of its forms. However, the world has now run out of time for casual hopes of a "vegetarian solution." It is doubtful that even ten percent of the world's people will choose to become vegetarians by 2030, given the current surge in world meat demand and the low numbers of vegetarians. Nor would that be enough to take the pressure off the wildlands since most "vegetarians" eat lots of eggs and dairy products, which take almost as many resources per acre as meat itself.

Only about 0.2 percent of Americans are vegan (eating virtually no animal products), and the percentages are similarly tiny in other affluent countries. We would need more than 80 percent of the world's 21st-century adults to be vegans to save the wildlands with dietary change. There is no current trend that will produce such large numbers of vegans. There is not even a financial commitment from the environmental movement to start a massive vegan campaign, let alone a persuasive rationale.

If we are truly concerned about wildlife conservation, we must assume that the world faces an enormous increase in its domestic livestock and poultry populations:

- The world's hog population is likely to expand from the current one billion to perhaps three billion (depending on how much we can raise the productivity of each sow).

- The number of cattle in the world may top three billion, more than double the current 1.3 billion.

- More and more of the world's fish consumption is likely to come from farmed fish, grown in ponds and cages, and fed primarily on grain and oilseeds. This will ease the pressure on the world's wild fish stocks, but increase feed grain requirements.

There are three major challenges in further expanding meat and protein production: feed supplies, land requirements of the creatures, and constructive use of the huge amounts of manure that will be produced. We've already talked about the need for higher crop yields (and for research, fertilizers and biotechnology) to support the feedstuffs expansion on existing cropland.

Fortunately, the world also has a similarly constructive approach to minimizing the livestock land requirements and using the manure: confinement production of beef, pork, poultry meat, and fish. The confinement meat production systems have been made possible by the same sorts of medical advances that tamed so many of humanity's epidemic diseases: vaccines, antibiotics, and good hygiene. Without these medical advances, livestock, birds, and farmed fish stocks would either have to be widely dispersed (raising costs and land requirements) or the meat industry would have to start with two or three times as many creatures to make up for huge death losses.

Confinement rearing raises questions in many people's minds about the rights of the livestock and the safety of the meat production. Such questions need to be examined realistically. But always understand that without domestic meat production the people of the world will hunt down virtually all of the wildlife for the stewpot, fish the wild stocks of marine creatures to collapse, and clear vast amounts of forest for outdoor production of cattle, hogs, and chickens.

NORTH CAROLINA HOG HOUSES—Provide clean, safe shelter for the animals. (Courtesy Murphy Family Farms.)

Excerpted from a column published in the *Des Moines Register*, July 7, 1995

The Coming Blue Revolution:
Why Corn and Soybeans Will Have to Feed Tomorrow's Fish
By Alex Avery

Canadian warships recently fired on a Spanish fishing trawler in international waters, then cut and seized its fishing nets. The trawler was fishing for turbot, once considered a trash fish, because the cod and halibut have been fished out of the once-rich Grand Banks fishing grounds.

The world's ocean fishers are running out of fish to catch.

Eco-activists say this is just one more sign that the world is running out of food and natural resources. It's not. We should be able to feed 10 billion people a better diet than they eat now—and one with lots of fish and seafood.

In *State of the World, 1995*, doomsayer Lester Brown cites the condition of the world's ocean fisheries as one of the "imminent limits" to a sustainable human population. Instead, fishing is the last vestige of our hunting/gathering history. Far from being a harbinger of mankind's collapsing future, the declining ocean fisheries are a final farewell to an unsustainable portion of our past.

Tomorrow's fish will not come from the wild, they will come from fish farms.

Oceanic fisheries are in trouble and fishing regulations are wholly inadequate, but we can hardly blame those who fish. They are only acting rationally in an unregulated "commons." Individual fishers will always try to take the last fish; if they don't, the next person will.

Blame the governments. They have done very little to protect the fisheries. Few fish are taken in international waters. More than 90 percent are caught within some country's 200-mile limit. (The Spanish trawler was fishing the only portion of the Grand Banks that extends beyond Canada's tightly regulated 200 miles.) Instead of using good stewardship, most countries have made things worse. They built larger fishing industries trying to increase income, only hastening the collapse. Now many are confronting the declining stocks and catches by pouring more subsidies into their fishing industries. The European Union increased fishing support from $80 million in 1983 to $580 million in 1990, with one-fifth of that going to build new boats or improve old ones—hardly a sound course for rebuilding fish stocks.

Today, the most prominent source of seafood is still ocean fisheries, with more than 80 percent of the world catch. We will probably always harvest some from the oceans. But we won't ever again have to place the fisheries in peril.

Let's look at an important trend: Inland and coastal aquaculture has gone from 12 percent of the world catch in 1981 to 17 percent in 1990 (78 percent worldwide growth) and the percentage is rising. We're producing fish in ponds and cages and feeding them wheat, corn, and soybean meal.

This points to a much larger role for fish farming in the future. Norwegian production of farmed Atlantic halibut in 1997 is expected to exceed wild catch of the species by their fishermen. Presently, our trout and catfish industries produce far more than we could catch in the wild. Since 1982, farm-raised shrimp production has increased more than sevenfold, and tops 610,000 metric tons—23 percent of the total world shrimp production.

The best news is that aquaculture stands now where livestock agriculture stood several centuries ago, before breeding and high-energy feed produced huge increases in carcass weight, feed conversion efficiencies, rate of weight gain and milk and egg production.

Aquaculture is entering the modern age through private enterprise and public efforts such as the International Center for Living Aquatic Resources Management. ICLARM began a tilapia improvement program in 1988 in the Philippines. Tilapia (called an "aquatic chicken" by its proponents) is an African freshwater fish. By 1992, using a classic livestock breeding approach that meant tagging tens of thousands of tiny fish, the "Super Tilapia" was growing 60 percent faster and with 50 percent greater survival than the most commonly farmed strains.

Dr. Rex Dunham, of Auburn University, estimates that commercial carp and catfish production in the United States has increased two- and threefold over the last 20 years because of better management and feed. Even bigger gains will come through biotechnology. Dunham's lab has achieved another 50 percent increase in yields through genetic improvement, resulting in more meat and less fat per fish. This type of technology will soon be reaching commercial producers.

By developing the management skills, feeding strategies and improved strains of fish and seafood, there is every hope that the Blue Revolution will match the productivity achieved in the Green Revolution of the 1960s.

We are well on the way.

The Environmental Virtues of Confinement Meat Production

For the sake of the environment, future expansion of the world's flocks and herds should be keyed to confinement production:

- Confinement raising of hogs and poultry takes very little land per bird and animal, achieves lower death rates and produces higher growth rates. Confinement helps minimize the land needed for the creatures and their feed production.

- More and more of the world's expanding hog herd will live in big indoor buildings where they are protected from heat and cold, suffer 10 percent lower death losses, and convert their feed to meat with 20 percent higher feed efficiency.

- More of the world's cattle will be finished, not on rangeland, but in feedlots.

- More of its dairy cattle will have their feed brought to them instead of grazing grasslands, or be moved daily from one small plot of grass to another, in intensive rotational grazing. This will produce more milk per acre from high-yield forage crops.

The alternative—traditional "outdoor" production—would be an environmental disaster. Dr. Gordon Surgeoner, of the University of Guelph, says that if the Canadian province of Ontario raised its chickens on free range, it would take 1.2 million acres of additional land. That's ten times the land area of Toronto, the province's largest city. By the same formula, producing all of America's 1.6 billion chickens and turkeys on free range would take wildlands equal to the cropland in Pennsylvania—more than 7,000 square miles.

Scaling up to produce a global flock of 50 billion chickens and turkeys in 2030 would take about 220,000 square miles of additional wildlands, equal to the total land area of Thailand.

Adding another two billion "outdoor" hogs would be far worse for the environment than outdoor chickens. The outdoor hog production system used recently in America required about one acre for each four adult hogs. Increasing the world hog herd from the current one billion to three billion might thus take an additional 780,000 square miles of land—the total land area of Indonesia or half the land area of the Amazon rain forest.

The additional two billion "outdoor" hogs would also trigger huge soil erosion losses. Hogs evolved digging a substantial part of their food—roots and tubers—out of the ground. I grew up next to an outdoor hog herd in Michigan; even with rings in their noses, the hogs created serious erosion in their pastures. They also virtually destroyed the ecology of the little stream that ran through the farm with mud and wastes.

Cattle present almost no ecological problem when grazing well-managed pastures, but the world is already using virtually all of its grasslands. Moreover, we have had more success in raising the yields on our best croplands than in boosting the productivity of rangeland. (In large part this reflects the scarce rainfall on most rangeland.) Expanding beef and milk production for the 21st century will mean moving well beyond the current capability of the grasslands.

Using Manure Constructively for the Environment

Manure is organic fertilizer. For centuries, it was humanity's only fertilizer. For decades, organic farmers have been telling us that fertilizing crops with manure produces even better food—and organic produce generally sells for a substantial premium. However, in outdoor production of hogs and chickens, much of the manure is flushed into the streams and rivers, or volatilized back into the air.

Confinement production makes it possible to collect virtually all of the livestock and poultry wastes and apply them effectively on growing crops. Since the world is short of organic fertilizers, this should seemingly be regarded as a major environmental plus. Surprisingly (or not), critics of intensive livestock production say that the manure from big hog and poultry houses is a serious threat to the environment.

The charge is that too much manure is being put into our lakes and streams from the big hog and poultry houses and the feedlot cattle; that more manure is being put on our fields than the soil can accept. They demand that pork be produced on small farms, preferably outdoors; and big hog farms be moved or shut down. There is a strident call for poultry houses to be shut down in places like the Eastern Shore of Maryland, which lies next to the Chesapeake Bay.

Environmentalists' desires have been intensified by the failure of some hog farms to build big enough lagoons, keep their premises clean, and spread their manure when it would offend their neighbors the least. However, most of the "bad actors" are apparently the smaller and least-well-financed producers.

In response to public worrying, North Carolina, which has expanded from 2.2 million hogs in 1984 to more than 9 million in 1997, declared a two-year moratorium in 1997 on further construction of large hog houses. In addition, it has ordered the State Department of Agriculture to find an alternative to applying the hog manure to fields. The Environmental Protection Agency, under a mandate from Vice President Gore, is "Federalizing" regulations for confined animal feeding, under the premise that state regulation has been inadequate.

Yet, virtually all of the big hog and poultry farms in the First World are governed by zero-discharge pollution requirements. They are not allowed to discharge any nutrients or other pollutants from their farms into streams, rivers, or groundwater. This is a very tough standard. Dams, lagoons, and manure storage tanks are subject to regular inspection by state and local governments. Any farm that violates its zero discharge responsibility by letting waste flow into streams, or leak from tanks or lagoons can quickly be stopped by the authorities.

Unfortunately, the opponents of big hog farms seem more interested in shutting them down than in the development of effective food production systems. For example, opponents of Premium Standard Farms in Missouri complained that the farm was putting too much manure on too little cropland. When the farm applied for the permits to build a 3.5-mile slurry pipeline, so it could be cost-effectively spread on more acres, the critics also opposed that.

AERIAL VIEW OF HOG FARM— Lagoon contains hog waste until it can be cost-effectively spread on feed crops as fertilizer. (Courtesy Murphy Family Farms.)

Cities Are Not Zero-Discharge

Ironically, big hog farms are taking better care of the environment, especially in terms of water quality, than America's cities. Our cities are not zero-discharge. Modern sewage treatment takes out very little of the nitrogen in human wastes; it mainly channels the nitrogen directly into our rivers. Nor do cities treat the nitrous oxide from auto exhausts, which is deposited on their paved surfaces. That washes into the storm sewers, and then downstream with every storm. In North Carolina, during 1995, dams on seven hog lagoons broke (out of more than 4,000 in the state). The biggest spill, however, was only an 8-acre pond. It spilled about 7 percent as much nitrogen in a onetime insult to the local river as the small city of Wilmington, North Carolina, puts into the Cape Fear River every year.

Eco-activists charge that big hog farms are overfertilizing the Albemarle and Pamlico Sounds which are important coastal waters in North Carolina.[1] But only a few big hog farms are located in the Albemarle-Pamlico watershed, and they're already under zero-discharge regulations. On the other hand, more than a million people live in Raleigh, Durham, and the other cities that drain into those waters. Many more of North Carolina's hog farms are in the Black River watershed—and the Black River is still designated as Outstanding Resource Water.

REALITY CHECK:

". . . during] the most unusual summer of 1996, when Eastern North Carolina was slammed into by two hurricanes and a tropical storm . . . data from [the Department of Water Quality] shows that 22 lagoons out of 4000 had some sort of malfunction—either running over or floodwater encroachment. In contrast, 122 municipal waste treatment systems (40 percent) malfunctioned, resulting in 35 . . . discharges . . . totaling 270 million gallons and another 87 reporting discharges as 'amounts unknown'. . ."

Garth Boyd, "Manure Management—No Easy Solutions,"
Journal of Soil and Water Conservation, September/October, 1997

Small Farms Not as Careful of the Environment

On average, the big hog and poultry farms have been doing a consistently better job of protecting the environment than small farmers. This probably should not surprise us. Big farms are always watched more carefully. Little farms have less technical expertise and less capital. And, their neighbors are more likely to give them the benefit of the doubt.

The Journal of Soil and Water Conservation, November/December 1997, carried an article by a team of sociologists and economists from North Carolina State and Auburn Universities, titled "Industrialization of Poultry and Swine Production: Implications for

Natural Resource Management." The article clearly sympathized with traditional small farmers, but also noted:

"Large livestock and poultry farms clearly benefit the farm economy. . . . Overall, rural counties where the industry has grown tend to have higher standards of living than before because contracting provides steady, predictable incomes to producers.

". . . Some independent farmers (especially small-scale and limited-resource producers) have been unable or unwilling to comply with recently enacted, more stringent, regulations. . . . Most of what the [hog] integrators have done voluntarily has become law. For example, they started using lagoon markers and initiated setback [requirements from property lines]. These were written into law.

". . . Independent producers do not have the same level of technical support that the companies provide to the contract producers. . . . Independent producers are caught up in the daily operation of the farm so they do not think as much about the larger context. . .

". . . After a major accident in 1995, a round of inspections revealed that about 5 percent of hog producers in North Carolina were 'bad actors.' These bad actors were identified mainly as independents.

". . . Most independents have older farm operations, with technical and economic problems that need to be overcome. Their lagoons tend to be too small. Many need improved irrigation systems and advice on the use of best management practices. Public agencies need to expand and target their technical assistance, financial assistance and education to help [small] producers understand and meet new regulatory standards. Older farmers may be unwilling to make the capital investments given their short time horizons and intended retirement."

Globally, the future health of the environment demands that additional hogs be raised indoors, and that their wastes be carefully and appropriately applied to cropland. Given the economic reality that the big, specialized hog operations produce more pork per man, at lower costs, it seems likely that larger hog operations will produce an increasing share of the world's pork. From the available evidence, this will be an environmental plus.

What About the Fate of the Small Meat Producer?

Today's world too often confuses nostalgia with conservation—and this puts too much of our wildlife at needless risk. I grew up on a small dairy farm. It was an extremely confining existence. We were tied to feeding and milking about 25 cows, morning and night, every day of the year, even though we didn't make much money from the cows. (Fortunately, my father had an off-farm income as well.)

In the early 1980s, I visited small dairy communities in Minnesota where the norm was a 40-cow herd. Incomes were low, and complaints about milk prices were widespread. Most of the farmers told me they couldn't handle more cows because of the cold climate. But then I found a young farmer with a 70-cow herd. He said, "You need a mechanized silo unloader, a mechanized manure-handling system, and the willingness to be shunned in church on Sunday." Those small Minnesota communities pretty much stuck with the 40-cow herds until the mid-1990s, as their young people left for more attractive jobs in the cities. Suddenly, in the 1990s, the region went from 40-cow herds to 400-cow herds, with no intermediate step.

Few young people today are willing to tie themselves to the daily needs of livestock for low pay and no vacations.

I have a neighbor who lives on a 100-acre farm. He sold his drug store, and wasn't able to find an appropriate job in the area. He was forced to consider moving, and leaving the 150-year-old farmhouse he and his wife had renovated and the community life they loved. Finally, they signed a contract with an integrated turkey producer, and built two turkey-breeding stock houses on their farm—instead of moving. The turkey manure will be used to increase the forage production on their remaining acreage. This is good for my community, and good for the environment. The story has been replicated hundreds of thousands of times across America.

We would not try to produce today's crops with a 1940 tractor that required dangerous hand cranking and lacked the hydraulics to carefully control its tillage machinery. Why then are we so emotionally wedded to the farm structure of the1940s?

What About Hog Odors?

Hog odors are a serious problem in a modern society. The farmers' old response, that "hogs smell like money" doesn't justify imposing intense hog odors on neighbors who don't have hogs and don't share the profits. But systems and technologies are available to control the odor.

I recently visited Denmark, where all the hogs are indoors. The manure is collected in big concrete tanks covered with floating chopped straw. It is spread early in the spring, when the weather is still cold enough to minimize odors (and most of the neighbors are still indoors). Too many hog farmers in the past have tried to get by with lagoons which were too small, or which needed second or third stages. More recently, some of the big hog farms have begun moving to tanks, because they can be covered more effectively with either straw or plastic covers.

Some of the larger farmers are using a new type of spreader that inserts the manure a few inches into the ground and scrapes a layer of fresh earth over it. Half an hour after the manure has been spread the only odor is the fresh earth. But this type of spreader may cost up to $25,000, so small farmers are not likely to buy them.

However, if the American hog industry is to help provide the pork for the world of the 21st century—and help use the world's manure assets to their maximum advantage on growing, high-yield crops—they will have to solve the odor problems to their communities' satisfaction.

What About Groundwater Quality?

Sampson County, North Carolina, is the largest hog-producing county in North Carolina and one of the U.S. counties with the most hogs per square mile. A study of 214 water supply wells in the county was performed in 1990 and repeated in 1996. In the 1990 study, 72 percent of the wells had less than background levels of nitrate-nitrogen, and 23 percent had levels above background levels but below the health-recommended ten milligrams per liter. Only five percent of the wells had nitrogen-nitrate levels above health-recommended levels. The 1996 test results were almost identical, even though the number of hogs in the county had doubled in the meantime.[2]

The wells with nitrogen levels above the health-recommended levels were *all* shallow (less than 50 feet deep) and/or poorly constructed (not grouted, not sealed, with wellheads below grade or down gradient from septic systems). Most were also within 100 feet of a livestock operation, a septic system, or a chemically treated lawn or garden.

Since midwestern groundwater levels tend to be deeper than those in the coastal regions of North Carolina, midwestern wells should be even safer from any groundwater impacts of confinement hog farming.

What About the Welfare of the Animals and Birds?

Some critics of modern agriculture believe that it is cruel to keep birds and animals confined. However, the case for "freeing" the birds and animals is very much weakened by the reality that "outdoor" birds and animals suffer much higher diseases and death losses.

Chickens: Battery production protects chickens from the sharp beaks of the other hens trying to enforce the "pecking order." It also minimizes the chances for disease epidemics. Sweden has mandated that laying hens kept in battery production facilities be allowed out for a "recess" each day. However, this offers the worst of both worlds for the hens; they are subject to all the fear and injuries of the pecking order, and all of the disease risks of dense populations, without really being free.

My wife has a dozen chickens given to her by a friend. They roam freely around our farm. (Obviously, we have more a rural residence than a farm.) Ironically, the chickens spend much of their time in an unused dog run where the chain link fence surrounds them, or perched on the barn rafters, both of which provide a feeling of security from hawks and foxes.

Maybe that shouldn't be too surprising. We also have two large white geese that live on our pond. We used to have Muscovy ducks; but they disappeared or were killed. Being smaller, they were carried off by foxes, killed by owls, and caught by visiting dogs. Their ducklings suffered terrible losses from the snapping turtles and large-mouth bass in the pond itself. Many more ducklings would have survived if we'd kept them penned. It may well be that our chickens are more realistic about the dangers of freedom than some "animal rights" advocates.

Hogs: Any hog producer who fails to pay close attention to health and happiness among his animals is likely to fail economically as well as morally. Hogs are extremely sensitive creatures and easily stressed. One farmer found that cleaning the empty hog houses between litters, caused a pounding in the hot water lines that ran through all of the hog houses, this upset his sows—some to the point of miscarrying.

Hogs are also sensitive to the weather. They have a difficult time with heat, because they lack sweat glands. (That's why they are historically associated with mud; they lie in it to cool themselves in hot weather.) However, raising sizable numbers of pigs in muddy conditions is an invitation to diseases and health problems. Modern

A WELL-RUN HOG HOTEL—Provides a clean, roomy, heat-controled living environment. (Courtesty of Murphy Family Farms.)

confinement operations generally use insulation, fans, and heaters to keep their pigs at comfortable temperatures. Slatted floors are often used to let urine and manure fall into pits below the houses, while the pigs stay dry and comfortable in their pens.

Some animal welfare critics want producers to stop using farrowing crates, which confine the sow's movements. However, farrowing crates were invented to keep the sows (who are very large and not very agile) from accidentally rolling over on their piglets. The freedom of the sow comes at a risk to her offspring.

Still another "welfare" proposal advocates that farmers should be barred from cutting pig's tails. But this ignores the reality that a favorite fighting strategy of pigs is to grab another pig's tail in their teeth. The pain and the bloodshed of such struggles can be avoided with a simple and nearly painless docking of the tail.

Cattle: The beef cows and calves in my pasture have a nearly idyllic existence. The calves nurse from their mothers' udders and frolic with each other. They have ample pasture in the summer, with clean water readily available. They are protected from predators by fences, dogs, and human efforts to control coyotes. Their health is protected by vaccinations, medicines, and—if necessary—visits by the veterinarian. (Unlike family physicians, large-animal veterinarians still make house calls.)

Most of the calves eventually move on to a feedlot, where they are surrounded by other calves. Cattle are herd animals and do not seem to feel stressed in feed yards—so long as they have water, food, comfortable humidity, and the ability to stay in reasonably close contact with one another. They get all of the rich feed they want and are still defended from diseases and predators. When the end comes, it is swift and humane.

In the wild, most of these calves would be dead soon after they were born, from diseases or predators. If too many of them lived, the feed supply for the herd would run short, and large numbers of them would starve. If there were no market for meat or milk, I would not spend $25,000 to fence my 100 acres. Instead, I would plant fast-growing trees. Farmers who lack my off-farm income would be even less able to afford cows, hogs and chickens as "pets."

The most extreme animal rights activists think there should be no domestic meat or dairy animals at all. They regard all meat and milk production as exploitation of the animals.

However, it is a puzzle to me that this movement calls itself "animal rights." The logical outcome of their recommendations would be that the billions of cattle, hogs and chickens thriving in the world today would disappear. Their offspring would never be born and their species would become nearly extinct.

Now that is a strange definition of "animal rights."

Notes

1. Nowlin, Janet, "Manure Management—No Easy Solutions: PointCounterpoint," *Journal of Soil and Water Conservation,* vol. 52, no. 5, September /October 1997, pp. 314–317.

2. Kimley-Horn & Associates, Professional Opinion Regarding the Impact of Hog Farming on Drinking Water Quality in Eastern North Carolina, December 16, 1997.

Bob Lang, Reprinted from CFGFI Quarterly, Spring 1998

7

Do the Rat Tests Mean Anything?

MYTHMAKERS SAY:

". . . [I]n 1969, Dr. Malcolm M. Hargraves of the Mayo Clinic said: 'Since the advent of pesticides in 1947, I've seen and taken inquisitive personal histories on 1,200 cases of blood dyscrasia and lymphoid diseases. Every patient at some time or another had great exposure to a pesticide, an herbicide, a paint thinner, a cleaning agent, or the like."

Quoted by Frank Graham, Jr., *Since Silent Spring*, Boston, Houghton Mifflin, 1970, pp. 147–48

"I read with amazement Dennis Avery's . . . article in the *Star* November 25. His bashing of worldwide efforts to make environmental progress is unconscionable. . . . To say don't worry about use of pesticides and chemical fertilizers is to step back 30 years before Rachel (Carson's) alarm. We do not really know how many pollutants we can spread over the globe and what the cumulative effect will in the next 75–100 years—the years in which our children and grandchildren will live."

William F. Steinmetz, letter to the editor of the *Indianapolis Star*, November 27, 1993

"When I read the extensive list of organochlorines, organophosphates, dioxin-containing sprays and other chemicals used as weedicides, fungicides, herbicides, pesticides and fertilizers (in cotton production) I was shocked. . . . Then someone told me that women with tiny babies are advised to wash new cotton garments three to four times to remove chemicals adhering to the cotton fibers before they use the clothes."

Helen Caldicott, antiwar and environmental activist, *If You Love This Planet*, 1992

REALITY SAYS:

"The standard carcinogen tests that use rodents are an obsolescent relic of the ignorance of past decades."

Science editorial by Dr. Philip Abelson, "Testing for Carcinogens with Rodents," vol. 249, September 21, 1990, p. 1357

"The current practice of feeding animals massive doses of chemicals to determine whether they might cause cancer in people has been attacked as misleading by two reports in today's issue of the scientific journal *Science*."

Larry Thompson, "High-Dose Chemical Tests on Animals Overestimate Cancer Risk, Critics Say," *Washington Post*, August 31, 1990

"We feed rodents 'all-you-can-eat' buffets every day, yet we know that caloric intake is the single greatest contributing cause of cancer. In fact, we found you can modify the cancer-causing impact of one of the most potent carcinogens from 90 percent down to 3 percent just by cutting rodent caloric intake 20 percent."

Dr. Ronald Hart, Director of the National Center on Toxicological Research, 1990[1]

". . . the early claims relating to health and environmental risks posed by pesticides, herbicides and food additives were based on plausible scientific theories backed up by semi-plausible scientific data. . . . Using more comprehensive data, it has been possible to show that many of the hypothesised risks are not worthy of concern. The more recent scares, however, are based upon scientific theories that are only semi-plausible and are backed up by almost no data at all."

Julian Morris, *Fearing Food,* Butterworth-Heinemann, Oxford, 1999, p. xxvii–xxviii

Rat tests mislead the public. They may communicate some useful information to the experts, but they have misled the public on the real cancer risks in our lives. For the last 20 years, we have tested compounds for cancer by exposing rats to what we call the Maximum Tolerated Dose (MTD). In other words, we see how much of the stuff we can cram into their diets without either killing them or triggering a dangerous tumor. Some of the rats have successfully tolerated 100,000 times the maximum expected human exposure (MHE) to certain chemicals. If they developed tumors at 101,000 times the MHE, however, the compound went on the "cancer list." The case of the sweetener saccharine illustrates how the high-dose tests can mislead. In the saccharine tests, male rats developed bladder tumors after eating a daily lifetime dose of saccharine totaling a relatively huge five percent of their *total diets*.

We now know that at such high exposures, saccharine crystals formed in the rats' bladders and caused constant irritation. Female rats didn't get the cancers—because their bladders don't form the crystals.

In this case, the cancer clearly seems to result from the crystals, not the chemical compound. Nevertheless, saccharine went on the cancer suspect list.

Typical MTDs (Maximum Tolerated Dose) and Human Equivalents

Chemical	*Lab Animal Dose*	*Comparable Human Dose*
cyclamates	5% of diet	138–522 12-oz sodas per day
saccharin	5–7.5% of diet	500 times typical consumption of sweeteners
Alar	0.5–1% of diet	28,000 lbs. of apples daily for 10 years

Source: *From Mice to Men: The Benefits and Limitations of Animal Testing in Predicting Human Cancer Risk*, New York, American Council on Science and Health, New York, 1991

Half of all the compounds tested by MTD produce tumors. Is it really possible that half of all the chemical compounds in the known and natural worlds cause cancer? Two-thirds of the compounds now on the National Toxicology Program cancer listing are there because of the MTD tests. Virtually all of the public pesticide scares have likewise come from MTD tests. The first was the original Thanksgiving cranberry panic (over a compound named aminotriazol) in 1959. The latest was the Alar scare in apples.

Now, we're finding that the rat tests didn't tell us anything accurate about the real human risks from the chemicals. Instead, they frightened us about safe compounds and foods. They pushed our priorities in the wrong directions. As an example, the Alar scare reduced fruit consumption—especially among kids—when we should have been raising it. In cancer as in almost everything else, the dose makes the poison.

MYTHMAKER:

"Environmentalists argue that since it is not known whether cancer-causing substances have a threshold, i.e., whether a certain amount need be in the body before it triggers uncontrollable cell proliferation, it is foolish to try to find a 'safe' exposure level."

"Priorities," *Sierra*, Sierra Club, January/February 1994, p. 43

VOICES OF REALITY:

"We believe that the MTD (Maximum Tolerated Dose) should be converted to the minimally toxic dose . . . or the highest subtoxic dose."

C. Jelleff Carr et al., "A Critique of the Maximum Tolerated Dose in Bioassays to Assess Cancer Risks from Chemicals," *Regulatory Toxicology and Pharmacology*, 1991, vol. 14, pp. 78–87

"Environmental policy has too often evolved largely in reaction to popular panics. . . . As a result, many scientists and health specialists say, billions of dollars are wasted each year in battling problems that are no longer considered dangerous, leaving little money for others that cause far more harm . . . for instance, thousands of regulations were written to restrict compounds that had caused cancer in rats or mice, even though these animal studies often fail to predict how the compounds might affect humans."

Keith Schneider, "New View Calls Environmental Policy Misguided," *New York Times*, March 21, 1993, p. 1

Just to put the dietary skewing of the rat tests into perspective, a big recent Finnish study of beta-carotene in human diets may have been thrown off by a fifteenfold increase in dosage. A $43 million test, in which beta-carotene was added to the diets of 29,000 smokers, was supposed to reduce cancer rates. Instead, it raised the lung cancer rate by 18 percent. The researchers were astounded—and suggested as one possible cause that they had the respondents take too much beta-carotene. They ate three times as much beta-carotene as they would have gotten from a diet rich in fruits and vegetables. The synthetic beta-carotene they consumed was also five times more readily available to the respondents' systems than normal beta-carotene. The resulting levels were enough to produce skin yellowing in more than one-fourth of the men.[2]

And then we assume that giving rats 100,000 times a "normal" exposure will give us a useful measure of toxicity!

The dose certainly causes the cancer in the case of gray asbestos. Canadian asbestos workers have had a higher risk of lung cancer than the general population. Their wives, living near the mines, do not have more cancer. Thus, the fibers of gray asbestos (the type used in virtually all building insulation applications) are toxic—but only at the high doses suffered by asbestos miners and installers working in air filled with floating fibers. There are no cases of passive non-workplace asbestosis.

Our state and local governments apparently have spent some *$20 billion* of public money to remove the gray asbestos from schools and public buildings—with no public health gain! (Blue asbestos, used in World War II ships, is far more deadly—but it was never used in buildings.)

DIOXIN DOUBTER:

"You can imagine my horror when [the Agent Orange] story, the biggest of my life, began to slip through my fingers. . . . Nothing checked. . . . Scientists, all off the record, were telling me the whole thing was hysteria. . . . Ranch Hand was the code name for the squadron that sprayed Agent Orange. Well, flyers are pretty macho. . . . They developed an initiation rite in which all new arrivals had to drink a cup of Agent Orange. They were supposed to wear protective clothing; they wore boxer shorts and tennis shoes, and were commonly covered with

Agent Orange. . . . Well, these guys were obviously prime targets . . . and their health, ten and fifteen years after exposure, was very normal. No excess cancer, heart disease, alcoholism. . . . Most newspapers didn't run this story."

Jon Franklin, journalist and two-time winner of the Pulitzer prize, in "Poisons of the Mind," keynote address to the Society of Toxicology, March 16, 1994[3]

What About Dioxin and Other Poisons in Our Food?

In 1999, the European Union ordered Belgium to test all of its beef exports for traces of the dreaded chemical dioxin. The entire Belgian government was forced out of office after dioxin traces were discovered in some feed on Belgian farms and, presumably, some dioxin-bearing pork and chicken got into supermarkets.

European consumers reacted in horror, thinking the dioxin was a cancer risk. Ralph Nader once said that three ounces of dioxin could kill a million people with cancer—but he was wrong. The United States government bought up and evacuated the entire town of Times Beach, Missouri, because dioxin-contaminated oil had been unwittingly sprayed on the roads to keep down dust. The official who evacuated the town now says it was "an overreaction."

At the same moment that Europeans were firing their leaders for not preventing the dioxin exposure, many of them were accepting much larger food risks. As "natural food" devotees, they cheerfully ate alfalfa and bean sprouts (high-risk carriers of the deadly E. Coli 0157:h7), and drank unpasteurized "natural" fruit juices (likely to carry salmonella and other dangerous bacteria).

We now know that dioxin is hugely toxic—to guinea pigs. It seems to have virtually no impact on hamsters or people. The carefully monitored former residents of Times Beach show no unusual pattern of ill health or disease. Neither do the residents of Seveso, Italy, who were exposed to very high levels of dioxin after a chemical plant exploded in 1976.

The U.S. National Institute of Occupational Safety and Health (NIOSH) examined the deaths of 5,000 U.S. workers whose jobs had brought them into contact with dioxin. (Dioxin is not manufactured; it is a by-product of high-temperature combustion. Most of it comes, ironically, from forest fires; the United States generates about four pounds of dioxin per year.)

NIOSH found a slight risk elevation, among workers who had more than 500 times the normal accumulation of dioxin in their tissues. However, the elevated risk was for lung cancer, and NIOSH has no data on how many of the workers smoked. NIOSH has found no elevated cancer risk among forestry workers.

The International Agency for Research on Cancer still classifies dioxin as a human carcinogen, but admits it has no convincing evidence. It assumes that dioxin can cause human cancer because guinea pigs and humans have similar dioxin-receptor proteins. (But then again, so do hamsters who are unaffected by dioxin.)

Belgium spent $1.5 billion in the frenzy to collect and destroy the feed and meat containing dioxin traces—with no discernible health benefit. That may be the most expensive spasm of cancer fear since the United States spent $20 billion to remove inert gray asbestos from its schools and public buildings.

Organic proponents say, "It doesn't take a genius to understand that if we put poisons on our crops, we'll be consuming poisons on our plates." The problem is that there's no way to avoid the world's poisons. Remember what Paracelsus, one of the world's first real scientists told us 500 years ago, "The right dose differentiates a poison from a remedy." Even grass, water, and sunshine will poison us at high enough doses.

Table salt can cause cancer at high enough doses. The limonene in your orange juice can cause cancer in lab rats, but you'd drown before you ingested enough orange juice to get cancer from it.

The medicines we take are certainly poisons; we take them because they kill things, like dangerous bacteria. Even too much aspirin can kill us, but an aspirin a day helps to fight off heart attacks.

Instead of fearing toxic chemicals, because virtually everything is a toxic chemical, we should fear toxic doses. "The dose makes the poison" is especially valid in a world where modern scientific wizardry can detect chemicals at parts per trillion. (A part per trillion is one aspirin in a million gallons of water.) The one additional thing we could reasonably do? If Paracelsus were alive today, he'd say, "Skip the cancer scares and irradiate your food."

DNA Testing Is Also Testifying Against the Rat Tests.

Researchers say the examination of frozen DNA sections from the test animals give another good secondary test for suspicious symptoms. With it, researchers can sometimes distinguish between problems brought on by the heavy dosage and those triggered by the compound itself. The DNA tests say saccharine is safe—unless you're a male rat consuming 100,000 times the expected maximum human exposure of that compound.

Twenty years ago, we weren't sure that the dose *did* make the poison in cancer. We opted for the most drastic test. Today, we can develop a more realistic set of cancer risk tests. We can figure out the maximum expected human exposure to the test compound, and test at a safety multiple of that. (One hundred times Maximum Expected Human Exposure? One thousand times? Certainly not 100,000 times.)

Western Europeans use a more realistic test. They look for the level where the animals begin to develop symptoms of stress such as weight loss, or other symptoms which show up in autopsies.

A more realistic set of animal tests would be truly more useful in guiding our lives and our choices—instead of being used primarily as a scare tactic to drive us away from our most healthful foods.

The OECD/European Union guideline for carcinogenicity studies states the following:

> *The highest dose level should be sufficiently high to elicit signs of minimal toxicity* without substantially altering the normal life span due to effects other than tumors. Signs of toxicity are those that may be indicated by alterations in certain serum enzyme levels or slight depression of body weight gain (less than 10 percent).

Ranking Cancer Risks

This is a ranking by a top cancer expert of the relative risks posed by pesticides and food additives *if the rat tests were an accurate guide to human risk:*

Source and daily exposure	*Risk factor*
wine (one glass)	4,700.0
beer (12 oz.)	2,800.0
cola (one)	2,700.0
bread (two slices)	400.0
mushroom (one, raw)	100.0
basil (1 gram of dried leaf)	100.0
shrimp (100 grams)	90.0
brown mustard (5 grams)	70.0
saccharin (in 12 oz. diet soda)	60.0
peanut butter (one sandwich)	30.0
cooked bacon (100 grams)	9.0
water (one liter)	1.0
additives and pesticides in food other than bread and grains	0.5
additives and pesticides in bread and grain	0.4
coffee (one cup)	0.3

Source: Bruce Ames, et al., "Ranking Possible Carcinogenic Hazards," *Science*, vol. 236, April 17, 1987, p. 271

In contrast, the 1987 EPA position document on MTD from the Office of Pesticide Programs states the following:

> The highest dose to be tested in the oncogenicity study should be selected below a level which resulted in significant life-threatening toxicity in the subchronic study. *The level should not be selected too far below a life-threatening level. . .* [emphasis added]

Europe has had no epidemic of cancers from using its milder form of testing. In effect, we've had a 30-year experiment in whether the dose makes the cancer or not, using a milder test in Europe and a more radical test in the United States. The MTD has flunked. It has not delivered lower cancer rates. Worse, the MTD results have been widely used to frighten consumers away from the one major positive behavior change they can make to cut their cancer risks substantially—eating more fruits and vegetables.

The MTD results also identify so many "carcinogens" that the public simply shuts off the results. If everything causes cancer, then nothing causes cancer. This is surely a dangerous attitude, because some things *do* cause cancer, and some other things help *fight* cancer.

The EPA is currently being sued by the government of Peru for classifying chlorinated drinking water as a carcinogen. Peruvian officials saw the classification, removed chlorine from their drinking water—and caused a cholera epidemic which cost an estimated 5,000 lives.

The clincher: Researchers have found evidence that most of the known human carcinogens can be detected in animals at dose levels well below the MTD. Thus, reduced doses would show up the important health hazards without flagging the trivial.[4]

Science editorial, September 9, 1994

"Risk Assessment of Low-Level Exposures"

"In one example, 11 chemicals known to cause cancer at high doses were administered at low levels. With 8 of 11 substances . . . Instead of damaging the rodents' livers, the low doses were apparently beneficial to them. . . . In the above instances, safe (diminished cancer) levels of exposure exist for substances known to cause cancer at higher doses. . . .

"The use of linear extrapolation from huge doses to zero implies that 'one molecule can cause cancer.' That assertion disregards the fact of natural large-scale repair of damaged DNA. . . . Adult humans are internally exposed to about 500 [grams] per day of oxygen—a relentless known destroyer of DNA. . . . Creatures ranging from microorganisms to mammals could not survive if they did not have mechanisms to respond to challenges from their environments. . . .

"The current mode of extrapolating high doses to low-dose effects is erroneous for both chemicals and radiation. Safe levels of exposure exist. The public has been needlessly frightened and deceived, and hundreds of billions of dollars wasted. . . . "

Source: Philip H. Abelson, *Science*, vol. 265, p. 1507

What About the Circle of Poison?

The Circle of Poison is another of those brilliant public relations ploys invented by the environmental movement.

We've already seen that pesticides don't kill people, unless farmers or farm workers are careless with the few compounds that are truly dangerous to handle. We've already noted that currently approved U.S. pesticides don't threaten wildlife except when misused, and that the dangers are declining rapidly as we get more of the safer, low-volume pesticides. We've just detailed the exaggerations in human risk introduced by the rat tests.

But all of that rational thought pales beside a menacing, provocative phrase like "Circle of Poison."

That phrase vividly conveys the impression the environmental movement wants to create: an America beset on all sides by foreign farmers using vile chemicals banned in the United States to grow unsafe food that will then be smuggled into U. S. supermarkets past the unseeing eyes of the Food and Drug Administration.

In the first place, few countries dare to use much in the way of pesticides that aren't legal here to grow food for the United States market. Any country thinking about slipping us Circle of Poison pesticides on their exports has surely noted what happened a few years ago when two grapes from Chile were found to contain small amounts of cyanide. Chile's huge fresh fruit and vegetable trade with the United States was virtually shut down for the rest of the season. The country, its farmers and farm workers lost hundreds of millions of dollars. (Later, tests indicated that if the cyanide had been put in the grapes in Chile, the grapes would have rotted before their arrival in the United States. It is almost certain that the cyanide was injected on the docks in Philadelphia. That would also explain the telephone tip-off to FDA inspectors.)

But what if we can't detect the residue? If we can't detect it, given today's gas chromatography and parts-per-trillion technologies, then it is almost certainly because it isn't there.

Of course, if the dose makes the poison, then residues that can't even be detected at parts per billion will be far too weak to overwhelm our natural defenses.

POLITICAL MYTHMAKER:

"If we are going to have tougher pesticide standards at home, it makes no sense to allow American companies to use a loophole in current law to dump unsafe pesticides abroad. . . . I am pleased the Administration is willing to do more to close the Circle of Poison loophole."

News release from the office of Sen. Patrick Leahy, D-VT, then chairman, Senate Agriculture Committee, January 25, 1993

REALITY:

"First, there may be situations in which a hazardous pesticide is essential for control of a major pest, as in the case of fungicides used to control late blight in Ecuadorian potato production. . . . There are no effective substitutes for fungicides and farmers who understand the health risks of fungicide exposure may choose to use these materials because their (food production) benefits are commensurately high."

John Antle and Susan Capalbo, invited paper for the 1994 annual meetings of the Allied Social Sciences Association, Boston, January 4, 1994[5]

Reality Comment: Note that Antle and Capalbo are talking about pesticide risks to Third World farmers, not First World consumers. There is no evidence that particular pesticides are being exported to foreign countries and coming back (past our FDA inspections) to afflict U.S. consumers.

Are We Poisoning the Rest of the World?

Then there is the matter of making pesticides in the United States that aren't needed here. We don't grow bananas, for example, or coffee. Should American companies be forced to make good coffee and banana pesticides overseas rather than providing the jobs here? (The banana pesticides, considered relatively dangerous, are a problem to the applicators, not to consumers or factory workers.)

Asia is also using some harsher pesticides (as measured by applicator risk) on their rice paddies, but they don't sell rice to the United States. Moreover, they are making their own decisions about the relative risks of applicator health versus running out of rice. What about a grass herbicide that was approved more rapidly in Argentina than in the United States? Should Argentina be denied the right to use it? Or again, should the company have to relocate its production and jobs outside the United States because our chemical approval process is too slow?

NEEDED—A CANCER TEST THAT TELLS THE TRUTH:

". . . High dosing may falsify the experiment in one of two ways: It can either poison the cells and tissues so severely as to prevent a carcinogenic response that might otherwise have been found, or it can 'overload' and change metabolic processes so as to cause a carcinogenic response which would not normally occur . . ."

Edith Efron, *The Apocalyptics*,
New York, Simon and Schuster, 1984, p. 248

America needs a better cancer testing system than the current high-dose rat tests. The system needs to incorporate these facts that we've learned about cancer over the last 30 years:

•Cancer is fundamentally a degenerative disease of old age.

•The doses and repetition of the insult are critically important to the risk.

•Smoking and our own heredity are the biggest risk elements in cancer, not the external environment.

The Alar scare was perhaps the most vicious swipe at America's health the environmental movement has ever perpetrated. It led parents and school administrators to take apples away from their kids, pour healthful apple juice down the drain, and ban apples from our schools. By implication, every mother and father in the country was forcibly reminded of the message the Natural Resources Defense Council wanted to convey—that nonorganic fruits and vegetables were dangerous.

Nothing could be further from the truth or from the message we need to send to kids, parents, and grandparents:

> The EPA, the FDA, and the USDA are doing their jobs, and making sure there's no harmful pesticide on that attractive apple or potato. You can eat produce in full confidence that it will add to your health, not subtract from it.

Virtually every cancer scare we've ever had has come from the high-dose rat tests. If there were no high-dose rat tests, there would be no scare material. We could then use the billions of dollars wasted on the high-dose tests on things that *would* improve public health. It is past time to make the tests a more realistic reflection of actual human risks. If left as they are, they will continue to be used for blatant fearmongering which will undermine our health.

Notes

1. Dr. Ronald Hart, quoted in Warren T. Brookes, "Pesticide Phobia, a Dangerous Health Threat," *Detroit News*, April 16, 1990, p. 7A.
2. Kathleen Meister, "Antioxidants and Lung Cancer: What the Conflicting Reports Mean," *Priorities for Long Life and Good Health*, New York, CAST, vol. 6, no. 3, 1994, pp. 7–11.
3. Jon Franklin is currently professor of Journalism, University of Oregon, Eugene, Oregon.
4. A. Apostolou, "Relevance of Maximum Tolerated Dose to Human Carcinogenic Risk," *Regulatory Toxicology and Pharmacology*, vol. 11, 1990, pp. 68–80.
5. Antle and Pingali, "Pesticides, Productivity and Farmer Health: A Philippine Case Study," *American Journal of Agricultural Economics*, August 1994, vol. 76, no. 3, pp. 418–430.

8

There Is Much Less Hunger Than We've Been Told

MYTHMAKERS SAY:

"A lifetime of malnutrition and actual hunger is the lot of at least two-thirds of mankind."

Lord Boyd-Orr, director-general,
UN Food and Agricultural Organization, 1950

"If present trends continue, the world in 2000 will be more crowded, more polluted, less stable ecologically and more vulnerable to disruption than the world we live in now. Barring revolutionary advances in technology, life for most people on earth will be more precarious in 2000 than it is now. . . . [T]he number of malnourished people in the LDCs could rise from 400–600 million in the mid-1970s to 1.3 billion in 2000. . . . In the developing world, the need for imported food is expected to grow."

Major Findings and Conclusions, *Global 2000 Study*,
Carter White House, 1980

"Does human society want 10 to 15 billion humans living in poverty and malnourishment or 1 to 2 billion living with abundant resources and a quality environment?"

David Pimentel et al., "Natural Resources and an Optimum
Human Population," *Population and Environment*,
Human Sciences Press, 1994, p. 348

REALITY SAYS:

"Approximately 600 million people in the developing countries do not have access to enough food to meet their nutritional needs. Only a small portion of those people are clinically malnourished. The preponderance of them are mildly to moderately undernourished. Others are poorly nourished on a seasonal basis. Still more are at the margins of adequacy such that major illness, increases in food prices or decreases in real income could force them into nutritional deficit."

USAID policy statement, 1984

There is a great deal of good news on population and hunger in today's world. In the first place, we now know that we don't have an upward population spiral. Thanks to affluence, contraceptives, and TV, we have a fighting chance to peak the world's population at 8.5 billion rather than 10, 12, or 25 billion.

Furthermore, the Third World countries that have done best in raising their crop yields, have also done best on bringing down birth rates! We don't have to watch people starve in order to restabilize the world's population at a sustainable level.

Bread for the World, a charitable organization, is even telling the public that we have the capacity to eliminate hunger in the world.[1] They are probably correct.

To eliminate hunger quickly, however, we will have to press forward on two fronts with more effort than we have mustered thus far.

We must, first, make larger investments in high-yield farming for the Third World. Those investments are not currently being made, in large part because eco-activists have been crusading against the high-yield seeds, the fertilizers, and (yes) the pesticides that will need to be part of the high-yield packages.

Second, we must simultaneously upgrade the skills of Third World workers and instill the necessary concern for honesty and human rights in the governments of Third World countries. Virtually all of the world's famines since World War II have been due to "mistakes of government" such as government grain monopolies and shooting wars (especially civil wars).

Even *with* these limitations, the world has made enormous progress against famine. The biggest factor in that achievement has been high-yield farming. The resulting high food production has been backed by rapidly spreading public health services such as clean water, waste treatment, and vaccinations. These cut down diarrheic and other diseases that prevent people from taking full nutritional advantage of their food supplies.

With the notable exception of Africa, the world has come perhaps 90 percent of the way to eliminating famine since the Rockefeller Foundation started the Green Revolution some 30 years ago. Consider just a few dramatic facts:

- The threat of severe hunger no longer stalks perhaps two-thirds of the world's population on at least a periodic basis, as it did within living memory. Today, the threat of severe hunger affects perhaps 5–7 percent of the world's people in any year when Africa is *not* having one of its big continent-wide droughts.
- *Per capita* calories in the Third World have risen by about one-third since 1960.[2]
- Asia, with about three-fourths of the world's population, has raised its per capita food intake by about one-third since 1960, in a region that many experts thought would suffer mass starvation

Figure 8.1 demonstrates graphically how sharply increased food productivity and rising consumer incomes have reduced hunger and malnutrition in recent years. Again, only in Africa has added food production *not* begun to break the back of the malnutrition problem.

Astonishing Gains Against Hunger

In 1970, the world gave the Nobel Peace Prize to Dr. Norman Borlaug—the plant-breeder who developed the Green Revolution wheat varieties. Back then, the Green Revolution's "miracle" wheat

Figure 8.1

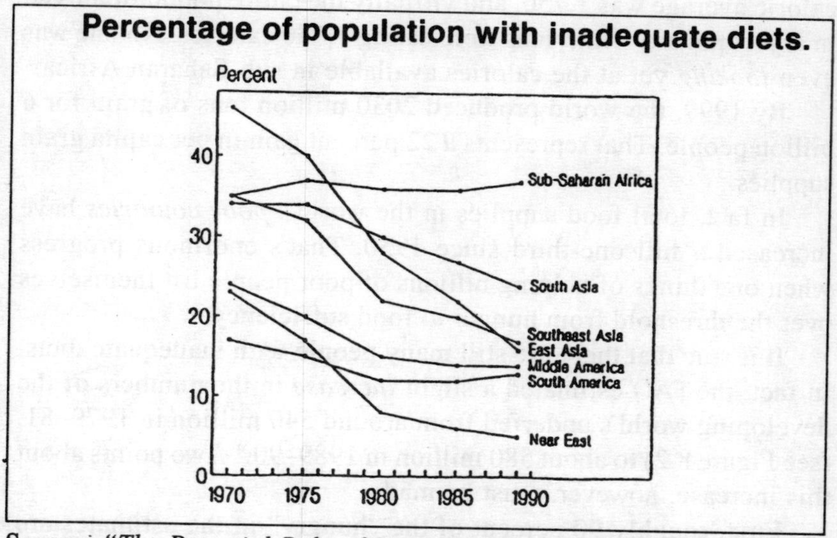

Source: *"The Potential Role of Biotechnology in Solving Food Production and Environmental Problems in Developing Countries."* Presented to the ASA-CSSA-SSSA Annual Meeting, Cincinnati, Ohio, November 1993.

and rice varieties seemed one of the outstanding achievements in human history. In the years since, however, some people have been having second thoughts. They worry that saving 500 million people from famine might just give the world a bigger famine—with even more deaths—later on.

But famine is not inevitable. (Even the famous Dr. Thomas Malthus came to realize this later in his life, and the tone of his later writings was far different than the Malthusian gloom for which he is famous.)

In 1950, the world produced 692 million tons of grain. This represented the key food supply for what was then a population of 2.5 billion people. That was the period when the FAO director-general was estimating that two-thirds of the world's population suffered from food deprivation. The world had just suffered a major famine in Bengal (1943) and was about to suffer another in China.

In 1950, Americans were eating around 3,200 calories per day. The average resident of China was getting about 2,100 calories per day. In India, the average was estimated at 1,700. In Indonesia, the caloric average was 1,750, and virtually the entire population lived in abject poverty. Bolivians were eating 1,760 calories. No one was even *looking* yet at the calories available in sub-Saharan Africa.[3]

By 1999, the world produced 2030 million tons of grain for 6 billion people. That represents a 22 percent gain in per capita grain supplies.

In fact, total food supplies in the world's *poor countries* have increased a full one-third since 1950. That's enormous progress when one thinks of helping billions of poor people lift themselves over the threshold from hunger to food sufficiency.

It is true that there are still many people with inadequate diets. In fact, the FAO estimated a slight *increase* in the numbers of the developing world's underfed from around 540 million in 1979–81, (see Figure 8.2) to about 580 million in 1989–90.[4] Two points about this increase, however, must be made.

First, roughly 90 percent of the "hungry" in the estimates are within 10 percent of having fully adequate calories for good health.[5] In fact, many of the so-called hungry lack "adequate" calories because they prefer to spend part of their food budgets for higher quality calories such as milk and fruit, rather than increasing their consumption of low-cost calories such as cassava flour.[6]

Second, world population has grown by more than one billion

Figure 8.2

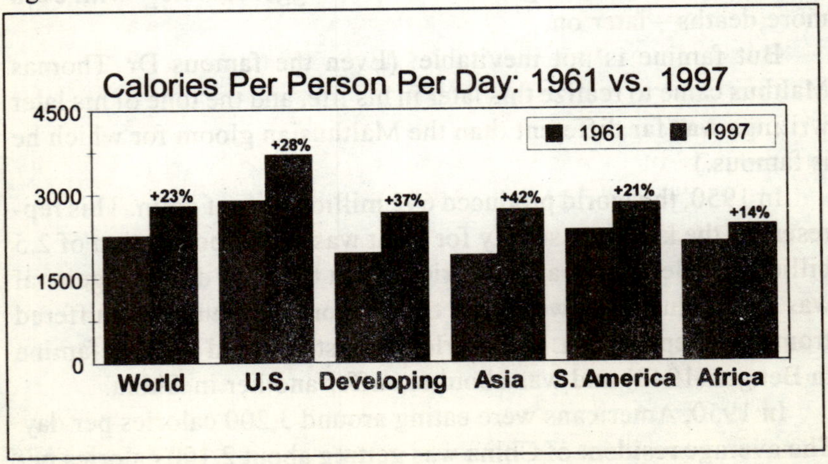

Source: *FAO Production Yearbooks*

Figure 8.3

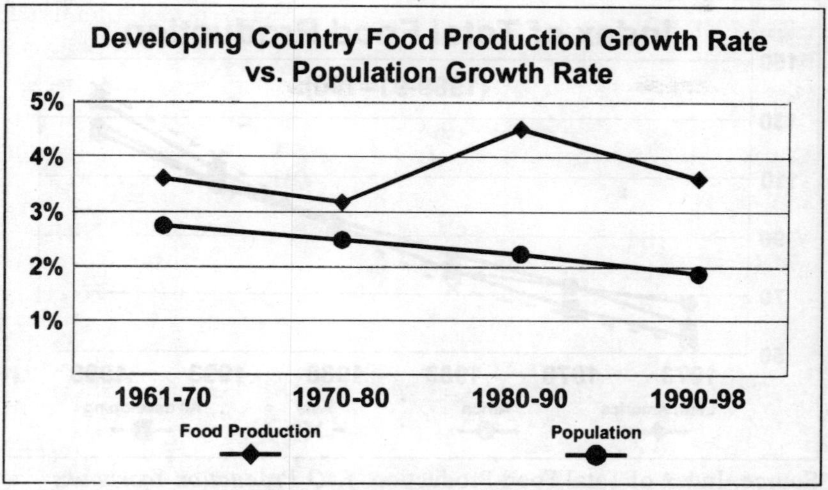

Source: Index of Total Food Production, All Developing Countries, *FAO Production Yearbook*, 1992, pp. 43–44

people during those years, so the increase in per capita food supplies represents a major productivity triumph.[7] Meanwhile, further progress beckons as better seeds and farming systems reach into such remote regions as Ethiopia, Ecuador, and Mongolia.

Per Capita Food Gains Accelerating Now

In fact, now that population growth rates are tapering off, progress against malnutrition is *really* picking up speed. (See Fig. 8.3.)

The rate of increase in the Third World's food production in recent years has been more than double its population growth rate. Progress in the *next* decades should be even more rapid, since the Third World population growth rate over the next decade will be declining from 1.6 percent. Nor has there been any *slackening* in the rate of food production gain, contrary to some highly publicized reports. (See Fig. 8.4.)

Per capita food supplies, as well as production, have continued to gain (again, with Africa as a modest exception). (See Fig. 8.5.)

It is important to remember that sub-Saharan Africa, where the remaining hunger is concentrated, contains only 7 percent of the world

Figure 8.4

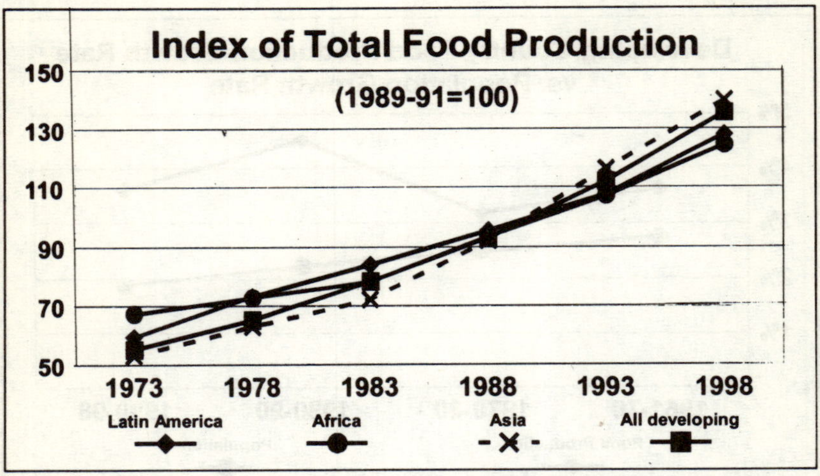

Source: Index of Total Food Production, *FAO Production Yearbooks*

population. Asia, where the success has been concentrated, represents the huge majority of the Third World's population, and *three-fourths* of the total world population. Latin America, which has also registered increases, accounts for 6 percent of the world's people.

The African Exception

Africa is today the outstanding exception to the good news on hunger. There are several reasons why this is true. Moreover, Africa should be able to do far better at food production in the future.

- First, Africa was until recently a sparsely populated continent with ample room for a low-yield/low-cost farming system called bush fallow. As a whole, the continent has probably never planted even 25 percent of its good cropland in any one year, because most of it has been in bush fallow, recovering its fertility without benefit of fertilizer.

- Second, the early Green Revolution scientific efforts were targeted at Asia, because our fear in 1960 was that billions

Figure 8.5

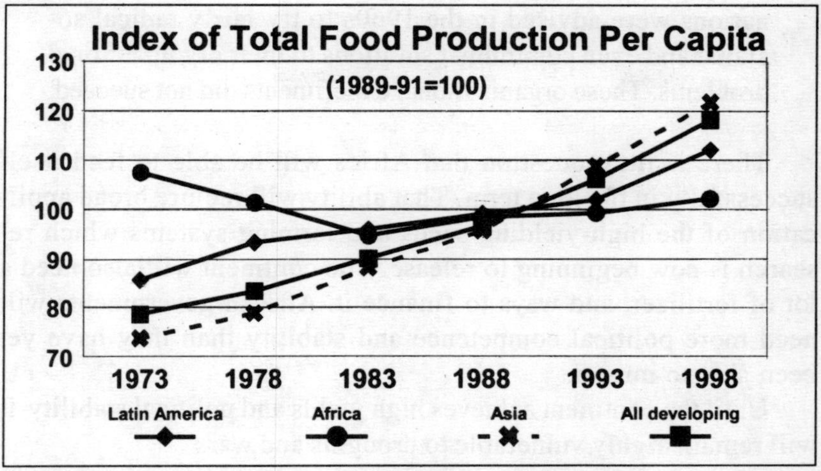

Source: Index of Total Food Production Per Capita, *FAO Production Yearbooks*

of people would starve there first. Africa was considered a distant second in urgency.

•Third, when attempts to raise African farm productivity did begin, researchers were startled to find that almost none of the productivity results from other regions could be transferred successfully to Africa. The research on Africa had to start from square one, and that delayed the results by many years. Now, however, high-yield seeds and systems are beginning to flow from the African research stations.

•Fourth, Africa's dramatic drop in food security since 1970 reflects a real decline in living standards. But it also reflects the decline of African *crop reporting* systems. Many postcolonial governments have lacked the jeeps and gasoline to send their crop estimators out into the countryside. What is not seen is not counted.

•Finally, African countries had little experience with self-government in 1960 when most of them began to get their independence. As an additional handicap, many African nations were advised in the 1960s to try fairly radical socialist and even communist solutions to their organizational problems. These organizational experiments did not succeed.

There is little question that Africa will be able to feed itself successfully in the long term. That ability will require broad application of the high-yielding seeds and farming systems which research is now beginning to release. The continent will also need a lot of fertilizer, and ways to finance it. African governments will need more political competence and stability than they have yet been able to muster.

Until the continent achieves high yields and political stability, it will remain highly vulnerable to droughts and wars.

Recent Famines Were Due to Shooting, Not Food Failure

Most of the world's recent severe hunger has been due to shooting wars, with wrongheaded policies of the hungry countries' own governments ranking as the second most frequent cause.

The last major famine in the non-Communist world was in Bengal in 1943, when 1.5 million people died of starvation after severe flooding destroyed their crops. Communist China suffered a much larger famine during Mao Tse-tung's Great Leap Forward in the late 1950's. Outside estimates range from 16 to 30 million deaths during Mao's Great Leap disaster.[8]

Civil wars have been the key in the recent hunger events in Somalia, Ethiopia, Liberia, Angola, and most of Africa's other recent famines. Thus, in a real sense, the world's recent famines could have been avoided.

They have also afflicted only a tiny fraction of the world's population. The civil strife and resulting hunger in Somalia in 1992 involved only 0.001 percent of the world's total population and a few thousand deaths. The hunger pattern of Sudan in the 1990s involved perhaps 0.002 percent of the world population—and is occurring because the Sudanese government is actively trying to prevent food from reaching their rebellious South. In both Somalia and Sudan, there was enough food available to prevent deaths from famine.

The fact that our famines now are small doesn't make them unimportant. But we must not let our anguish over small-scale suffering blind us to the fact that we have found successful strategies to stop the large-scale famines. Famine caused 20–25 million deaths in the last quarter of the 19th century. Realistically, the famine death toll for 1975–2000 is probably less than one million, rather than 50 million or more.

Virtually all of the recent famines have featured the same deadly combination:

First, *they all occurred in African countries* that had not yet developed high-yielding agricultures. Africa got a late start on agricultural research, in large part because it was land-rich (or sparsely populated) until recently. As mentioned earlier, high-yielding seeds and farming systems are only now beginning to become available to African farmers, 30 years after the Green Revolution began in earnest and 60 years after U.S. farmers got hybrid corn.

World Vision International Photo by E. Mooneyham

THE FAMINE-STRICKEN FEW—Newspapers are fond of publishing photos of famine victims like these Ethiopians beset by civil war and drought in 1981. Fortunately, rapid gains in the Third World's per capita food production are making it harder and harder to take such photos.

Second, *All of the famine countries' governments had agricultural policies that discouraged their farmers from planting high-yielding seeds and trying to meet rising food needs.* Such policies as low fixed prices for grain, government grain monopolies, and export taxes amount to powerful disincentives, which drive farmers back into the subsistence mode they know so well.

Most important, *shooting wars,* which make it impossible for farmers to raise and market food normally. Combat turns problems (like ethnic tensions, poverty, and drought) into full-blown disasters.

Has there been any recent famine attributable to lack of food production potential? Only in the Sahel, the "edge of the desert" region south of the Sahara Desert, which suffered severe and extended drought both in 1973–74 and in 1983–84. But the entire population of the Sahel is about 25 million people, *one-half of one percent* of the world's population. The Sahel can support most of them most of the time, but not during the deep, extended droughts that occur there. Even high-yield farming research has been unable to break the grip of aridity on the Sahel to date. Archeologists say that, in the past, Sahel has been unpopulated for centuries at a time apparently because it was too dry even for pastoral grazing.

Nor does all of the progress mean that the next big African drought will not produce almost as much suffering as the last one. We have not yet drought-proofed Africa the way we have India. But the solution is not to let more Africans die. The fastest, most humane solution is to breed more high-yielding seeds for African farms; build new fertilizer plants there; provide consistent, realistic incentives for farmers; and support them with political stability, roads, and storage silos.

Nearly All Countries Could Now Produce Adequate Diets

Other than the Sahel, there are only few countries or regions that could not readily produce their own base calories today, using their own natural resources and the high-yielding seeds and farming systems of the continuing Green Revolution. Base calories are the fundamental calories—usually root crops and cereals—needed to supply minimum nutrition requirements in a poor country. Wealthy countries substitute quite a few "affluence calories" from meat, milk, eggs, and fresh fruits and vegetables.

Meanwhile, of course, the affluent countries of the world are struggling with surpluses and deliberately *trying* to limit food production gains; their low rate of increase, combined with troubles in the agriculture of the former USSR, kept the world average of grain production increase down to less than one percent in 1986–91.[9]

Ironically, Lester Brown of the Worldwatch Institute has been claiming that the world is now headed for famine, because per capita grain stocks have been declining since 1984. What he does not mention is that the slump has all been in the First World—and has no hunger implications at all.

This does not fully answer the question of how densely populated countries can provide their consumers with high-quality diets rich in fruits, vegetables, meat, and milk. Generally, however, when countries can afford such high-quality diets, they can afford to import foodstuffs, meat, and the other items they prefer.

USDA Photo Library

A MOUNTAIN OF EXTRA GRAIN—For the past 25 years, we have been told to expect massive famine, but we had surplus grain piled on the ground instead of famine victims in the streets.

Reprinted from BridgeNews Forum, June 5, 1998

THE WORLD'S LAST BIG FAMINE

A lesson to remember: 30 million Chinese died to preserve a politically correct myth.

We're approaching the 40th anniversary of the world's last big famine. It occurred in China, between 1958 and 1961, when some 30 million people starved in the only big famine that didn't start with a food shortage.

China's Communist Party leader, Mao Tse-tung, had read Russian press reports about new farming systems, such as "deep plowing" and "close planting," that supposedly were producing record yields on Russian state farms. Mao was busily forcing his millions of peasants into collective farms, reneging on his promise to give them their own land. He hoped ample harvests would keep farmers content even if they didn't own their land.

But the Russian reports were lies: Plowing too deep turns up subsoil with few nutrients, few soil bacteria, and little capacity to nurture crops. And planting ten times as many seedlings per acre, without fertilizer, starves all the plants. (Mao refused to spend scarce capital on chemical fertilizers.)

Rural officials didn't dare oppose the Chairman, so they followed his instructions. They didn't want to tell him about failure, so they claimed his new methods had tripled the harvests.

When Mao toured model collective farms, they showed him flourishing rice fields so densely planted that small children stood atop the growing grain stalks! (Huge numbers of rice plants had been laboriously transplanted just for his visit, and the children were standing on hidden benches.)

He was driven past miles of piled-up vegetables. Officials said the communes had more food than they could eat, even with five meals a day.

Historically, the government had taken about 30 percent of the harvest for the cities and the Army. And since the countryside was reporting three times as much grain production, Beijing demanded three times as much grain too. But that left only 10 percent of the harvest to feed the farm workers and their families.

Jasper Becker, author of *Hungry Ghosts: Mao's Secret Famine*, says the officials in one county of Henan Province reported a 1959 harvest of

239,000 tons, though the farms had actually produced only 88,000 tons.

Naturally, officials found that they couldn't collect the entire harvest. They ended up with only 63,000 tons, but the inflated production statistics meant that the county still owed a grain levy of 75,000 tons.

Local officials began beating peasants, hoping to force confessions of hidden grain. Some farmers were buried alive or left outdoors to freeze. But no one could find grain that had never been grown.

The state granaries were still full, but the kitchens on the communal farms had nothing but soup made from tree bark, wild grass seeds and wild vegetables. Party cadres smashed household cooking pots to prevent families from cooking hidden food secretly.

After the first year, weakened peasants began to die in huge numbers. The elderly were the first to go. Families stopped feeding young girls. Cannibalism became widespread.

At first, the famine didn't touch the 90 million privileged people living in the cities. Their grain was "guaranteed" by the state. But when the farmers got too weak to farm, the state's grain ran out too, and the famine spread to the cities. The government began distributing food substitutes: ground-up banana tree roots and stems to be mixed with steamed rice; cakes made of rice husks, sugar cane fiber and turnip tops. People hunted rats and sparrows.

Finally, in 1961, Mao's senior lieutenants assembled conclusive evidence of the massive death toll. Mao was forced to reverse his farming policies. Peasants were allowed to plant grain for themselves, but only for a while, and the Chinese government began to make big investments in chemical fertilizer. In 1962, the Chinese government even bought grain—from the hated capitalists in Canada and Australia. People were encouraged to write overseas relatives for food parcels.

However, Mao hated to face his farming failure. Most of the realists were soon to be purged in his Cultural Revolution. Den Zhaio-ping, Mao's eventual successor and the author of China's economic miracle after 1978, was exiled to a remote village for the third time in his career.

Looking back, it seems hard to believe that Mao and his colleagues would be taken in by the Russian pseudo-science, which could have been checked easily with a few test plots.

It is also hard to believe that government officials would rather starve their own people than admit to the world a policy mistake had been made.

China's politically correct famine should stand as an urgent reminder that objective reality must remain the foundation of successful society, especially for its food production.

The Mismeasure of Food—and Famine

Food-providing agencies like CARE and Bread for the World, no matter how well-meaning, are staffed by activists who want more food to distribute. Organizations like the United Nations cite very large numbers of "hungry people"—but many of these organizations have both a legitimate concern for hunger and a vested interest in maintaining hunger donations.

Dr. Thomas Poleman of Cornell University pioneered the effort to achieve more accurate food estimates. His work suggests that we have typically underestimated food supplies by 10–15 percent, especially in the poorest tropical countries.[10] His on-the-ground research found a whole series of mistakes in food estimation which have led us to the erroneous (and rather astonishing) conclusion that most of the world's people have failed to provide themselves with their own first requirement—adequate food. For example:

- Few studies take into account hunting and gathering that are particularly important for people in marginal economies—and which often supply 15 percent of their calories.

- There is a strong tendency to under-report food output. Farmers look at the crop reporter as a tax assessor and governments often choose to justify more aid rather than less.

- Food surveys typically leave out the "street foods," which are part of virtually every culture. Street foods are particularly important for protein, because much of the frying in poor countries is done on the street. (It seldom pays to heat frying fat for one family, whether it is for French-fried potatoes at an American fast-food restaurant or fried ants in Zambia.)

- In the tropics, you can't tell what's growing in the intercropped fields unless you crawl through the fields virtually on hands and knees.

• The poorer the country, the less of its food moves to market; most of it is eaten where it's grown. Few Third World countries spend their scarce dollars on good crop reporting systems.

Death That Mimics Starvation

Most of the world's so-called "hunger" deaths—especially among children—are actually caused by untreated water, lack of sewage systems, lack of vaccinations, and contaminated food.

In fact the most effective method ever found for cutting childhood malnutrition and death doesn't involve food at all. It's a cheap, simple treatment for diarrhea. The UN Children's Fund (UNICEF) says one-third to one-half of all infant deaths in the world are due to diarrhea. UNICEF is saving millions of lives a year by distributing packets of *oral rehydration salts* that quickly restore body fluids and also help the victims digest some of the food available to them. By contrast, food distribution programs seldom have any lifesaving capability, except in genuine famine situations.[11]

Meanwhile, despite such evidence, many in the affluent West cling firmly to our belief in widespread hunger. I constantly find people who have traveled in Third World countries who "know" there is hunger because they have seen emaciated people, and even occasional wasted corpses. They have simply assumed that emaciated people were emaciated because they were hungry. In most places, however, it is far more likely that they are ill.

My wife, Anne, contracted both typhoid and amoebic dysentery while living in Ethiopia, and became a walking skeleton before the diseases were diagnosed—and she was an American attached to our embassy. (While working at a clinic, Anne also had tiny children die in her arms after repeated bouts of diarrhea due to local infection sources.)

When We Don't Know—We Overestimate

Most of the nutrition studies in the world have been done with college kids in rich countries—a practice that helps guarantee that the world's "nutritional requirements" are overestimated. There is little research on the *real* food needs of small-but-otherwise-healthy people whose ancestors have been coping with hunger for generations.

Dr. David Seckler, who has worked with USAID and the Ford Foundation, suggests that a child's body responds *first* to a food shortage by limiting its own growth. Seckler followed up by checking a sample of young Indian men who had been medically screened and found healthy. When he surveyed their food consumption, more than 90 percent of them were "malnourished.[12]

Seckler's idea seems to be strongly backed by a USAID–sponsored nutritional assessment of children in 14 countries: The studies found lots of "malnutrition." Ninety percent of it, however, was kids who were short for their ages (by the standards of the World Health Organization). They had normal weight for height. They were short but not malnourished.

It would certainly be better if these children had fully adequate diets—but being short certainly beats being dead from starvation or mentally retarded by malnutrition.

The World Hunger Problem Doesn't Involve Much Food

The rich countries have donated food aid liberally in recent decades. More often than not, so much food aid arrives in the hungry countries that some of it sits in storage and depresses the local farmers' price for the *next* crop.

Even this ample amount of food aid has required a smaller share of the world's rising grain production. The total has averaged about 10 million tons per year, well under one percent of the world's grain production. Meanwhile, the world's annual grain carryover has ranged between 200 to 400 million tons.

MYTHMAKER:

"It is unconscionable in a world with so many going hungry that we aren't doing more to prevent unwanted children from being born."

Jane Fonda, appointed by President Clinton as U.S. special goodwill ambassador to the UN Population Fund, 1994[13]

Reality Comment: Most of the children born in the Third World *are* wanted. And relatively few of them are urgently hungry, except where there are shooting wars and government mismanagement.

Have We Solved Hunger in America?

What about conditions in the United States itself? It shouldn't be too surprising that few of our citizens are really hungry. The federal government alone is spending more than $33 billion per year on such feeding programs as food stamps, school lunches, women with infants, and special programs for pregnant mothers. Since 1961, federal spending on the food stamp program has expanded from $825,000 per year (for 50,000 recipients) to $22 billion per year, with the free food being offered to more than ten percent of the population of the richest nation on earth.

This expansion has not occurred because of increasing hunger in America, but because of congressional logrolling. Congressmen from urban areas voted more money for farm price supports, and the Agriculture Committees—in a direct exchange—budgeted more money for food stamps.

By the 1970s, we had essentially conquered America's hunger problem. In addition to the federal effort, there were and are hundreds of well-organized and effective feeding and food-providing programs run by organizations like Second Harvest, the Salvation Army, churches, cities, and local charities.

- Childhood anemia, one of the diseases most closely linked with malnutrition, has been cut in half in American children during recent years. At the same time, the percentage of poor children considered underweight has dropped below that in the general population.[14]

- A high proportion of our homeless have been found to be substance abusers and/or suffering from chronic mental illnesses such as schizophrenia. Hunger for these people is rarely a food problem but more accurately a consequence of substance abuse or illness.[15]

But in America, we count missed meals as hunger. Or the kids having to eat rice and beans rather than meat. This is not hunger, though it certainly is not affluence either.

Why So Much Mythology About Hunger?

Hunger is one of our strongest emotional "hot buttons." All of us have *felt* hungry. The emotion triggers age-old frantic responses, like a hungry kitten pouncing on anything that moves.

Hundreds of organizations have found it to be a key fund-raising appeal and many charities tie their campaigns to hunger themes.

High-yield agriculture is a huge success against hunger. Without it, we would have seen people die of starvation by the billions, and/or destroy most of the wildlife habitat in Asia and Latin America already. With high-yield farming, we have had small, terrible, and unnecessary hunger spasms in a few remote corners of the world—most of them caused by civil strife.

Notes

1. Bread for the World Institute, 802 Rhode Island Ave., N.E., Washington, D.C., 20018.

2. "Food Supply: Calories Per Capita Per Day," *FAO Annual Production Yearbook Series*, FAO, Rome.

3. "Calories Per Capita," *FAO Annual Production Yearbook Series*, FAO, Rome.

4. UN Administrative Committee on Coordination, Subcommittee on Nutrition, and International Food Policy Research Institute, *Second Report on the World Nutrition Situation*, World Health Organization, Geneva, 1992.

5. Thomas T. Poleman, *Quantifying the Nutrition Situation in Developing Countries*, Cornell Food Research Institute Studies 18, no. 1, 1981.

6. Cheryl Gray, *Food Consumption Parameters for Brazil and Their Application to Food Policy*, International Food Policy Research Institute Research Report No. 32, Washington, D.C., 1982.

7. Urban and Trueblood, *World Population by Country and Region, 1950–2050*, U.S. Department of Agriculture, Washington, D.C., 1993.

8. Ansley Coale, *Rapid Population Change in China, 1952–82*, National Academy Press, 1984; and Ashton et al., "Famine in China, 1958–1961," *Population and Development Review*, vol. 10, December 19, pp. 613–645.

9. "Cereal Production," *FAO Annual Production Yearbook Series*, op. cit.

10. Poleman, op. cit.

11. Taylor and Greenough, "Control of Diarrheal Diseases," *Annual Review of Public Health*, 10: 221–44, 1989. See also *Proceedings of the Third International Conference on Oral Rehydration Therapy*, sponsored by the U.S. Agency for International Development, The UN Children's Fund and the World Health Organization, Washington, D.C., 1989.

12. Dr. David Seckler, "Malnutrition," *Western Journal of Economics* 5, no. 12, December 1980, pp. 219–26.

13. Jane Fonda, "High Time for Some Population Intelligence," *E, The Environmental Magazine*, vol. 5, no. 1, January/February 1994, pp. 22–24

14. Carolyn Lochhead, "Data Don't Back Claims of Activists," *Washington Times*, July 27, 1988, p. F5.

15. Christopher Jencks, *The Homeless*, Harvard University Press, Cambridge, 1994. Jencks finds that about 25 percent of the U.S. homeless are suffering from mental illness, 33 percent from alcohol abuse, and many more from other dysfunctions such as schizophrenia. He says "the spread of homelessness among single adults was a by-product of five related changes: the elimination of involuntary [psychiatric] commitment, the eviction of mental hospital patients who had nowhere to go, the advent of crack cocaine, increases in long-term joblessness, and political restrictions on the creation of flophouses. Among families, three factors appear to have been important: the spread of single motherhood, the erosion of welfare recipients' purchasing power, and perhaps crack." Cited in a book review by Douglas Besharov, "Book World," *Washington Post*, July 10, 1994.

9

Organic Farming: More Problem Than Solution

MYTHMAKERS SAY:

"Shouldn't we . . . having concluded that we are being asked to take senseless and frightening risks . . . no longer accept the counsel of those who tell us that we must fill our world with poisonous chemicals; we should look about and see what other course is open to us."

Rachel Carson, *Silent Spring*, p. 278

". . . [T]he choice we have as consumers is not between chemically treated fruits and vegetables and no food at all, as some would have us believe. The choice before us is between chemically treated food and produce grown without the use of toxic chemicals. Alternative agriculture is a promise for the future. We have excellent evidence that it is efficient, productive and profitable."

Susan Cooper of the Coalition Against the Misuse of Pesticides, "Do Farm Chemicals Pose 'Unnecessary Risks?'" *Global Food Progress*, Hudson Institute, 1991

"We need a second Green Revolution that will focus on the needs of the Third World's poor, increase the productivity of small farms with low input agricultural methods, and promote environmentally sound policies and practices."

Vice President Al Gore, *Earth in the Balance*, New York, Houghton Mifflin, 1992, p. 322

REALITY SAYS:

"When the first organic wheat is harvested next year, Mr. Lister expects a yield of around 4 tonnes a hectare. Conventionally grown milling wheat . . . yields 8–9 tonnes a hectare."

David Blackwell, "Green Field Site in Essex," *Financial Times*, December 5, 1991, Commodities and Agriculture page

"The United States has about 28 percent of the *organic* nitrogen needed to sustain current farm production."

Conclusion from Gilbertson et al., *Animal Waste Utilization on Cropland and Pastureland,* U.S. Department of Agriculture, 1979

The secret is out. Organic farming is unsustainable on any kind of broad basis.

Environmental activists have focused worldwide discussion on the issue of "sustainable" food production. They have charged that high-yield farming's high-powered seeds are more susceptible to pests than the traditional "landrace" varieties; that irrigation water supplies are running out; that soil erosion is stealing fertility from the fields; that pesticides cannot continue to cope with the insects and diseases; and that chemical-based farming will ruin soils and increase cancer rates among consumers.

USDA

PEST DAMAGE—Few of today's consumers have ever seen the sort of damage that pests can inflict on crops. After this corn borer weakens the cornstalk, it falls over, and the ear of corn that it would have produced is lost.

None of these charges against high-yield farming is true. That's fortunate for the world, because organic farming offers no solution to the world's food or environmental problems.

Organic farming uses no man-made chemicals. It not only does without synthetic pesticides, but also foregoes man-made fertilizers. The organic producers believe that pesticides are dangerous to humans and the environment. They say man-made nitrogen is bad for soils—though all nitrogen is elemental and chemically identical.

Organic farmers *do* use pesticides. They allow themselves the use of "natural" pesticides such as sulfur, a natural biopesticide called *Bacillus thuringiensis*, and pyrethrins (a pest killer that is the natural product of a plant flower). They use copper sulfate, which is toxic to people as well as to insects. Most organic farmers spray more pesticide, more often, than nonorganic farmers.

Mainstream farmers use large amounts of both pesticides and man-made fertilizers. In the United States, pesticide use has increased from a little over 300 million pounds of active ingredient in 1964 to more than 800 million pounds per year in the latter 1980s.

Insecticide use has actually dropped in recent years, from more than 100 million pounds in 1964 to about 70 million pounds in the 1980s. New low-volume compounds and relatively lower crop prices have both had an impact in the decline.

The big recent increase in farming's chemical intensity, however, has been due to the increased use of herbicides (chemical weed killer) in new conservation tillage farming systems, as discussed in Chapter 2.

The use of herbicides expanded from only 70 million pounds in 1970 to nearly 500 million pounds per year in the late 1980s.[1] Since then, the level of pesticide active ingredients has remained roughly stable.[2]

Degrading the Environment with Organic Farming

The day may come when we'll understand biology and ecology well enough at the level of cells and molecules to make organic farming a high-yield success. The new science of molecular biology is beginning to peel away some of the layers of mystery now. But that in-depth knowledge is still at least decades away.

Until then, organic farming will produce far lower and far more erratic yields of many crops than science-based high-yield farming. Organic farming will thus force tillage of more crop acres to produce a given quantity of food. With present knowledge levels, no responsible authority or organization should recommend either organic farming or traditional low-yield farming systems as a broad-gauge alternative to high-yield agriculture.

In fact, slashing farm chemical usage is likely to produce more soil erosion, more human cancer, and less wildlife habitat. At present, organic farming could not even sustain the fertility of our existing cropland, or protect it effectively from erosion.

Nor do the organic farming boosters offer *any* plan to feed the expanded world population of 2050. That alone makes it a non-starter, because the rest of the world definitely plans to feed itself one way or another.

Lower Yields on Organic Farms

Organic farmers and their advocates often claim to get "yields as good as their neighbors." In fact, the yields from an individual field of organically grown crops *can* be high—if productivity has been "borrowed" in the form of rotation with green manure crops (like clover) or spreading animal manure. I say "borrowing," because the high-yield organic acres are either lowering the cropping intensity or taking manure from feedlots or pasture acres. One famous "low-input" farm uses 18 tons of manure and municipal sludge per acre—plus some commercial fertilizer!

A set of organic and low-input farms was presented in *Alternative Agriculture*, (National Research Council, 1989) Two points were particularly illustrative. First, the book's Iowa corn-soybean farm (The Thompson Farm) had corn and soybean yields comparable to those of other farms in the area. But two years out of five each field was in oats or meadow and thus virtually out of effective crop production. They also had severe yield losses in wet years when they could not get into the fields to control weeds with rotary hoeing. Second, the California rice farm had high yields, but could produce them only every other year.The off year the land was put in a combination of legumes and fallow.

A South Dakota Organic Farm

I recently talked with an organic farming consultant who stated that he could point me to a number of organic farms with high yields. The first farmer I reached was in South Dakota. He reported a number of telling points:

- His first year of normal rotation would be a crop of oats interseeded with red clover. He says his oats yielded about 60 to 100 bushels per acre. (Surprisingly, the state's average oat yield average was only about 50 bushels.) Mainstream farmers would rotate corn and soybeans, regarding oats as too low-yielding and low-value to claim field space.

- His second year, he harvested two cuttings of hay, and plowed down the third cutting to provide nitrogen. (Mainstream farmers would take three cuttings of hay and might add some chemical nitrogen the next year.)

- The third year, he planted popcorn, getting nearly as much popcorn per acre as the neighboring farms (about 50–60 bushels). He gave up growing field corn because his yields were only about half as high as his neighbors.' (Of course, the world needs only a small acreage of popcorn.)

- The fourth year was the most profitable in the rotation. The organic farmer grew soybeans, got yields comparable to those of his neighbors—and got double the price (from a maker of organic tofu).

- The farm produced "natural" beef (meaning the farmer did not use drugs to treat any animals that got sick). He did not produce organic beef, he said, because that would have forced him to raise his own low-yielding organic field corn. This way, he could feed his cattle the cheaper chemically supported corn and soybeans from his neighbors' farms.

Over and over, we find organic yields, measured over time and over the whole farm, run far less than the yields from mainstream farms.

Organic Farming Needs a Crop Yield Breakthrough

If, as organic farming advocates sometimes claim, we could grow high yields of the crops the world demands without expensive off-farm inputs, all farmers would want to do it. They would save money and increase their profits. Presently, however, this is not the reality.

It makes little sense to applaud an organic farmer who gets county-average yields by importing huge amounts of animal manure and/or urban sewage sludge onto his acres. We already don't have enough organic nitrogen to go around.

If that organic farmer isn't doubling county-average yields, he isn't part of the wave of tomorrow, because the best mainstream farmers are doubling county averages. Or the new-wave organic farmer could be matching the county average with half the organic nitrogen. Otherwise, he's not part of the organic breakthrough we need.

Of course, organic farmers *can* make money. They get vastly higher prices. Their chemical costs are lower than those of mainstream farmers (even though, as mentioned above, they do use sprays, and quite a lot of them). However, their labor and management costs per acre are almost certainly higher, and often their labor constraints confine them to low production volumes. But we're not worried here about organic farm incomes.

I freely concede that mainstream farmers have often used chemicals more heavily than was really necessary. They were encouraged in their prophylactic pesticide use by a poorly conceived set of government subsidies. That's not the key point either.

The key point is that mainstream farmers can safely step up their intensity and yields as the world needs it. Can organic farmers?

To be part of the solution, organic farmers need a productivity breakthrough. They need to produce more food with fewer natural resources and less erosion than organic farming would today produce if it was extended to the less ideal lands.

Organic farmers haven't thought yields were that important to date. To many organic producers, the important thing is how few chemicals they can use and still produce *something*. To many others, the size of the price premium they get from consumers is the key.

But the amount of land we use to produce food in the world governs how much land is left over from farming for forests, wildlife, and other nonfarm uses. That's why minimizing the amount of land needed for crops is *far* more important to maintaining wildlife, and the natural ecology than eliminating farm chemicals.

The Shortfall in Organic Plant Nutrients

Low yields are only one disappointment about organic farming. Its biggest shortcoming is the global shortage of "natural" nitrogen fertilizer.

Experts at the U.S. Department of Agriculture have calculated that the available animal manure and sustainable biomass resources in the United States would provide only about one-third of the plant nutrients needed to support current food production.[3] Most of the world has far less pasture and manure per capita than the United States. Globally, we may have less than 20 percent of the organic plant nutrients needed to sustain current food output.

This shortfall in plant nutrients should come as no surprise. Organic farming deliberately does without the off-farm inputs used in most high-yield farming. Instead, organic farmers deliberately put a heavier burden on farming's "natural" resources.

There is no precise way to calculate the food shortfall or the wildlife encroachment which organic farming would force on the world, but it would be massive—hundreds of millions of tons of grain per year, and/or millions of square miles of wildlife.

Questions That Need Answers

Organic fans reassure us that "lots" of organic fertilizer is being wasted. But they apparently don't realize the magnitude of the challenge they face. Where, for example, might we find additional organic fertilizers?

Is there any animal manure being stored away in silos?

No. Hundreds of millions of tons of manure are produced in a year, but it's already being spread on fields either by farmers or by the animals themselves.

What about all the crop biomass that's left on the fields; couldn't that be turned into high-quality compost?

Yes, but that biomass is already being incorporated back into the soil with the current farming systems. Incorporating it into the soil probably captures more of the biomass's value than would composting (because nitrogen generally escapes over time and during the composting process). Composting and redistributing would not maintain all of the soil erosion benefits we currently gain from incorporating the crop residue under conservation tillage systems.

What about using urban sewage sludge more broadly on crops?

It would take 50 times the nation's current total of sewage sludge, all dedicated to farm application, to replace the nitrogen currently provided by chemical fertilizers. All of the urban sewage in the country equals only 2 percent of the nitrogen currently being applied in commercial fertilizers.[4]

The EPA's National Sewage Sludge Survey for 1988 indicates that perhaps 25 percent of the nation's sewage sludge is already being used for agricultural fertilizer.[5] Perhaps another 20–40 percent could be retargeted for agriculture, but at a cost.

First, many big cities are not near major farm production centers and transportation costs can be very high for wet sludge.

Second, sludge is perishable, the nitrogen quickly dissipates into the air if it isn't handled quickly.

Third, unless the city has a strong industrial pretreatment program, the sludge will carry high concentrations of heavy metals. Liming the fields correctly can prevent much of the heavy metal uptake, but this takes careful management and a good deal of cost. Otherwise the food grown could carry harmful levels of heavy metals into consumers' systems.

It remains a puzzle how the eco-activists can be so fearful of pesticides and so casual about the use of human waste as fertilizer. Some organic farms are reportedly using as much as 50 tons per acre of sewage sludge *on vegetable crops*! When both the disease potential and the heavy metals are taken into account, sludge looks, if anything, more dangerous than pesticides. (Neither, of course, ranks with highway accidents as a life threat, but both need to be handled properly.)

Environmentalists claim they can gain 30 million tons of compostable materials from current urban landfill waste. Won't that help?

Thirty million tons of compost sounds like a lot—until you compare it with the 2 billion tons of manure, corn cobs, and other farm waste already being spread annually on American farms.[6] Then the compost just looks like an expensive way to add 1.5 percent to our agricultural biomass.

A recent New Jersey recycling report indicates this "free" compost will cost perhaps $100 per ton just for gathering the compostable material (with the heavy metals still in it) and getting it to the recycling site. We then have to take out the heavy metals (hand sorting?), make it into compost, and haul it to the field. (Separation alone adds $44 to $260 per ton to New Jersey's recycling costs.)[7]

Nor are the costs much offset by not using landfills. Even in densely populated New Jersey, the cost of building and maintaining a modern landfill that meets all standards is often less than $10 per ton.[8] Neither New Jersey nor the rest of the United States is short of rough wasteland on which to put landfills. The only problem is political—the "Not in My Back Yard" syndrome.

It's hard to see how this major increase in costly off-farm biomass would make us environmentally better off than using commercial fertilizers.

More Soil Erosion from Organic Fields

Why would organic farming mean a major increase in soil erosion? Organic farmers have always claimed they care *more* about soil preservation than the average mainstream farmer. Unfortunately for the organic enthusiasts, however, organic farming now carries huge and inevitable soil erosion penalties.

Due to its low yields, organic farming would force us to plow millions of additional acres that are generally more fragile than those where our farm production is currently concentrated.

During the 20th century, mainstream American farming has radically increased its crop yields. Average U.S. corn yields, in fact, have increased about sixfold. For most crops in prime farming regions, crop yield potentials have at least tripled. This tripling of yields has let us take those millions of acres of steep, erodible land out of crop production.

The environmental movement fails to grant credit for these erosion reductions. If we shifted to organic farming systems as the bases for all of our food production, big tracts of erodible soils would be put back in crops to replace the lost productivity.

Organic farmers' second erosion disadvantage is that they deliberately ignore modern weed-killing chemicals. Organic farmers don't use conservation tillage or no-till farming, the soil-safest farming systems ever devised, because both systems depend on herbicides for weed control.

They leave themselves only mechanical tillage for weed control. Hoeing and cultivating can kill weeds, but they inevitably leave the soil open to erosion. These are "bare-earth" farming systems. Organic farming leaves far more of its soil bare to wind and water more often than do the best modern high-yield farming systems.

The Thompson low-input farm in Iowa uses ridge tillage to minimize weed competition and selects soybean varieties that are tall and fast-emerging to get canopy shading as quickly as possible. Still, the farm has to rotary hoe its crops at least twice, which means a long window of opportunity for soil erosion.[9]

Organic farmers are trying to surmount these obstacles. They have always paid close attention to soil tilth, which has helped modestly to reduce their erosion rates. They also tend to use conservation practices like contour farming and strip cropping to minimize their erosion losses.

Try as they will, however, organic farmers have nothing to offer that is nearly so powerful against soil erosion as the new conservation tillage and no-till farming systems. Organic farmers' mainstream competitors, armed with herbicides, have made a quantum leap in sustainability. A 90 percent reduction in erosion on about 50 million U.S. acres farmed with no-till represents a *huge* cut in soil risks. So does a 50–65 percent cut in soil losses on another 200 million American acres farmed with conservation tillage.

The Long-Term Record of the Morrow Plots

The Morrow Plots at the University of Illinois have been farmed since 1976. The University says, "The Morrow Plots are the oldest agronomic research plots in the United States, and include the oldest continuous corn plot in the world.

They have been farmed with a variety of crop rotations and fertilizer regimes. The plots received no fertility treatments between 1888 and the mid-1930s, and their yields gradually declined. In 1955, most of the plots began to receive annual applications of liquid nitrogen, phosphorus and potassium. Corn yields have roughly tripled since then.

The continuous corn plot which has received no fertilizer is still producing just about the 50 bushels per acre today that it produced in 1888, despite the higher yield potential of the modern seeds now being planted on it. The nutrient constraint is holding firm.

Another plot recently grew 200 bushels per acre because it received adequate fertilizer—and despite the fact it has now grown corn continuously for over a century. Putting it another way, the yields quadrupled during a century of continuous intensive cropping.[10]

Fruits, Vegetables, and Organic Farming

Organic farming's final big shortcoming is in the production of fruits and vegetables.

Produce crops look just as succulent and delicious to pests as to people. A wider variety of insects, disease organisms, fungi, and microorganisms attack fruits and vegetables than any other crops. Most organic fruit and vegetable production is barely able to cope with these pests, even though virtually all organic growers spray their crops with "organic pesticides."

Organic producers manage quite well with *some* produce crops. Red raspberries, for example, have relatively few pest problems. Some grape growers apparently manage effectively most of the time with mainly organic methods. Some Florida citrus producers, whose fruit is headed for processing, rather than the fresh market, use organic methods to cut production costs.

Organic techniques can also be used by dedicated home gardeners.

THE MYTH OF ORGANIC PEST CONTROL

"As steward of a self-sustaining garden, your first job is to recognize that the forces in your garden will never be in 'perfect' balance. There will always be some plant damage. . . . Your garden should be able to defend itself from severe damage from most pests most of the time. . . . Such natural defense is promoted by four major factors: sun, water, soil, and air circulation"

Pest	*Organic Remedy*
Colorado potato beetle	1. Use row covers in early season. 2. Handpick immediately when sighted and crush adults and egg masses. Very effective control. 3. Apply thick organic mulch to impede movement of overwintered adults to plants. Beetles walk more than fly during early season.
cabbage maggot	1. Use row covers early in season. 2. Maggots don't like alkaline environment. Circle plants with a mixture of lime and wood ashes, moistened to prevent blowing.
aphid	1. This may be a symptom of too much nitrogen or pruning. 2. Use row covers. 3. Control aphids . . . with sticky bands, sticky yellow traps, or yellow pans filled with water.

Tanya Denckla, *Gardening at a Glance: The Organic Gardener's Handbook on Vegetables, Fruits, Nuts, and Herbs,* Franklin West Virginia, Wooden Angel Publishing, 1991

Reality Comment: Squashing potato beetles by hand is, indeed very effective in that the bugs are certainly dead; but there can be a million beetles and egg masses per acre of potato field.

Poisoning Topsoil?

Dr. Sharon Ingham, an associate professor in botany and plant pathology at Oregon State University is one of those who claims that farm chemicals poison soils. She recently said, "We can keep dumping fertilizer and chemicals on farm land, but at what cost to the rest of the ecosystem?" (She is trying to set up a "sustainable" farming research program at her university.)

She says both her own research and that of other scientists has shown that the concentration of nutrients in chemical fertilizers kills many of the bacteria and other organisms in soils. Such biologically "dead" soil, she says, is less able to hold agricultural chemicals that are applied to it later. Fertilizers, herbicides, and pesticides then leach more quickly into surrounding water and land.

She says organic fertilizers would be preferable to chemical ones. In her view, just focusing on yields is "simple-minded."[11] Dr. Ingham is apparently worried that we will have too few soil microbes and earthworms under our crop fields. But the problem is self-limiting. If too much fertilizer kills too many useful soil organisms, crop yields will decline. Expensive fertilizer will be wasted. Farmers will change their systems.

In fact, one reason why conservation tillage and no-till farming systems are sweeping the mechanized agricultures of the world is that they *encourage* more soil microbes and more earthworms. These farming systems don't disturb the lower soil profile with plowshares, and provide year-round supplies of nutrients from decomposing cover crops and crop residue. The killed sod in no-till fields makes especially fine habitat for earthworms and soil microbes. The farmers get yield gains as a result.

Oddly, Dr. Ingham seems not at all worried about a much more serious problem: The fact that the world has far too little of the organic fertilizers she favors to support world food needs.

If yields in chemically supported fields were going down, Dr. Ingham would have a powerful argument.

If high-yield farming was creating serious dangers for our groundwater or wildlife populations outside the crop fields, she would have an important argument.

But hers is a relatively minor quibble, not an argument.

What About Biological Controls?

The world has been searching for biological controls to suppress noxious pests for more than 100 years. (The boll weevil was an early target.) Some successes have been achieved:

- The biggest success to date has been in West Africa, against the cassava mealybug and the green spider mite. The two cassava pests emigrated from South America and became serious threats to Africa's key root crop. Researchers from the International Institute for Tropical Agriculture (IITA) in Nigeria led the attack, searching out, propagating, and distributing a set of natural predators to suppress the mealybug and mite. The effort won the King Badouin international agricultural research prize in 1990. It was particularly important in Africa because cassava growers there could not afford to use pesticides, even to protect one of their most important food sources. It was relatively easy to do because there were highly developed predators of the two pests in their South American home territories.

- In America, biological parasites have been introduced to control the alfalfa weevil. They work, but only well enough to *reduce* the amount of spraying the growers must do, not eliminate it.

Unfortunately, it is not likely that biological controls can substitute for very much of the pest control currently provided by pesticides. Biological controls are generally too narrowly targeted to give broad, cost-effective pest control, according to Leonard Gianessi of Resources for the Future. Gianessi recently addressed that problem in the National Academy of Science quarterly, *Issues in Science and Technology*.[12]

Gianessi notes that a pathogenic fungus is likely to be effective on a single weed species, while an herbicide can sometimes control hundreds of kinds of weeds.

East Coast apple growers can use one fungicide to control nine significant diseases of apples. Even if successful biological controls were developed for eight of these diseases, the apple growers might still have to use the fungicide to control the ninth disease.

Gianessi says that just in the United States thousands of pest species infest 80 to 100 crops in many different regions. Thus, there are hundreds of thousands of possible combinations of pests, crops, and regions that researchers would have to target to replace any particular deployment of one of the roughly 200 active pesticide ingredients now in use.

There is also a problem with releasing and encouraging new biological controls. They too can have side effects on the local ecologies. Dr. Francis Howarth of the J. Linsley Grassit Center for Research in Entomology believes that introduced biological control organisms have themselves led to the extinction of nearly 100 species worldwide.

Being "natural" is obviously not enough.

MYTHMAKER:

"The Hyper-Expansionist scenario seems to be the route our society is now traveling. HE is a path by which the agri-food industry could further increase its control over the natural world and one where nature's long-term reliability would be replaced by clever technologies promising unparalleled abundance at merely the cost of computerized vigilance. This path implies systems that will demand the monitoring of fallible sensors by bored humans serving the giant corporations that control the food system. . . . We are not feeding the world, we are robbing the world. *I believe that localized agriculture is needed in order to make agriculture more sustainable, but also because I simply can't imagine how we can teach people to protect the resources that produce their food unless somehow those resources are closer to home.*" [emphasis added]

Dr. Joan Dye Gussow, professor of nutrition at Columbia University in New York City, interviewed in *Safe Food News*, Winter 1993, pp. 8–10

Reality Comment: Dr. Gussow may be a very competent nutritionist. She does not, however, seem to understand very much about how food is produced.

First, she implies that big corporations produce our food, when the truth is that ninety-plus percent of America's farm output is produced by family farms. (Many of the family farms have incorporated under the subchapter S rules for small businesses, mainly for inheritance tax reasons.)

Second, (and more embarrassing for a nutritionist) Dr. Gussow's local-production option would severely restrict our access to fresh fruits and vegetables for about 90 percent of the year.

Third, her "local and organic" production preference would mean famine for billions, stunted children for the survivors, and the loss of huge tracts of wildlife.

Fourth, she gives no indication how having city folks observe their neighborhood farms is going to result in better farm resource use.

Finally, how is her concrete island of New York going to feed itself on her model? Will New York use its own sewage to fertilize former parking lots, re-plowed for vegetable gardening? Will multistory parking garages be converted to feedlots, with the waste collecting in the subbasements for recycling?

ARS photo by Keith Weller

HIGH-YIELD PRESERVATION—The Germplasm Enhancement for Maize (GEM) project works to increase genetic diversity of U.S. corn by combining exotic germplasm from Latin America with domestic corn lines.

Biosphere Reality

The crew of the Biosphere 2 emerged from two years in their self-contained ecosystem near Tucson, AZ. They had originally planned to spend half their time raising their own food and the other half on scientific experiments.

They never got the chance to do much science. Insects and plant diseases found their way into the biosphere, and the crew had pledged to fight them without man-made chemicals. Mites ate the beans and potatoes. Powdery mildew shriveled the squash. The natural oils that were supposed to control insects attracted cockroaches, which also infested the living quarters. The sweet potatoes did well—and the crew ate so many that the beta carotene gave some of them orange-tinted skin. Even though 20 percent of the food they ate had been stored in the biosphere before the mission began (for emergencies) the crew lost about 25 pounds each. Hunger was a constant companion, they reported, dominating their thoughts.

Viability of High-Yield Versus Landrace Seeds

Opponents of high-yield farming claim that modern high-yielding seeds are such tender, exotic plants that they will be wiped out by natural disasters, leaving mankind to wish it had hung onto the old seed varieties.

The truth is that the high-yielding varieties are tougher than the old seeds. Much of their yield advantage comes from having pest resistance and stress tolerance that the old seeds lack. (American farmers have harvested good corn crops recently in years which would have been too dry to produce good yields before World War II, because of the added drought tolerance of the seeds.)

The old farmer-saving-the-best-seeds selection process had its good points. It might even have come up with the breeds we have now—but not very quickly. The real impact of modern seed breeding has been to speed the selective breeding process. Instead of

While environmentally friendly, pesticide-free crop dusting has proven to be extremely labor-intensive.

tripling yields over 1000 years, we have done it in 35. That has been rapid enough to prevent the massive famines that the skeptics have been predicting for so long.

It is also true, however, that we need to preserve the genes in the old seeds. And we can't expect them to survive in the wild; they're just as dependent on man as the high-yielding varieties. That's why there is a major effort underway to save the landrace varieties in seed banks around the world. (And these efforts, too, need more funding!)

We also need a modest amount of gene farming, especially for livestock and poultry breeds.

The Global Perspective on Organic Farming

It may be difficult for America's urban residents to understand why organic farming could not feed the current human population, let alone the one expected in 2050. They are continually hearing farmers complain about big crops and low prices.

The United States is one of the few countries in the world that could provide a quality diet for its expected population in 2050 *with* organic farming. But even the land-rich United States would have to rein in the 40 percent of its farm production that is now exported to countries less richly endowed for agriculture.

The answer is that we can spare a *small* amount of land for organic farmers and chemical-free experiments—if we continue to pursue high-yield research and technology at the same time.

By way of analogy, we can also spare *some* farming resources to produce a few ranch-mink coats. (The mink eat lots of fishmeal and soy protein which we would otherwise need for human nutrition.)

With present knowledge, neither organic farming nor mink coats can be supported except as part of a much larger, more powerful *chemically assisted* farming system.

AFRICAN REALITY:

"Mr. Norman Borlaug, a prominent agriculturalist . . . told a meeting of the Overseas Development Institute yesterday: 'Some people say that Africa's food problems can be solved without the application of chemical fertilizers. They're dreaming. It's not possible.' He said that the environmentalists advocating traditional farming methods failed to recognize the rapid growth in population expected in the continent. . . . Sub-Saharan Africa had the lowest use of fertilizer in the world and soil nutrients were so low that other efforts to raise crop productivity would not be successful until fertility was improved."

Financial Times, June 10, 1994, p. 26

MYTHICAL REALITY:

"In Germany, 10,000 farmers have converted to organic methods and 2.4 percent of the country's agricultural land is cultivated without the use of artificial fertilizer and pesticides. However, the German government pays the equivalent of [$400] a hectare for farmers to produce in this way, which costs up to 30 percent more than conventional methods."

Financial Times, May 20, 1994. p. 26

Reality Comment: Western Europe, ironically, is paying for this shift to organic farming because it lowers production, thereby reducing the farm surplus induced by its high price supports!

Reprinted from BridgeNews Service, July 24, 1998

What You Don't Know May Hurt You:
The Organic Risk in Your Foods

People who eat organic foods are apparently at least eight times as likely to be attacked by the deadly new E. coli bacteria as people who eat mainstream foods. Organic consumers are also at increased risk from natural toxins produced by fungi—some of which cause cancer. Organic foods also carry far more of the dangerous bacteria, such as salmonella, campylobacter and listeria, which kill thousands of people every year.

But you'd never know of these dangers from reading the draft of a new food-safety leaflet that the Environmental Protection Agency will start distributing in U.S. supermarkets. Instead, the EPA's new leaflets talk about undocumented "dangers" from pesticide residues in our food. The leaflets warn mothers "some pesticides have been shown to cause health problems such as birth defects, nerve damage, cancer, and other toxic effects in laboratory animals. In addition, infants and children may be more vulnerable to pesticides because their bodies are not fully mature."

To combat these perceived chemical dangers, the EPA urges homemakers to wash, peel, boil, and skin their food to remove pesticide residues—or to buy organic foods. You'd never know from the EPA leaflet that the U.S. National Research Council has recently said publicly that pesticide residues pose no significant danger to consumers (or their kids). Nor would you learn that:

- The most deadly of the health threats in our food today is a virulent new strain of E. coli bacteria—O157:H7. It can attack kidneys and liver, causing permanent internal damage and possibly death. Animal manure is the primary reservoir of these bacteria—and organic farmers rely heavily on manure to fertilize their crops.

- The Centers for Disease Control says it confirmed 2,471 cases of O157 in 1996 and estimates at least 250 deaths. But CDC says this was only a small fraction of the poisonings that occurred. Organic food made up only 1 percent of the U.S. food supply—but organic/natural foods were implicated in at least 8 percent of the confirmed O157 cases.

•*Consumer Reports* recently found that free-range chickens are morelikely to be infected with dangerous bacteria than those raised in poultry houses. Free-range chickens pick up almost anything from each other's wastes, including parasites and bacteria.

•The U.S. Food and Drug Administration says organic crops have higher rates of infestation by natural toxins—including aflatoxin, one of the most violent cancer agents known to man. Organic foods suffer more damage from insects and rodents and the resulting surface breaks lets toxin-causing fungi get into crops.

•"Natural" (unpasteurized) milk and juices may claim to be "fresher," but are potentially contaminated with dangerous bacteria and such diseases as tuberculosis. More than 1500 companies sell such products in the United States.

•Organic producers too often compound the animal manure risk by refusing to use chemical washes, disinfectants—or even chlorinated water—to prepare their products for market.

"I was really horrified that something supposed to be so safe for my kids could really almost kill them," says Rita Bernstein, a Connecticut woman whose two daughters became seriously ill from E. coli after they ate triple-washed organic lettuce in 1996. Her younger daughter, Halee, still suffers severe vision problems and reduced kidney function. Also in 1996, an 18-month-old Colorado child, Anna Gimmestad, died after drinking Odwalla brand non-pasteurized apple juice. The Odwalla Company now pasteurizes its juice.

Dr. Dean Cliver of the University of California–Davis says most organic farmers compost manure rather than putting it directly on the crops. However, the growers' guideline has been to compost wastes for two months, at temperatures of 130 to 160 degrees. Cliver found that O157 E. coli could survive at least 70 days in a compost pile, and that it takes at least 160 degrees of temperature to kill it. (Nor do organic farmers routinely check their compost piles with thermometers.)

Why hasn't the public been warned about the dangers of organic food? Apparently, it's too politically correct to be criticized. The CDC has never even officially released the data on the high proportion of O157 cases involving "natural" foods.

Organic farmers are fond of claiming they use "no chemicals whatsoever." But Dr. Robert Tauxe, Chief of the Foodborne Diseases Branch of the Centers for Disease Control, says "organic means the food was grown in animal manure."

Dr. Tauxe, is one of the few government officials who dared to be quoted implicating organic foods as a significant health hazard. However, even Tauxe has since backed away from his statements, apparently under pressure from superiors or activists. (Vice President Al Gore is an ardent supporter of organic food production, and EPA administrator Carol Browner formerly worked for Gore.)

Ironically, technology has already produced a way to make organic food safe: irradiation. But when the U.S. Department of Agriculture recently proposed that irradiated foods be permitted within its new organic food standards, it was flooded with thousands of protests from organic believers.

The EPA's new leaflets are a danger to the public. They will frighten people about nonexistent cancer risks in safe fruits and vegetables—while pushing them to buy the most dangerous foods in the marketplace.

Why is the EPA recommending organic food and farming? Why does the EPA not take the Food and Drug Administration's recommendation to also warn of the dangers of organic foods? Why are government agencies keeping quiet about the need for irradiation to protect consumers against virulent bacteria?

Excerpted from BridgeNews Service, February 11, 2000

Are Organic Foods Really Better for You?

Natural-Grown Killers in Organic Food Make It No Safer Than Produce Grown With Pesticides

By Dave Juday, former White House agricultural adviser

LEESBURG, Va.—Stop the presses: America's top organic food executive has admitted organic foods are neither safer nor more nutritious than mainstream foods.

Meanwhile, food safety tests commissioned by ABC News found no pesticide residues on either the organic or mainstream produce and up to 100 times as many dangerous pathogens on the organic vegetables.

On the Feb. 4 edition of the ABC newsmagazine "20/20," John Stossel questioned Katherine DiMatteo, director of the Organic Trade Association, which represents organic farmers and retailers.

Also interviewed was Les Crawford, former director of food safety for the Food and Drug Administration, and Dennis Avery, my colleague at the Hudson Institute's Center for Global Food Issues.

He also spoke to Rita Bernstein, whose young daughter, Hayley, was attacked by E. coli O157:H7 after eating contaminated organic lettuce. Hayley nearly died and remained on a respirator for months. She suffers from impaired vision.

Here's an instructive excerpt of Stossel's interview with DiMatteo:

Stossel: "Is (organic food) more nutritious?"

DiMatteo: "It's as nutritious as any other product."

Stossel: "Is it more nutritious?"

DiMatteo: "It is as nutritious as any other product on the market."

Stossel: "There's a sales campaign to dream about. The organic industry admits organics are no more nutritious than other food, but the customers think it is."

DiMatteo: "Organic agriculture is not particularly a food safety claim. That's not what our standards are about."

Stossel: "But your customers think it's better for you."

DiMatteo: "I don't know that that's true."

Stossel: "Sure it is. (We) did a poll on organic foods and found 45 percent of the public think organics are more nutritious. Half the people said healthier, and they're not."

Avery told Stossel organic food is potentially more dangerous because much of it is fertilized with manure, a known reservoir of bacteria dangerous to humans.

Crawford, now of Georgetown University, warned viewers that 76 million Americans get sick from food-borne bacteria every year. Health authorities are especially concerned about the new strain of E. coli, O157, which is deadly enough to kill even healthy people and leaves many of its survivors with internal organ damage.

This new E. coli is heat-resistant microbe, making it uncertain whether routine composting done by organic farmers will consistently kill it.

Stossel found no comparative tests had been done on the safety of organic food compared with mainstream food, and commissioned the University of Georgia to test samples of both.

The Georgia tests found no pesticide residues in either the organic or mainstream produce. Mainstreams farmers avoid using pesticides when the residues would persist after harvest.

The most dramatic finding, however, was that the organic produce had 100 times as many pathogens as their mainstream counterparts. Most of these bacteria were strains of E. coli, an indicator of fecal contamination.

Notes

1. Osteen and Szmedra, *Agricultural Pesticide Use Trends and Policy Issues*, USDA Agricultural Economic Report No. 622, Washington, D.C., 1989.

2. Arnold Aspelin, *Pesticide Industry Sales and Usage: 1992 and 1993 Market Estimates*, Washington, D.C., EPA, June 1994.

3. Van Dyne and Gilbertson, *Estimating U.S. Livestock and Poultry Manure Nutrient Production*, U.S. Department of Agriculture, ESCS-12, March 1978; and *Animal Waste Utilization on Cropland and Pastureland*, Washington, D.C., EPA, 1959, EPA–600/2–79–059. The preliminary results of an update indicate that the new CAST findings will be similar to those of the earlier studies.

4. S. Graef, director of technical services, Western Carolina Regional Sewage Authority, Greenville, South Carolina, based on "Estimated Mass of Sewage Sludge Disposed Annually," Table I–1, *Federal Register,* vol. 58, no. 32, February 19, 1993, p. 9257.

5. Graef, op. cit. from *Regulatory Impact Analysis of the Part 503 Sewage Sludge Regulation*, Table I–2, EPA Office of Water, EPA 821–R–93–006, March 1993.

6. James DeLong, *Wasting Away: Mismanaging Municipal Solid Waste*, Washington, D.C., Competitive Enterprise Institute Environmental Studies Program, May 1994, p. 4.

7. Grant Schaumberg and Katherine Doyle, *Wasting Resources to Reduce Waste*: *Recycling in New Jersey*, CATO Institute, Policy Analysis Report No. 202, January 26, 1994, p. 13.

8. Ibid, p. 11.

9. "The Thompson Farm," *Alternative Agriculture*, Washington, D.C., National Research Council, National Academy Press, 1989, pp. 308–323.

10. *The Morrow Plots: A Century of Learning*, Agricultural Experiment Station Bulletin 775, University of Illinois, Champaign, Illinois, 1974, Updated through 1992.

11. David Wheeler, "Expansion of Agricultural Research Said to Have Fueled Dramatic Increases in Yields of Corn, Rice and Wheat," *The Chronicle of Higher Education*, September 22, 1993, p. A10.

12. Leonard Gianessi, "The Quixotic Quest for Chemical-Free Farming," *Issues in Science and Technology*, National Academy of Sciences, Fall 1993, p. 31.

10

Who Has a Soil Crisis?

MYTHMAKERS SAY:

". . . [E]very ton of fertile topsoil unnecessarily washed away, every hectare claimed by desert sands, every reservoir filled with silt further drains world productivity and spells higher costs for future gains in output."

Erik P. Eckholm, *Losing Ground: Environmental Stress and World Food Prospects*, New York, Norton, 1976, p. 181

"The long-term social threat posed by uncontrolled soil erosion raises profound questions of intergenerational equity. If our generation persists in mining the soils so that we may eat, many of our children and their children may go hungry as a result."

Lester Brown and Edward C. Wolf, *Soil Erosion: Quiet Crisis*, Washington, D.C., Worldwatch Paper 60, Worldwatch Institute, September 1984

REALITY SAYS:

". . . If current rates of (U.S.) cropland erosion prevail for 100 years, crop yields will be from 3 to 10 percent lower than they would be otherwise. Yield increases (resulting from technology) that are modest by historical standards would much more than compensate for such a loss."

Pierre Crosson, one of the world's top soil erosion experts, reported in a special issue of *Scientific American* on "Managing Planet Earth," September 1989[1]

". . . [P]ractices such as mulching, manuring, low tillage, contour cultivation and agroforestry can frequently reduce surface runoff of water, sediment loss and erosion by 50 percent and more. These techniques are not yet widely used."

World Bank, *Development and the Environment*, 1992 World Development Report

"Satellites Expose Myth of Marching Sahara"

Heading of article in *Science News*, July 20, 1991

High-yield farming is producing a triumph over man's age-old enemy, soil erosion.

For centuries, man has accepted relatively high levels of soil erosion as the price of growing food. That trade-off is no longer necessary. Modern farming systems are radically cutting soil erosion on farms around the world—by raising yields so less land has to be open to erosion, and by using herbicides to kill weeds instead of plows and cultivators. These new farming systems preserve topsoil, encourage soil microorganisms and build soil organic content better than any previous mainstream farming systems in history.

Research is also providing new low-erosion farming systems like tied ridges and alley cropping for the Third World as well.

An impressed agronomist said it well: "Do you realize that now, for the first time in ten thousand years, people can grow crops without destroying the land?"[2]

These technical breakthroughs will permit the world to sustainably produce record-high yields of food and fiber from its best and safest cropland—with record-low losses to soil erosion. In the future, we should even be able to increase topsoil depths on our best farmlands, even while carrying forward intensive high-yield farming.

Remember, soil erosion is an inevitable, ongoing natural process. It has already shrunk the elderly Appalachian Mountains into a tiny vestige of the towering peaks they once were. Over millions of years, it will turn the Rocky Mountains into a set of foothills. And it will take all of today's topsoil off our fields.

What no one is telling our children is that the same unstoppable processes that *erode* topsoil also *create* topsoil. Sunlight, bacteria, moisture, and earthworms are just as powerful as wind and water.

They're just more subtle.

The question is not whether erosion is occurring. It is. But topsoil *creation* is also occurring.

The real questions about erosion?

First, is topsoil eroding faster than it is being created on a particular surface? Second, are we creating problems downstream from the erosion that could be solved by reducing soil erosion upstream (such as silting up lakes or reservoirs)?

High Yields Cut Soil Erosion

High yields cut soil erosion in and of themselves.

A field of corn that produces 150 bushels of corn per acre with high plant populations—protected by pesticides from weed competition and corn borers—will probably suffer less than half as much erosion as two acres of organically grown 75-bushel corn. Not only will the organic corn present double the land area to erosive forces, but it will have less leaf canopy to moderate the effects of thunderstorms and high winds. The organic field will also have a much longer window of erosion potential; the organic farmer will need to combat weeds with mechanical tillage well into the growing season, while the chemically assisted farmer uses herbicides that do not disturb the soil surface.

After we factor in lower topsoil losses and the redoubled number of bushels, the *soil loss per bushel* on the 150-bushel field should be perhaps one-third of the soil loss per bushel on the low-yield field. This soil erosion differential may not be a serious problem so long as organic farming is concentrated on top-quality, level land. If organic farming were extended into more difficult conditions, however, even its present yields would prove far harder to sustain.

Reduced-Tillage Systems for Mechanized Farms

Now, in addition to their high-yield advantage in soil conservation, the chemically assisted farmers now have another enormous advantage—conservation tillage systems that use chemical weed-killers instead of bare-earth farming systems like traditional moldboard plowing.

"Conservation tillage" cuts soil erosion *by 65 percent* compared to moldboard plowing.[3] It leaves a heavy layer of crop residue in the upper soil profile that radically slows runoff and soil loss. Nor does it disturb the microbes and earthworms below the top 3–5 inches of earth as plowing does. (The formal definition of conservation tillage is at least a 30 percent residue cover on the soil after planting.)[4]

Conservation tillage swept across the crop production areas of America after the 1970s and early 1980s, and is now cutting erosion on more than 100 million acres in the United States alone. Reduced-tillage systems were used on another 100 million acres—and on low-erosion fields the reduced-till can be enough to control erosion effectively. Thus more than 50 percent of America's cropland is being farmed under low-erosion farming systems.[5]

"No-till" farming is an even better soil-saver. It keeps a layer of sod on the field virtually throughout the entire year, cutting erosion by as much as 98 percent compared to plowing.[6] No-till is now displacing conservation tillage all across America—the second sweeping wave of revolutionary farming systems in as many decades. No-till was used on more than 47 million U.S. acres in 1998—up from 21 million in 1991.

Both conservation tillage and no-till are also being extended in such places as Western Europe, Brazil, Paraguay, Kenya, Australia, and New Zealand. It seems likely that low-till farming systems will become the global norm.

Lowest Erosion Rates in Farming History

If you doubt the triumph of new low-erosion farming systems, remember how primitive farming was—and is—carried out:

- Ancient farmers controlled weeds in grain fields with "clean fallow"—keeping half of their land bare of all vegetation for the entire year. Any weed seeds that sprouted were uprooted before they could set new seed. But the price was enormous in terms of high wind and water erosion rates, lost soil moisture, and lost productivity per acre.

- Row crops (such as the turnip) came to Western civilization about the 12th century. They represented a big gain in food production, because farmers could then produce crops from their fallow land—by hoeing the turnips to control weeds. (The extra food and feed from turnips and other root crops helped support the emergence of medieval cities.) However, row crops permitted almost as much soil erosion as clean fallow.

- As recently as the 1960s, modern American farmers plowed their land in the fall and left it open to erosion throughout the winter's storms. Plowing was such a slow process that it couldn't wait until spring. (That would have sacrificed too much of the growing season and cut yields.)

Soil-Saving Tillage
United States, 1998 Summary

Crop	Conservation Till*	Reduced Till**	Total Low-Till
Corn	39%	25%	64%
Small grain	32%	34%	66%
Soybean (dbl crop)	54%	22%	76%
Cotton	12%	13%	25%
Sorghum	34%	34%	68%
Forage crops	24%	26%	50%
Other crops***	18%	26%	44%
Total planted	37%	27%	64%

* Plus 30% residue,
**15–30% residue
***vegetables, tobacco, peanuts, sugarbeets

Source: *1998 National Crop Residue Management Survey,* Conservation Technology Informaton Center, West Lafayette, Indiana.

•Organic farmers still plow and cultivate mechanically. Contour plowing and strip cropping help reduce their erosion, but they cannot match the current low erosion rates of mainstream farmers.

The University of Minnesota Soils Department has done one of the most thorough and complete studies of U.S. soil erosion, based in part on the 1982 national soils inventory. The study concluded that Corn Belt soil erosion was far lower than the doomsayers (and the Soil Conservation Service) had been telling us.[7]

They estimated that America's 1982 soil erosion rates meant productivity losses of only about 0.03 to 0.1 percent per year. That means that if the 1982 soil erosion rates continued unabated for 100 years, soil productivity would decline only 3 to 10 percent. With crop yield rising by 3 to 5 percent annually, such productivity losses are almost trivial.

Today's reality is even better than that, since low-till farming systems have continued to spread—and further reduce erosion rates—in the years since 1982.

Frank Lessiter, a pioneering no-tiller and founder of *No-Till Farmer*, says the tillage revolution is succeeding because the new systems accomplish all the following goals:

•cutting soil erosion and cutting surface water pollution
•cutting air pollution (dust)
•conserving water
•improving soil tilth
•increasing earthworm numbers
•providing more cost-effective weed control
•adding organic matter to the soil
•cutting soil compaction
•saving time and saving fuel
•cutting machinery costs
•increasing farmers' management flexibility

John Deere

FARMING WITHOUT THE PLOW—The crop residue stays in place to prevent soil erosion. At planting time, the tough no-till planter slices right through the old cornstalks to plant the new seeds.

How Conservation Tillage Works

Farmers had long known they could prepare a seedbed faster and with far less fuel using a disc plow. (A disc plow is a gang of concave steel discs dragged across the ground at an angle).

Disc plowing also halved soil erosion by leaving most of the crop residue on the surface of the soil and/or in its top few inches. However, disc plowing had never controlled weeds effectively.

Herbicides came into wide use in the 1960s. But they didn't deal effectively with the whole range of broad-leafed weeds and grasses until the 1970s. Also in the 1970s, the Organization of Petroleum Producing Countries (OPEC) sharply raised the price of oil (and tractor fuel). New herbicides and the opportunity to save on fuel costs combined to drive a rapid tillage revolution.

How No-Till Works

Under no-till, the farmers plant a heavy grass sod in the fall. In the spring they spray it with a vegetative killer and plant right through the killed sod. Later in the growing season a new crop of grass is interseeded to protect the soil after the harvest removes the crop canopy. It will be killed in its turn the next spring.

No-till eliminates soil erosion almost entirely. Soil is virtually never exposed. No-till is especially useful on steep slopes and problem soils.[8]

No-till ought to gladden the hearts of conservationists and environmentalists all over the globe. Not only does it reduce erosion to near-zero levels, but due to the year-round sod cover creates an especially hospitable home for earthworms and soil microbes. The improved tilth validates the tilth emphasis of traditional organic farmers—but boosts yields far higher than they have been able to do.

No-till farmers start out with yields comparable to plowed fields—and talk about a further yield gain after about the fifth year of the system.

Purdue University agronomist Peter Hill documented this longer-term yield gain in the histories of Indiana fields signed up in the 1992 MAX record-keeping program. Note that the yields *slumped* after the second year, but then moved up to a higher plateau than they reached the first year.

No-Till Corn Yields[9]

Years	*Yields*
1–2	167 bu.
2–5	158 bu.
6 and up	172 bu.

ELECTED MYTHMAKER:

"My earliest lessons on environmental protection were about the prevention of soil erosion on our family farm, and I still remember clearly how important it is to stop up the smallest gully. When I was a boy, there were plenty of examples elsewhere in the county of what happened when gullies got out of control and cut deep slashes through the pasture. . . . Unfortunately, little has changed. . ."

Vice President Al Gore, *Earth in the Balance*, pp. 2–3

Reality Comment: According to the Tennessee Agricultural Statistics Service, the use of conservation tillage in Mr. Gore's home state increased from 17 percent in 1987 to 64 percent in 1992. Happily, this cut Tennessee's soil erosion close to 10 million tons per year.

Fortunately for the vice president and the rest of us, a great deal *has* changed.

Soil-Conserving Systems for the Third World

Alley Cropping: Alley cropping is a farming system designed for tropical regions like West Africa.

The farmer interplants rows of leguminous trees with his rows of crops. The trees shade the young seedlings for part of the day, hold down soil temperatures, and reach deeper than the crop plant roots to bring up nutrients and moisture. Later in the growing season, the trees are pruned, and the nitrogen-rich leaves are used to mulch the crop plants, raising yields.

Alley cropping is the first fully stable farming system ever available to much of West Africa. The farmers who use it do not have to practice shifting cultivation, and they do not have to buy fertilizer. The system was developed under the auspices of the International Institute for Tropical Agriculture in Nigeria. It is spreading significantly where population pressures have shortened the fallow periods (and cut yields) for the traditional bush fallow system.[10]

Kudzu Cropping for the Rain Forest: The Amazon rain forest currently has about 3 million people attempting to subsist on shifting cultivation in its harsh environment. The subsistence farmers burn an acre or so of rain forest in the dry season, and the ash from the burned trees provides enough fertility for perhaps two years of crops before the farmer has to move on to another plot and let the first one rest. Over time, this means large areas of the rain forest are impacted by even a small population of farmers.

Recently, researchers have discovered that one crop of a legume called kudzu does as much for soil fertility as 14 years of bush fallow! (Kudzu is a noxious weed in the southeastern United States because it develops a storage root that is almost impossible to kill; it does not develop the storage root under Amazon conditions.) This means that the 3 million people who are currently attempting subsistence farming there could be supported on about 10 percent as much rain forest land as they now occupy with their slash-and-burn system.

However, kudzu is a double-edged sword. If Amazon subsistence farming can actually *succeed*; and, if the Brazilian government's economic policies continue to act to prevent the creation of off-farm jobs, more and more discouraged people might migrate to the Amazon.[11]

Tied Ridges in Burkina Faso: In Burkina Faso, much of the annual rainfall runs off before the hard crust on the soil softens and lets the rains soak into the soil. A technique called "tied ridges" which uses animal traction to till a grid of soil ridges about a yard apart across the fields before the rains start is being extended. These interlocking or "tied" ridges trap the early rainfall. That extra moisture makes fertilizer a paying proposition. The combination of more water and fertilizer gives a fourfold yield increase.

The adoption of tied ridges will be slow, however, because it means the adoption of animal traction in regions that have no animal tradition, and a restructuring of traditional village land allocations and farming systems. Still, it holds important promise for both raising food production and reducing soil erosion.[12]

MYTHMAKER AS HUMANIST OF THE YEAR:

"Is an economic system that leads to the loss of 24 billion tons of topsoil from our cropland each year environmentally 'sustainable'"?

Lester Brown and Werner Fornos, "The Environmental Crisis: The 1991 Humanists of the Year Speak to One of Today's Most Serious Concerns," *The Humanist*, November/December 1991, pp. 26–30

Reality Comment: Here is Lester Brown at his best—or worst. He takes a phony number on soil erosion—and misuses it. There is *no* solid estimate of worldwide soil erosion, but Brown has "constructed" one. Then he uses it as though modern capitalism was to blame for a problem that has been threatening since man scratched the first seedbed with a stick. In reality, the world's severe erosion problems are in the poorest

Third World countries and regions, where *tribalism*—not capitalism—is the dominant system of government.

Lester Brown has been forecasting famine, desertification, and soil erosion crisis nearly every year since 1965. The predicted famines, as we saw in earlier chapters, have never appeared. The soil erosion crises, for their part, have been largely confined to poor countries extending low-yield farming onto fragile soils.

During the big African drought of 1983–84, Brown claimed that the African climate had been changed by cattle grazing and the clearing of forests. Climatologists, however, firmly rejected Brown's theory. They noted that the 1983–84 African drought was nearly continent-wide—too large an event to be triggered by a relatively modest set of regional factors. Moreover, Africa has always been subject to periodic major droughts.

Meanwhile, the years since Brown's African calamity forecast have brought more moisture rather than less, and Africa's great Sahara Desert has been receding since 1985, not expanding. In fact, most of Africa's severe grazing and deforestation problems have been confined to the Sahel region that contains few of its people.

The World Bank 1992 Development Report, *Development and the Environment*, states rather flatly, "Desertification in the form of advancing frontiers of sand that engulf pastures and agricultural land, as so often depicted in the media, is not the most serious problem in dryland areas, although it occurs locally."

But then, Lester Brown had already blown *his* credibility on soil erosion in 1980. He announced in that year that the United States was in a "soil erosion crisis" just when the new conservation tillage was being adopted on millions of acres, and reducing erosion to all-time lows.

Lester Brown and his Worldwatch Institute have been among the most prominent soil erosion crisis-mongers. They have frequently warned that the rich rolling fields of the Corn Belt will eventually look like the mountain slopes of Nepal if we don't stop population growth.

However, Brown doesn't command much respect among soil erosion professionals. Pierre Crosson, one of the top world experts on soils and erosion wrote a 1993 paper on "Future Supplies of Land and Water for World Agriculture." He noted the following:

> Brown provides no evidence to support his 24 billion ton estimate (of world soil erosion), nor for his assertion about the cumulative yield effects of the erosion. . . . [Brown and Wolf's] global estimate of 25.4 billion tons of excessive erosion annually is in fact based for the most part on erosion estimates for the United States and on a rule of thumb . . . adopted for [Soil Conservation Service] technicians to use in advising farmers which land should be treated with soil conservation measures . . . I know of no soil scientist in or out of SCS who considers the standard to have scientific merit as an indicator of the land to which erosion in fact poses a significant threat to long-term productivity.
>
> For China, Brown and Wolf note that although the drainage basin of the Yellow River is much smaller than that of the Ganges, the Yellow carries a much heavier sediment load. . . . They infer from this that erosion in China must be . . . 30 percent higher than in India. From this they derive an estimate of excessive erosion in China.
>
> (They) assume that excessive cropland erosion per hectare in all countries is the same as in the "big four."
>
> The boldness of Brown and Wolf in developing their global estimate commands a certain admiration. But that the estimate should be taken seriously is highly dubious.

THE WORLD'S MYTHICAL GRAIN CRISIS

"Between 1950 and 1984, world grain output climbed from 624 million tons to 1,645 million tons, a prodigious 2.6-fold gain. Since [1984], world output [of grain] per person has declined each year, falling 14 percent over the last four years. In part, this fall measures the unsustainable use of soil and water. . . . This grim process of eroding soils is leading the world into a period of agricultural retrenchment."

Lester Brown, "Reexamining the World Food Prospect,"
State of the World 1989, Washington, D.C.,
Worldwatch Institute, pp. 41–58

Recently, Brown has been claiming the world has been suffering a grain production "slump" since 1984. This is his latest turning point. *Now*, he says, the famines will occur.

Unfortunately for Brown, and fortunately for the rest of us, most of the slowing in world grain production gains in the mid–1980s has been in rich countries, not poor ones. It has been the result of the United States, the EC, and Japan *trying to suppress surplus grain production*. In addition, the grain output in the former USSR has declined sharply with the collapse of its old command-style economy. That has nothing to do with soil erosion or Third World famine.

True, India did have a monsoon failure in 1987. But India *always* has a monsoon failure about one year out of five. This time, the country had some 26 million tons of grain in storage from record crops in the early 1980s. Thus, India only had to import about 2 million tons—to make up for its worst monsoon failure of the 20th century. (And India has since rebuilt its grain stocks from still-larger record crops.)

No reputable agriculturists agreed with Brown's forecast of food production retrenchment in 1989, and no such retrenchment has in fact appeared. Instead, world grain production potential has continued its escalation. For example:

- Grain yields in the developing countries rose 32 percent in the 1981–92 period, outstripping their population growth of 26 percent.

- As a result of the recent food output gains, per capita calories in the Third World rose to an all-time high of 2,473 in 1990. That compares with 2,117 calories in 1960 and 2,129 in 1980.

- Grain yields in the industrialized countries rose 25 percent from 1981 to 1992.

- World grain production in 1993 was 108 million tons higher than in 1984, with most of the gain in developing countries.

- *Total food production* in the developing countries rose 3.9 percent annually during the 1984–1992 period, nearly double their population growth rate of 2.1 percent! Lester Brown's "grain crisis"depended on some carefully selected numbers.

Since 1992, the World Bank stopped offering its statistics on the basics of "developing countries," but world food output continued rising by 1.3 percent—the rate of global demand growth.

If soil erosion rates haven't visibly slowed the gains in world food production, how can the anguish about soil erosion be so high?

In part, the soil erosion anguish has been manufactured—as an excuse to demand population suppression and/or reorganization of the world's "wasteful" socioeconomic systems.

Reprinted from BridgeNews Forum, September 3, 1999

"Soil Archeology" Finds Modern Farming Sustainable

For at least 25 years, critics have labeled modern farming "unsustainable." One of their major indictments is that aggressive cropping generates too much soil erosion. Over and over, they have predicted the return of the Dust Bowl days of the 1930s, and even global famine driven by topsoil losses.

Now, an industrious erosion expert from California has literally dug up the topsoil history of the Coon Creek watershed in Wisconsin—to find that its steep, intensively-farmed slopes are losing only 5 percent as much topsoil as they did in the Dust Bowl era. Dr. Stanley Trimble, a professor of geography at the University of California at Los Angeles (and a part-time farmer in Tennessee) used historic records and 20 years of his own soil surveys to prove the critics wrong.

Trimble found that erosion in the Coon Creek watershed rose strongly after the prairie was plowed in the 1850s and skyrocketed during the droughts of the 1920s and 1930s. However, erosion was sharply reduced by the contour plowing, strip cropping and other techniques introduced by the U.S. Soil Conservation Service in the late 1930s.

Coon Creek soil losses continue to decline, he says, as farmers improve their systems to reduce runoff from the fields. Now, Trimble says, the creek's waters again support the brook trout that heavy sediment loads eradicated before 1900.

Trimble says the Coon Creek farmers are not even yet making much use of the most powerful soil conservation systems—conservation tillage and no-till. Coon Creek was chosen for intensive study in the 1930s by the SCS. Trimble rummaged that SCS data out of the National Archives. Instead of relying on models or small-scale studies, he then resurveyed the soil profiles in nearly 100 soil cross-sections across the basin's profile in the 1970s and again in the 1990s.

Trimble even dug down to find the old roads, railroads and building foundations marking soil levels in 1900. At lower depths, he found the dark organic soils of the pre-1850 prairie.

Soil erosion has long been the biggest threat to the sustainability of the world's expanded population, and traditional low-yield farming systems have long suffered high erosion rates. Modern farmers claim they are achieving both high yields and soil protection; but official USDA soil erosion estimates have suggested that modern farming has reduced soil erosion only slightly.

Lester Brown, of the Worldwatch Institute, has claimed huge worldwide soil erosion losses—26 billion tons per year—and specifically warned in 1980 that the topsoil of the U.S. Corn Belt was rapidly disappearing. David Pimentel, a Cornell entomologist, published a 1995 article in *Science* magazine claiming the world had lost one-third of its arable land in the last 40 years, basing his conclusions mainly on his own estimate of 74 billion tons of annual soil erosion.

Asked about the soil erosion history in Coon Creek, Pimentel claimed that Trimble has studied sediments, not what happened in the fields. "Trimble has a good imagination," he said. However, Pimentel has little physical data to back up his own soil loss estimates. The world has gathered little systematic data on soil erosion outside the United States, and Trimble's Coon Creek data call into question even the U.S. soil loss estimates.

Why would Brown and Pimentel invent huge soil losses? Maybe because both have built global reputations by declaiming against overpopulation. Brown has predicted massive world famines for the last 25 years, claiming that recent world food production gains were "an illusion of progress."

Pimentel, in a 1994 book titled *Population and Environment,* asked, "Does human society want 10 to 15 billion humans living in poverty and malnourishment or 1 to 2 billion living with abundant resources and a quality environment?" (Actually, global population is slated to reach 8.5 billion and then decline slowly.)

Modern high-yield agriculture is one of the cornerstones of today's high-tech science-based world. By suggesting a lurking threat to the topsoil, Brown and Pimentel struck at the core of modern food security and to our children's prospects for living in technological abundance.

The USDA's Natural Resources Conservation Service has a built-in conflict of interest. If soil erosion is not a big problem, then there is no reason for the country to pay a billion dollars per year to have a soil conservationist in every one of America's 3,000 counties handing attractive cost-sharing subsidies to soil-conserving farmers.

A few years ago, I met with a group of urbanites, at one of those weekend rallys for the political well-meaning, in northern Indiana. I was amazed to find that they urgently believed that the deep, level soils surrounding our meeting room were silently disappearing from beneath our feet. They were not convinced by the conservation tillage being practiced on their neighboring farms. They were upset when I suggested that the best thing they could do to combat soil erosion would be to send their local soil conservation agent to some backward farming region like Indonesia.

When asked about high soil erosion estimates that imply the need for radical global depopulation, Trimble says, "They owe us physical evidence."

The Real Soil Erosion Crisis

The history of such long-term farming experiments as the Rothamstead test plots in England and the Morrow Plots at the University of Illinois proves that we *can* keep cultivating farmland for hundreds of years with *rising* crop yields.

The rice paddies of Asia, in fact, have been producing high and rising yields of rice for 4,000 years. Agricultural researchers have found no reason why they should not continue to do so. Much of the world's agriculture could achieve similar gains in the long term, if given continuing investments in research, inputs, and infrastructure.

Development and the Environment, the World Bank Development Report for 1992, goes out of its way to politely put the soil erosion crisis into perspective. Essentially, the Bank says that modern farming systems have reduced the soil erosion problem to manageable proportions in the United States and throughout the temperate-zone agricultures of the whole world. For example:

> The few comprehensive analyses of soil erosion that have been done in temperate areas indicate that the consequences are not large for aggregate agricultural productivity, although they are a concern locally for susceptible soils . . . . [S]tandard measurements of gross soil erosion from test plots typically overestimate the consequences for productivity, since the eroded soil can remain for decades elsewhere in the farming landscape before it is delivered to the oceans.

The Bank report also offers strong hope for erosion reduction in the rest of the world; as the quote from the Bank's 1992 World Development Report noted at the beginning of this chapter, most of the world's farms are not yet using the mulching, contour farming, and low-till systems that can cut erosion by more than 50 percent.

Pierre Crosson, of Resources for the Future, notes that the scary erosion estimates for other parts of the world haven't held up under close examination, even without the best farming systems:

> The United States is the only country in the world that has reasonably accurate and comprehensive estimates of soil erosion and its effect on productivity. . . . Global estimates must be taken with more than a few grains of salt. . . . Experts who have examined these estimates . . . concluded these evaluations have little scientific merit. There is no question that erosion and the resulting loss of productivity is significant in some regions, including Nepal, parts of India, the highlands of East Africa, and parts of the Andes.

As Crosson notes, however, these risky regions don't contain much of the world's cropland or people.

The world still has a potential soil erosion crisis, but as the World

Bank properly notes, it is in tropical developing countries, where the soils and rainfall patterns are different, and where soil conservation has so far relied mostly on bush fallow. With the population surge, tropical areas will now have to use more intensive farming systems. That means they will also have to use more aggressive soil conservation systems, such as kudzu crop rotations, alley cropping, ridge tillage, and mulch cropping. And eventually, they will use no-till.

Fortunately, there are only a few regions where the agriculture cannot be sustainably modernized to provide the base calories for the population growth expected. For example, even Nepal could provide its own base calories safely if it achieved the high yield potential of its Churai lowlands.

There are also some "edge of the desert" regions like Africa's Sahel where farming and herding will always be precarious. But the Sahel and the Nepalese highlands contain only about 0.01 percent of the world's population. If the ultimate answer in such regions involves people emigrating, it will produce only a small number of emigrants.

America's Soil Erosion History

Soil erosion has been an emotionally charged issue in the United States at least since the 1930s. The droughts of the 1930s coincided with—and aggravated—the Great Depression. The USDA's Soil Conservation Service was created after a dust storm from the Midwest blew over Washington and frightened the wits out of the Congress in session there.

Without question, American farming had extended crops onto some droughty and erosion-prone soils in its move westward. The experience of the 1930s made it clear that some of this land had to be returned to grass.

Since the 1930s, the Soil Conservation Service has spent $18.6 billion to encourage soil conservation on American farms, 55 percent of it since 1980. SCS has fostered such laudable techniques as contour farming, farm ponds, filter strips to protect waterways, and rows of windbreak trees to reduce wind erosion on the Great Plains.

Ironically, the 1930s also led to price supports for U.S. farmers—which worked in direct opposition to soil conservation programs. Price supports had the unintended side effect of stimulating many farmers to push more crops onto erodable soils, cut down woodlots, and drain wetlands.

In the 1990s, the United States essentially ended a 60-year-old policy that relied on farm price supports, government farmer payments, and cropland set-aside to improve farmer incomes. After 1996, farmers who had been getting crop subsidies began to get a declining set of payments ("golden parachutes" if you will) scheduled to end after 2002. Instead, farmers were encouraged to produce whatever would earn them the most profit, given their individual costs and prospective market demand.

There is no evidence that the subsidies helped slow the exodus of the farmers and farm kids who wanted to accept higher-paying off-farm jobs. The subsidies did succeed in raising land values—or rather the values of the land to which payments and allotments were attached. Unfortunately, after the programs were initiated, the higher land and "franchise" costs fell on the young farmers taking over from the original subsidy grantees. The cropland diversion also cut the number of off-farm jobs in farming areas by about one-third, because the land that wasn't planted didn't need seeds or machinery repairmen, or processing or storage.

The federal government also continued the Conservation Reserve Program. The CRP had originally been created in 1983, after the end of the commodity boom that began with big grain sales to the Soviet Union in 1974, and was further stimulated by the OPEC oil price boom. With commodity prices low and a surge of farm bankruptcies, Congress voted to take up to 36 million acres of land out of production on long-term contracts to "prevent soil erosion." The CRP was a thinly disguised attempt to raise farm prices and incomes.

In the ensuing years, conservation tillage sharply reduced the erosion problems on American farmland. The low-rainfall soils of the Great Plains, which had made up the bulk of the CRP lands, became not only safer, but more profitable to farm. (Conservation tillage often saved enough moisture for annual cropping, instead of the traditional grain-fallow rotation.)

After the passage of the 1996 farm bill (widely known as Freedom to Farm) the focus of the CRP was shifted to protecting water quality. The land put under CRP contracts was increasingly targeted at stream banks, wetlands, and other land that could help protect streams, rivers, and wildlife habitat.

Fighting the Good Fight

With high-yield farming, the world should be able to support its expected human population with less soil erosion than we have today. In fact, with high-yield farming extended to the best land, the world can *build* topsoil depth on many of the best farmlands.

Other putative "solutions" miss the mark. "Population management" cannot solve the Third World's soil erosion problems. Third World populations are already too high to be supported on the low-yield farming systems its farmers have used to date. Can we shoot the "extra" people? Can we forcibly sterilize young couples by the millions? If we leave people to starve in "benign neglect," how much of their local forests and wildlife would they destroy before accepting famine as their fate?

Obviously, we can accept none of these policies.

CGIAR Photo Gallery, ICRAF Project

EROSION CONTROL—Tree hedges in Uganda provide a good barrier against erosion on these steep slopes, wildlife habitat, and firewood for people, which helps hold back deforestation of the wild lands.

Resource reading: Charles E. Little, *Green Fields Forever: The Conservation Tillage Revolution in America*, Washington, D.C., Island Press, 1987.

Notes

1. Pierre Crosson and Norman J. Rosenberg, "Strategies for Agriculture," *Scientific American,* vol. 261, no. 3, September 1989, pp. 128–135.

2. Greg Schmick, agronomist and farm equipment executive in Spokane, WA, speaking of "no-till" agriculture in the mid-1980s. Quoted in the frontispiece of Charles E. Little's *Green Fields Forever*, Washington D.C., Island Press, 1987.

3. Jerry Hytry, executive director, Conservation Technology Information Center, West Lafayette, Indiana.

4. Ibid.

5. 1993 National Crop Residue Management Survey, Conservation Technology Information Center, West Lafayette, Indiana.

6. Flinchum, "Producing Soybeans Under a Highly Erodable Situation" *Proceedings of the 1993 Southern Soybean Conference*," St. Louis, Missouri, American Soybean Association, 1993.

7. F. Pierce, R. Dowdy, W. Larson, and W. Graham, 1984, "Soil Productivity in the Corn Belt: An Assessment of Erosion's Long-Term Effects," *Journal of Soil and Water Conservation*, vol. 39, no. 2, pp. 131–136

8. Flinchum, op. cit.

9. J. Walter, "Better with Experience," *Corn Farmer*, Harvesting Issue, 1993.

10. B. T. Kange, G. F. Wilson, and T. L. Lawson, *Alley Cropping, a Stable Alternative to Shifting Cultivation*, International Institute for Tropical Agriculture, Ibadan, Nigeria, 1984; see also IITA Annual Report 1989–90, Ibadan, Nigeria, 1991.

11. Pedro Sanchez, "Low-Input Cropping for Acid Soils of the Humid Tropics," *Science* 238, December 1987, pp. 1521–27; also see Pedro Sanchez, Testimony before the U.S. House of Representatives Committee on Science, Space and Technology, February 23, 1989.

12. John Sanders and Michael Roth, *Field and Model Results from Burkina Faso for Tied Ridges and Fertilization*, Purdue University Department of Agricultural Economics, April 1985.

11

Is High-Yield Farming Sustainable?

MYTHMAKERS SAY:

"Pesticides often leave the most resistant pests behind. . . . Then . . . the resistant pests multiply . . . soon, enormous quantities of pesticides are sprayed on the crops to kill just as many pests as were there when the process began. Only now the pests are stronger. And all the while, the quantity of pesticides to which we ourselves are exposed continues to increase."

Vice President Al Gore, *Earth in the Balance*, p. 52

"The second cause of slower food production growth is environmental degradation, which is damaging agriculture more than ever before."

Lester Brown, *State of the World 1993*, Worldwatch Institute

REALITY SAYS:

"The Food and Agricultural Organization reported Sunday that the percentage of people in the developing nations who are hungry fell to 20 percent from 36 percent between 1961–63 and 1988–90."

Paul Overberg, Gannett News Service, quoted from the Binghampton, N.Y., *Press and Sun-Bulletin*, September 21, 1992

". . . [P]ublic and private research institutions, commercial R&D enterprises and especially the various international agricultural (research) centers . . . are moving forward in concerted efforts to extend, redirect and fine-tune the original Green Revolution thrust. With the added impetus of biotechnology and other new scientific tools, we see clear indications that many of the problems and constraints of the 1960s and 1970s have been surmounted."

Dr. John R. Campbell, then-president, Oklahoma State University
and former dean of agriculture, University of Illinois,
Global Food Issues, Hudson Institute, 1991

"No doubt there is a [biological limit to yield] waiting for us somewhere, but the evidence says we don't seem to be there yet in farm yields."
Dr. Donald Plucknett, in his farewell lecture as senior science advisor to the International Network of Agricultural Research Centers [1]

"Sustainable farms cannot be constructed by going back to traditional farming systems. The success of traditional systems lay in their nonextractive use of natural and human resources. Under intensification . . . the systems collapsed. . . . [I]t is only in the unrealistic scenario of reduced population pressure that these systems can be made to work."
Lightfoot, Pingali, and Harrington, "Beyond Romance and Rhetoric: Sustainable Agriculture and Farming Systems Research"[2]

One of the key charges of the environmental activists is the claim that high-yield farming is "unsustainable." This has resonated with the public because it implies a lurking, hidden threat.

Every serious study of the world's food production potential has concluded that we have more than enough resources to support the current surge in population, and sustain the larger-but-restabilized population expected in the future. The latest work to confirm the world's huge "carrying capacity" is by Dr. Paul Waggoner, distinguished scientist at the Connecticut Agricultural Experiment Station. His study, entitled *How Much Land Can 10 Billion People Spare for Nature?* was published by the Council for Agricultural Science and Technology (CAST).[3]

Waggoner concludes that the current globe's cropland could provide a vegetarian diet for 10 billion people right now with current yield levels (if we converted the Third World land that is currently in pasture for draft animals to crops instead).

Is There a Yield Limit?

Perhaps the strongest endorsement of the Green Revolution's continuing importance and potential has recently come from Dr.

Don Plucknett. As senior science advisor to the Consultative Group on International Agricultural Research, the key international network for agricultural research, Dr. Plucknett may have the world's best overview of agriculural research potential.

Dr. Plucknett until recently was inclined to say that we didn't know where our next farm research breakthrough was coming from. But in recent years, he has reviewed statistics from the UN's Food and Agricultural Organization for the world and for individual countries around the globe. He also reviewed historic—even archeological—evidence of yields in past centuries.[4]

Food Supply Systems in the Tropics

system	*farming intensity*	*population density*	*tools used*
gathering	0	0–4	None
forest-fallow	0–10	0–4	Axe, machete, digging stick
bush fallow	10–40	4–64	Hoe, axe, machete, digging stick
short fallow	40–80	16–64	Hoe, animal traction
Annual crops	80–120	64–256	Animal traction, tractor

Source: Pingali, Bigot, and Binswanger, *Agricultural Mechanization and the Evaluation of Farming Systems in Sub-Saharan Africa*, Baltimore, Maryland, Johns Hopkins Press, 1987[5]

Plucknett says that yield takeoff marks a transition point when a farmer stops expanding his fields and starts getting more from the land he is already using by benefiting from scienific advice on how to boost yields. Plucknett believes once a country's farmers begin "yield takeoff," the potential for production *keeps going upward.*

Dr. Waggoner says we've been looking at agriculture through the wrong lens. We've assumed that crop yields are governed by the law of diminishing returns. But he points out that high-yield agriculture has been cheerfully violating the "law" of diminishing returns for decades—and is still getting away with it!

Diminishing returns would mean that after a certain level, more fertilizer wouldn't produce more yield. Better seeds wouldn't raise yields much higher. Instead, we find that more nitrogen, for example, can add just as much yield at higher levels as at low ones—assuming such constraints as moisture and trace minerals are dealt with. Artificial insemination keeps adding more milk per cow, as long as we keep improving the cow's nutrition, comfort, and health care. Improved seeds keep pushing yields higher.

Waggoner thinks what is happening is that farmers are *removing limitations* on crop yield rather than moving out on a diminishing returns curve. Let us review the past surges in farm yields:

- Mankind's first and most obvious crop yield limitation was weeds. We learned to do clean fallow, then plow and hoe, and now no-till and conservation tillage with herbicides has made a giant leap toward sustainability.

- Adding nitrogen (at first with animal manure) removed a constraint on the plants' nutrient requirements.

- Adding phosphate as well as nitrogen removed another nutrition constraint, and yields surged again.

- Later we learned that the plants need up to 26 trace minerals as well, and removing the constraint of mineral deficiencies raised yields again.

- Breeding shorter stalks on wheat and rice plants removed two constraints: It allowed more of the plants' energy to go into the grain heads; and the plants could support the heavier grain heads produced by heavy fertilization without falling over.

- Using supplemental irrigation on rain-fed land can produce still another yield surge in dry years, without even requiring much additional water or water delivery cost.

- Blocking insect and disease attacks with pesticides and pest-resistant breeding removes another yield constraint.

Confidence in the sustainability of yield-enhancing technologies continues to be increased by new developments on almost a daily basis. One good example is the USDA's newest pest-resistant soybean for hot climates. The new soybean resists nematodes, leaf-eating insects, and stem canker, all of which have taken a heavier toll in hot climates than in regions with cold winters. Its yields, as a result of the pest resistance, match corn belt levels (45 bushels per acre) and double the yields of current Southern commercial varieties (21 bushels).[6]

Waggoner asserts that science and technology should enable the developing countries to raise their crop yields fully as much as the rich countries already have, by removing their yield constraints over the long term. He concludes that 10 billion people should need no more land for food production than they use today, assuming the world continues to support agricultural research and permit freer trade in farm products.[7]

REALITY DATA:

"Data from China's National Network of Chemical Fertilizer Experiments (during 1981–83) revealed . . . 74 percent of China's cultivated land was deficient in phosphate, . . . about 40 percent was severely deficient in P, and about 23 percent was deficient in potash. . . . This scenario of imbalance is, outside of the large alluvial plains and deltas, typical of most developing countries, and is one of the reasons for stagnating yields, poor quality crops, increasing incidences of diseases, and soil degradation. . . . With few exceptions, organic manures alone will not support the yield increases in crop production that are required. . ."

World Bank staff members Richard Grimshaw, Christopher Perry, and James Smyle "Technical Considerations for Sustainable Agriculture," *Agriculture and Environmental Challenges*[8]

Excerpted from Hudson Briefing Paper, Number 190, May 1996

Farming to Sustain the Environment

by Dennis Avery and Alex Avery

The Main Sustainability Issues

The single greatest challenge for global agricultural sustainability is to save the world's remaining wild lands and natural areas from being plowed for food. The world depends on the integrity of local ecosystems, climates, and water cycles. Crucial in maintaining that integrity is the global conservation of forests and other wild lands, as well as their wild plant and animal species. Most of this conservation responsibility falls on agriculture's shoulders, since agriculture dominates world land use decisions.

The second biggest challenge is to ensure adequate funding for agricultural research at this critical moment of world population growth and rising affluence. We must be able to triple the productivity of the world's agriculture over the short span of fifty years to meet the needs of a larger and richer world population. Current levels of productivity and knowledge cannot sustainably do the job.

The third major sustainability issue is soil erosion. The ancient enemy, erosion is still a significant problem in key parts of the globe, especially the tropics. Obtaining the knowledge and capital to implement established solutions more broadly, and developing new techniques and practices for regions where current solutions will not work, must be a global concern. The current solutions include no-till, mulch cropping, and other soil conservation practices, along with farming the best and safest farmland to its fullest potential—wherever it is found.

Monocropping, pesticide use, and finite resource depletion dominate much of the sustainablity debate. Environmental quality issues, such as salinization, soil compaction, depletion of soil organic matter, surface water nitrification, and many others follow behind these major concerns. Some of these problems pose direct threats to future production; others affect future production only indirectly or slightly.

Sustainability Through Agricultural Research

Agricultural research is the most important sustainability component under humanity's direct control. Adequate research funding is urgent now for two reasons: (1) it can take years or even decades to develop new research thrusts and bring research findings into practice and (2) the next half-century is the most critical period—when the wildlife will be saved or lost to the production of food.

Cures for Soil Erosion

Soil erosion has always been the Achilles heel of agriculture. In relative terms, though, we've been doing pretty well against soil erosion for the last 50 years. We've doubled total world farm output by tripling the yields on the best land. When we triple the yields on an acre of land, we can get the same food tonnage even though we open only one-third as much soil to wind and water. In that sense, fertilizer should be considered a powerful soil conservation weapon.

What are some of the other strategies for addressing this sustainability problem? The most basic soil-conserving strategies are contour plowing and strip cropping. In contour plowing, the farmer plows his field along the contour of the slope so that the furrows act like mini-terraces, slowing the water and, to a certain extent, catching the soil before it leaves the field. Strip cropping involves planting two or more different crops in "strips." As one crop is harvested, the other crop is still growing and the slope is never totally bare. The steeper the slope, the narrower the strips need to be. If possible, the strips are situated along the slope contour.

These strategies, however, only mitigate the soil erosion problems which plowing creates.

Mulch crops and cover crops are strategies for limiting soil losses in fields. As the name implies, a crop is planted that acts as a mulch or cover for the soil when there isn't another crop growing, thus protecting it from wind and erosion. (Cover crops are especially effective when combined with no-till farming systems.

There are many benefits to using cover crops. In addition to the soil savings, many cover crops are legumes which add nitrogen to the soil. Soil organic matter is enhanced from the increase in plant residues. The cover also suppresses weeds, reducing weed control costs. The residues can insulate the soil and the young plants of the next crop from cold snaps. Some cover crops can actually yield a grain harvest before the next crop is planted. For the southern United States, where half of the rainfall and a good portion of the solar radiation occurs in the cool season, cover crops can be very effective.

Cover crops have been used for many years, but their use has increased significantly in recent years, in large part because of the development of herbicides that allow farmers to fully kill the cover crop so that it won't compete with the primary crop. Herbicides have also made no-till and conservation tillage possible.

Tillage means plowing or cultivating the soil. Conservation tillage is formally defined as any tillage system which leaves at least 30 percent of the soil covered by crop residue. Conservation tillage controls weeds with a combination of limited tillage and herbicides. No-till also leaves at least a 30 percent residue cover on the soil, but goes further than conservation tillage. No-till disturbs less than ten percent of the surface area in planting. As the name implies, there is *no* plowing, disking, or cultivating, and no-till keeps a grass cover on the soil in the winter. No-till relies completely on herbicides to control weeds.

Conservation tillage and no-till cut soil erosion an additional 65–95 percent over traditional tillage practices.[1] Using these appropriate strategies, farmers should now be able to create soil faster than it is lost through erosion on the best land, and be truly sustainable for this precious resource. However, the benefits of these tillage systems go far beyond reduced soil erosion.

Innovative solutions such as no-till are only the most recent developments of a constantly changing body of research and technology that enhance the sustainability of our agriculture. We should encourage the spread and further development of these established technologies and assist in the creation of new systems that can be used more broadly or address unique problems.

Alternative to Monocropping

The sustainability charge against monocropping is that planting large tracts of relatively identical plants makes us more vulnerable to catastrophic disease outbreaks and to overall disease spread and damage. This is a valid concern. In 1970, there was a devastating outbreak of a fungal disease in corn due to an overdependence on a single cell line for hybrid corn seed production. This cell line eliminated the need to remove the corn pollen tassels (time-consuming hand work that kept many high school kids employed in the summers but increased the cost of seed production). Almost 80 percent of hybrid corn was produced using this cell line because of the cost savings.

Many growers lost 100 percent of their crop to the disease and the country paid a heavy price for overdependence on a narrow genetic line. Estimates of crop losses totaled more than $1 billion. But farmers, breeders, and seed companies learned a hard and valuable lesson on the dangers of genetic uniformity and have applied that knowledge since. As an example, in today's corn breeding, careful and deliberate shuffling of the genetic base, multiple pollen sterility factors, and hand detasseling are all used to maintain genetic diversity and reduce the chance of another catastrophe. Competition between the seed companies also plays a real part in maintaining genetic diversity.

Nor should we in the United States forget one of the biggest culprits in continuous monocropping—the federal government. The structure of the federal farm subsidy programs has done more to limit crop rotation and cropping diversity than any other factor.

The federal government subsidy programs have required farmers to maintain a base level of production every year in their commodity in order to continue to receive subsidy payments. Production of crops without attractive price supports has been severely discouraged. Unfortunately, the need to maintain a payment base in their commodity has acted as a powerful disincentive to practice crop rotation and diversification.

If the government switched to a policy that did not shackle the farmer to specific crops, the amount of diversity in the fields would increase dramatically. In addition to rotating more mainstream crops, farmers would also grow more of such immediately available and highly useful alternative crops as canola, sunflower, meadowfoam, and kenaf. (Farmers would still be limited to whatever crops grow best in their regions.)

The Case for Pesticides

The use of agricultural pesticides has long been one of the most hotly debated issues in environmental and agricultural circles. The claim is that they are unsustainable due to their impact on the environment, wildlife, human health, and the development of resistance in the pests. But a careful and thorough look at these issues fails to support the claims. In fact, there is tremendous evidence that the judicious use of chemicals greatly *enhances* farming's sustainability.

Some antipesticide activists have asserted that agrichemicals have damaged the environment so severely as to decrease potential future production. This contention is refuted, however, by the ever-increasing yields on our fields. Every high-yield farm is a long-term testament to the viability and sustainability of today's agricultural practices, including pesticide and fertilizer use. This assertion is supported by the University of Illinois' Morrow plots and the Rothamstead Plots in England, both of which go back many decades, and other objective scientific studies of long-term farm productivity.[2]

The point is often raised that pest damage worldwide has increased dramatically despite pesticide use. But the real question is how high crop losses would have mounted *without* pesticides and how much extra land would have been needed to compensate for the losses.

(We must also note that organic producers also use large quantities of "organic" pesticides. Are organic pesticides exempt from the accusations of environmental impact leveled against synthetic pesticides?)

Pest resistance to pesticides is also raised in discussions of pesticides and sustainability. Pests will always adapt to overcome obstacles. Our only solution to pest adaptation is constant development of new pest resistant crops, pesticides with new or altered modes of action, and prudent approaches to slowing the development of resistance in the pests. There is every reason to believe that we can continue to stay ahead of pest and disease organisms, especially in light of the potential of biotechnology. We cannot sit still on any broadly used pest controls, even organically derived pyrethrins.

The claims that agricultural pesticides wreak havoc on natural ecosystems are unrealistic for today's highly specific, low-volume, and short-lived pesticides. In addition, today's chemicals are applied very carefully so that virtually all of their impacts are within the fields where they are applied. For example, less than one percent of the herbicides applied in the United States leave the root zone of the fields.[3] The fields are far from being pristine natural environments; we can hardly burden our farmers with the task of creating ideal wildlife habitat within their fields while still demanding they protect the real wildlands from the plow with high yields.

Concerns About the Health of Our Soil

Soil Quality: Although conservation tillage and no-till are practical solutions to soil erosion, some critics of high-yield farming have suggested new "problems" associated with modern farming systems.

One of the primary issues they have raised is soil quality. John Haberern and Anthony Rodale, of the Rodale Institute, phrased it in lurid, emotional language: "The soil provides us with life. The quality of that life depends on the quality of the soil itself. . . . If we feed the soil, the soil will feed us. It's nature's way."[4] They see pesticides and even chemical fertilizers as contrary to this goal and contend chemical use must be reduced or eliminated.

Fertilizers—plant nutrients—are essential to maintaining and restoring soil fertility, structure, and organic matter content. Fertilizers allow plants to produce more crop and crop residues, and crop residues are the critical factor in building the content of soil organic matter in our fields. Long-term studies have shown that soil organic carbon and nitrogen levels are highest when we combine conservation tillage, crop rotation, and adequate fertilizer.[5]

Plants absorb all nitrogen fertilizers in an inorganic form regardless of whether the nitrogen comes from organic or inorganic sources. What this means is that, from the plant's perspective, organic nitrogen fertilizers, such as manures, have no advantage over synthetic nitrogen.

What about pesticides? Are they degrading the quality of our soils and undermining agriculture's sustainability? Evidence from long-term field studies using pesticides says not. The productivity of the fields continues to increase as long as the soil structure and quality remains high, and these factors are most affected by tillage and nutrient management practices.

So far, there is little evidence that the quality of the soil is degrading past a critical point in the First World. In fact, herbicide-based low-till systems are now improving soil quality without the animal wastes or added plant manures which are so land and resource-costly.[6]

Soil Degradation: Soil degradation has been cited as another reason that current practices in farming are unsustainable. These prob-

Reprinted with permission from *No-Till Farmer*

LOW-PRESSURE TIRE—The huge footprint of this combine tire sharply reduces soil compaction, even with big machines.

lems include waterlogging, salinization, nutrient depletion, and soil compaction. Although some lands are indeed subject to these problems, these are localized, manageable problems with limited global impact.

Nutrient depletion occurs when farmers produce crops without adequate fertilization. All fields must be fertilized to replace nutrients that are removed in the harvested crops. Nutrient deficiencies are manageable simply through appropriate fertilization. Soil testing for nutrient presence and availability is relatively inexpensive.

Soil compaction occurs when external pressure from farm implements (tractors, harvesters, etc.) causes large soil pores to collapse. This condition causes slowness of water infiltration, poor drainage, and poorly aerated soils, all of which limit root growth and nutrient uptake. Solving soil compaction involves appropriate management practices and equipment. These can include no-till and conservation tillage, deep ripping to break up hard pans where they occur under the soil surface, permanent tracks for equipment to run on, and low-pressure tires, multiple axles, and tracked equipment, which reduce the pressure on the soil and prevent compaction from reoccurring.

Salinization and waterlogging result from irrigation and over-irrigation of cropland, particularly in drier areas, and areas with poor drainage. Salinization can also be caused by poor-quality groundwater. It occurs when excess water evaporates, leaving behind salts that concentrate in the topsoil. Although salinization and waterlogging are separate phenomena, they are often linked in susceptible areas.

Prevention should be our primary goal in treating salinization and waterlogging problems, and it is often straightforward: more frequent light irrigation, higher soil moisture regimes, and conjunctive use of surface and groundwater. The overuse of water is the biggest contributor to the problem. (Flood irrigation used by traditional farmers is also one of the major culprits.) Private ownership of water rights and real-cost pricing of water resources would encourage conservation and help avoid overuse.[7]

There are also after-the-fact solutions to existing salinization and water logging, such as leaching with water and simultaneous lowering of the water table. These, however, are more expensive than prevention.

Fear of Resource Depletion

Extremists have cited agriculture's dependence on resources that can't be replaced as evidence of the unsustainability of current approaches. This is a harshly oversimplified indictment of high-yield agriculture. Agriculture's current nonrenewable resource dependence is relatively small and only two inputs seem to raise sustainability issues, petroleum and phosphorus.

Petroleum: Petroleum is currently a cheap and readily available carbon and energy source. The percent of U.S. petroleum used in farming is tiny. Agriculture currently accounts for only two percent of total U.S. energy consumption for direct energy use *and* the manufacturing and transport of agrochemicals and fertilizers.[8] Moreover, agriculture will use whatever energy system the rest of our industries and homes use. When the United States was horse-powered, so was agriculture. When the rest of the country changes from fossil fuels to electricity, hydrogen, fusion, or whatever, agriculture will change too.

There is no sound rationale for forcing agriculture to change its energy system before of the rest of the economy does so. Agriculture could, but it would ultimately hurt our sustainability in other areas. For example, if we removed a percentage of crop residues for direct energy production (for example, by burning for electricity production), the extra land we would need and the reduced levels of residues on fields (which are needed to maintain soil quality and structure) would undermine any environmental benefits. It is pure folly to use our food-producing industry for energy experimentation.

Phosphorus: Some individuals claim that the world is short of phosphorus rock ore reserves. Phosphorus is a major plant nutrient, and virtually all of the world's phosphorus consumption is in agricultural feeds and fertilizers. Phosphorus is less abundant than most of the other natural elements. Some argue that this resource is *very* limited and predict we will exhaust our minable ore reserves in a relatively short time. For example, Herring and Fantel predict we will have depleted known phosphorus ore reserves in a short 100 years.[9] A closer look, however, reveals several major flaws in their analysis. Their first and most minor mistake is in underestimating known phosphate ore reserves by ten percent. Their estimate is 34.3 gigatons (Gt), whereas the U.S. Bureau of Mines number is 37.8 Gt. Second, they offer ridiculous population growth estimates of between 58 billion (lowest estimate) and 82.1 billion people by 2150.

Predictions of resource depletion dates are *always* highly suspect, because of economics and human nature. Mining companies do not like to admit to having very large reserves of ore. As more reserves are demonstrated, the value of known reserves falls.

Reasonable estimates of phosphate ore reserves extend at least 250 years, and that does not include vast reserves of lower-grade ores that could become available with future mining technology.[10] The United States has phosphate concentrations in 28 states, but only four large fields are currently competitive. Also, biological extraction systems are a likely prospect for producing large quantities of phosphorus from the world's extensive deposits.

Conclusion

Calling alternative agriculture "sustainable" does not make it sustainable. Alternative agriculture may rely less on chemical inputs, this is not a prerequisite for sustainability. It could actually make farming less sustainable by increasing land requirements. Today's modern, high-yield farming practices are the most sustainable we've ever had.[11] They will continue to change and improve in efficiency, safety, and sensitivity to the environment in direct proportion to our investments in agricultural research and technology.

Notes for "Farming to Sustain the Environment"

1. Personal communication, Jerry Hytry, Conservation Technology Information Center, West Lafayette Indiana; Flinchum, "Proceedings of the 1993 Southern Soybean Conference," St. Louis, Missouri, American Soybean Association, 1993.

2. University of Illinois, *The Morrow Plots: a Century of Learning*, College of Agriculture Bulletin 775, Champaign, Illinois, 1984.

3. Dr. Jerry Hatfield, Director, Soil Tilth Laboratory, Iowa State University, *Preserving Groundwater Quality*, presented at the Hudson Institute conference on "Saving the Planet with the 1995 Farm Bill," Washington, D.C., Feb. 7, 1995.

4. Rodale Institute press release, Sustainable Agriculture Network (SANET) on the Internet, August 8, 1995. sanet-mg@ces.ncsu.edu or scottrodin@aol.com

5. *Facts and Scientific Evidence: The Basis for a Fertilizer Use Policy in the United States*, pp. 10–11, Potash and Phosphate Institute and the Foundation for Agronomic Research, May 1994.

6. Conclusion, op. cit, 10.

7. Carter Brandon and Ramesh Ramankutty, *Toward an Environmental Strategy for Asia*, World Bank discussion paper 224, Washington, D.C., 1993, pp. 130–131.

8. Agricultural Resources and Environmental Indicators (AH-705), USDA-ERS, Washington, D.C., 1994.

9. James R. Herring and Richard J. Fantel, Phosphate Rock Demand into the Next Century: Impact on World Food Supply, Nonrenewable Resources vol. 2, no. 3, Fall 1993, Oxford University Press.

10. S. J. Van Kauwenbergh, "Overview of the Global Phosphate Rock Production Situation," *Fertilizers in Asia: What Hinders Acceptance and Growth*, presented at Muscle Shoals, Alabama, International Fertilizer Development Center, 1995.

11. Farming for a Better Environment, Ankeny, Iowa, Soil and Water Conservation Society, 1995, pp. 48–50.

Preserving Genetic Diversity with Gene Banks

The agricultural research institutions of the world are in a crash program to save the old landrace varieties of plants, along with their wild relatives, by collecting their seeds and preserving them in gene banks. The world really is in danger of losing these old landrace varieties of crop plants as they are being displaced in the fields by higher-yielding modern seeds.

The "Green" solution to this risk of gene loss would be to eliminate the new varieties. Essentially, they would turn the whole Third World into a gene museum, without the productive power to feed humans or protect wildlife from encroachment. Nor does having a primitive rice plant in a Burmese farmer's field make its genes available, say, to a rice breeder trying to defeat the brown planthopper in Indonesia.

The International Board on Plant Genetic Resources (IBPGR) is part of the Consultative Group on International Agricultural Research (CGIAR) whose researchers launched the Green Revolution. (The Green Revolution was probably founded by the Rockefeller and Ford Foundations when they established a plant breeding program in Mexico in the 1940s; it later became the International Maize and Wheat Improvement Center, now a key part of CGIAR.)

CGIAR encourages and coordinates national and multinational gene preservation efforts all over the world. These gene banks now have more than 500,000 plant "accessions" representing hundreds of plant species and thousands of varieties. The board has working relationships with more than 120 countries and 600 research institutes.[9]

The International Rice Research Institute (IRRI) rice collection alone has 41,000 seed samples of landrace rice from Asia, and another 6,000 from Africa. The CIP potato collection has more than 5,000 accessions of domesticated potatoes, plus 1,500 from wild potatoes, covering 90 percent of the variation in about 100 wild species.[10]

One payoff has been the discovery of resistance to the bean weevil. A search of *all* the field bean varieties in the world's gene banks failed to turn up such resistance. Now, bean varieties that can protect themselves from this voracious pest are being bred for the whole world. The world should also invest judiciously in some additional "gene farming," to preserve more of the old farmer-developed landrace strains of seeds, livestock, and poultry. This gene farming has become controversial, unfortunately, because some non-government organizations are demanding that First World governments make subsidy payments so the Third World's farmers will keep on farming with the traditional low-yield genetics. (It may be that the non-government organizations see themselves collecting the First World payments and becoming heroes to traditional Third World farmers.)

The world cannot afford to turn the whole Third World into a gene museum. The Third World has more than half of the world's arable land! But a few designated gene farmers *should* get payments for preserving the old genes.

Reprinted with permission from *No-Till Farmer*

QUAD-TRAC—It's not a lunar lander, but the latest effort to provide farmers with a cost-effective way to prevent soil compaction. It is pulling a tool that breaks up compacted subsoil layers.

SUSTAINABLE AGRICULTURE FOR AFRICA:

"In the late 1930s, Machakos District, a semiarid area of East Central Kenya . . . was considered . . . to be degrading alarmingly and to be rapidly approaching, if not exceeding, its capacity to support its inhabitants and their livestock. Today the area has a population five times as great and the value of agricultural output per head is estimated to be three times larger than it was then. At the same time food production in the area is less susceptible to drought than before . . . the rate of erosion has been sharply reduced. . . . More than 200,000 hectares has been terraced in some way, most without external support. There are more trees than before and they are being actively managed by farmers. Projections made in the 1950s, the 1960s and again in the 1970s all foresaw severe fuel wood shortages, but there is no evidence that such have occurred. Initial emphasis was on coffee and cotton, with a subsequent shift over the past decade into fruit and horticultural crops. . . . Staple food production appears to have stabilized at about the level required for basic subsistence, about 200 kg of maize per head per year."

John English, World Bank staff[11]

SOME THINGS THAT ARE NOT SUSTAINABLE:

"Keeping the poor in their misery while protecting the environment and promoting economic growth is *not* sustainable development."

Ismail Seragelden, World Bank vice president for environmentally sustainable development, 1993[12]

The Role of Integrated Pest Management

Integrated pest management is a vital element in maintaining the sustainability of high-yield agriculture. It will probably become even more important as more of the Third World intensifies its farming; Third World farmers will have the same need to avoid building up pest tolerances to pesticides, and they also have less cash to spend on pesticides in the first place.

However, integrated pest management is more important for ensuring that costs are kept low and that our pest control technologies continue to work well, than as a way to eliminate (or even reduce) the use of pesticides.

Analyzing even the successes of IPM points out how complex it is, as the following example shows.

The Case of Pakistan's Mangoes and Indonesia's Rice

Mangoes in Pakistan are attacked by four major groups of insects: mealybugs, fruit flies, scale insects, and leafhoppers. During the 1980s, the International Institute of Biological Control worked with the mango growers, who were spraying about five times per year and still not getting good insect control.

Research showed that the mealybugs laid their eggs in the soil at the base of the trees, and moved up to the leaves in the spring. Ladybird beetles preyed on them, but the ladybirds didn't build up their populations until late in the season. Research offered two inexpensive solutions.

First, the farmers were told to hoe around the bases of the mango trees in the winter, to expose and kill the eggs.

Second, the ladybirds were having to overwinter on other trees because the bark of the mangoes isn't rough enough to shelter them. Simple bands of rough sacking around the mango trunks let the ladybird beetles overwinter right on the trees and start effectively preying on the mealybugs sooner in the season.

Fruit flies were the reason for most of the insecticide use, because their maggots, laid in the mangoes, ruined the fruit. As an alternative to spraying, attractant traps were set up using imported fruit fly attractant (methyl eugenol). The traps cut fruit fly infestation rates from 35 percent to 3 percent.

Mango hoppers still needed spraying—but a study of hopper distribution on plants revealed that the growers could spray just the lower part of the trees (up to 5 meters) and still get control. This cut the amount of spray used.

Cutting the other sprays also eliminated the scale problem, because beneficial insects now controlled the scale. (They had been killed by the pesticides.)

Overall, research cut the mango growers' spray program from five applications to one, with a fourteenfold reduction in chemical costs. About 25 percent of Pakistan's mango growers currently use the IPM program.[13]

The biggest success for IPM to date is against the brown planthopper in the rice fields of Indonesia. The hoppers were increasingly attacking the miracle rice varieties of the Green Revolution during the 1970s and 1980s—despite subsidized pesticides and additional varieties bred for planthopper-resistance.

Research in the 1980s indicated that cutting pesticide applications didn't reduce yields. The pesticides had been killing many beneficial insects, and the hoppers had developed enough resistance to the approved pesticides that they offset the impact of the chemicals. (Too many farmers had also focused too long on a few favorite rice varieties that brought high prices.)

In 1986, a Presidential Instruction banned 57 insecticide formulations from the rice fields. It also ordered that only resistant varieties be grown in affected areas and that the rice industry be trained in IPM. Most farmers now spray once a year instead of four times. The amount of pesticide used has been cut by 50 percent, while yields have risen.[14]

Leonard Gianessi, a pest control expert with Resources for the Future, says the United States alone has thousands of pest species infesting 80 to100 crops in thousands of microclimates. As we saw earlier, there are hundreds of thousands of possible combinations of pest, crop, and region for researchers to struggle with in replacing the 200 active pesticides ingredients now in use. Yet only a small and declining fraction of pesticide uses appear to threaten human or environmental health.

Preserving Wetlands

Amazingly, the United States is now *gaining* wetlands. High-yield farming and the environmental movement share the credit for this development. The environmental movement gave wetlands a far higher priority in the public mind than they had ever had before—and a priority much more in keeping with their ecological importance. High-yield farming provided the high farm productivity so that we could spare these important ecological assets.

America has recently undergone a dramatic shift in its attitude toward wetlands, as Jonathon Tolman notes in his important report for the Competitive Enterprise Institute, *Gaining Ground: An Analysis of Wetland Trends in the United States.*[15] At the turn of this century, as Tolman writes, the U.S. Supreme Court characterized wetlands as "the cause of malarial and malignant fevers," and said "the police power is never more legitimately exercised than in removing such nuisances." It is no wonder that America has lost half of its original wetlands.

Currently, however, the conversion of wetlands has been slashed from roughly 450,000 acres per year in the 1960s and 300,000 acres per year in the early 1980s to a current conversion rate of perhaps 100,000 acres per year.[16] Farmers' conversion of wetlands has been slashed dramatically, from a rate of more than 350,000 acres per year in the 1960s to less than 30,000 acres annually today.[17] Meanwhile, nonfarm conversions of wetlands for development have remained constant at about 80,000 acres per year.

Tolman says more than 100,000 acres of wetlands have been gained in each year since 1991. These wetlands gains were recorded under the following auspices:

- The Partners for Wildlife Program of the Fish and Wildlife Service, which encourages private landowners to restore converted or degraded wetlands on their property (210,000 acres total).

- The Conservation Reserve Program of the farm subsidy structure administered by the U.S. Department of Agriculture's price-support agency (300,000 acres total).

- The Wetlands Reserve Program administered by the USDA's Soil Conservation Service, which secures permanent easements for restoring wetlands on cropland (125,000 acres).

These totals do not include additional wetlands that have been restored on both federal and private lands. The Fish and Wildlife Service, for example, restores 33,000 acres per year in its national wildlife refuge system, which is not included in the figures. It is highly unlikely that today's public opinion of wetlands would have become nearly so positive if America were facing severe shortages of key foods—or rampant epidemics of malaria.

MYTHMAKER:

"The relatively cheap and abundant supplies of fossil fuel have been substituted for human and draft animal energy. . . . Per capita use of fossil energy in the U.S. is . . . 14 times the level in China. . . . As our population continues to grow, we will inevitably experience resource shortages similar to those now being experienced by China and other nations."

Pimental et al., "Natural Resources and Optimum Human Population," *Population and Environment,* 1994[18]

MYTHMAKER STILL DOESN'T GET IT:

"The use of more land to produce food reduces the total energy inputs necessary for crop production and would lead to greater solar energy dependence and sustainability in agriculture. This of course assumes the availability of sufficient land, halving crop yields per hectare. . . ."

Pimental et al., op. cit.

Reality Comment: How does Pimental protect the wildlife if he needs twice as much land per person? He says we can only support 2 billion people on his system. Three billion people must be subtracted so we can give up the fossil fuels now used in agriculture. Who gets to choose the survivors?

Agriculture is the crucial industry for people *and* wildlife. It should not be the place where we casually *start* our energy-change experiments. Rather, it should be the place where we implement tested-and-successful strategies. Maybe we should listen to Dr. Pimental after his University (Cornell) has banned autos for its students and faculty.

Notes

1. *Chronicle of Higher Education,* September 22, 1993.
2. Lightfoot, Pingali, and Harrington, "Beyond Romance and Rhetoric: Sustainable Agriculture and Farming Systems Research," *NAGA the ICLARM Quarterly*, International Center for Living Aquatic Resources Management, Manila, Philippines, January 1993, pp. 17–18.

3. Dr. Paul Waggoner, *How Much Land Can Ten Billion People Spare for Nature?* Council on Agricultural Science and Technology, Task Force Report no. 123, Ames, Iowa, February 1994.

4. Dr. Donald Plucknett, "Science and Agricultural Transformation" IFPRI Lecture, September 9, 1993. International Food Policy Research Institute, Washington, D.C.

5. Reprinted in Keck et al., *Population Growth, Shifting Cultivation and Unsustainable Agricultural Development: A Case Study in Madagascar*, Washington D.C., World Bank, 1994, p. 7.

6. U.S. Department of Agriculture, Agricultural Research Service, *Quarterly Research Bulletin*, December 1993, Washington, D.C.

7. Waggoner, op.cit.

8. Grimshaw, Perry, and Smyle, World Bank, "Technical Considerations for Sustainable Agriculture," *Agriculture and Environmental Challenges, Proceedings of the 13th World Bank Agriculture Sector Symposium*, Washington, D.C., 1993, p. 19.

9. *Partners in Conservation: Plant Genetic Resources and the CGIAR System.* Consultative Group on International Agriculture Research, International Board for Plant Genetics Resources, IBPGR Secretariat, Rome. See also IBPGR Annual Report for 1992, Rome, Italy, 1992.

10. See Erich Hoyt, *Conserving the Wild Relatives of Crops*, International Board for Plant Genetic Resources, International Union for Conservation of Nature and Natural Resources, World Wide Fund for Nature, Rome, Italy, 1988.

11. John English, "Does Population Growth Inevitably Lead to Land Degradation?" *Agriculture and Environmental Challenges, Proceedings of the 13th World Bank Agriculture Symposium,* Washington, D.C., 1993 pp. 46–47.

12. Ismail Serageldin, World Bank vice president for environmentally sustainable development, *Agriculture and Environmentally Sustainable Development: Thirteenth Agriculture Symposium*, Washington, D.C., 1993, p. 6.

13. Jeff Waage, Director, International Institute of Biological Control, "Making IPM Work: Developing Country Experience and Prospects," *Agriculture and Environmental Challenges, Proceedings of the 13th World Bank Agriculture Sector Symposium,* Washington, D.C ,1993, pp 119–133.

14. Waage, op. cit.

15. Tolman, Jonathan, *Gaining Ground: An Analysis of Wetland Trends in the United States* Washington, D.C., Competitive Enterprise Institute, May 1994.

16. Dahl and Johnson, *Status and Trends of Wetlands in the Coterminus United States, Mid-1970s to Mid-1980s*, U.S. Department of Interior, Fish and Wildlife Service, Washington, D.C., 1991.

17. USDA Soil Conservation Service, *1991 Update of National Resources Inventory, Wetlands Data for Non-Federal Rural Lands*, Iowa State University Statistical Laboratory.

18. Pimentel, Harman, Pacenza, Pecarsky, and Pimentel, "Natural Resources and an Optimum Human Population," *Population and Environment*, vol. 15, no. 5, May 1994, pp. 347–369.

Bob Lang

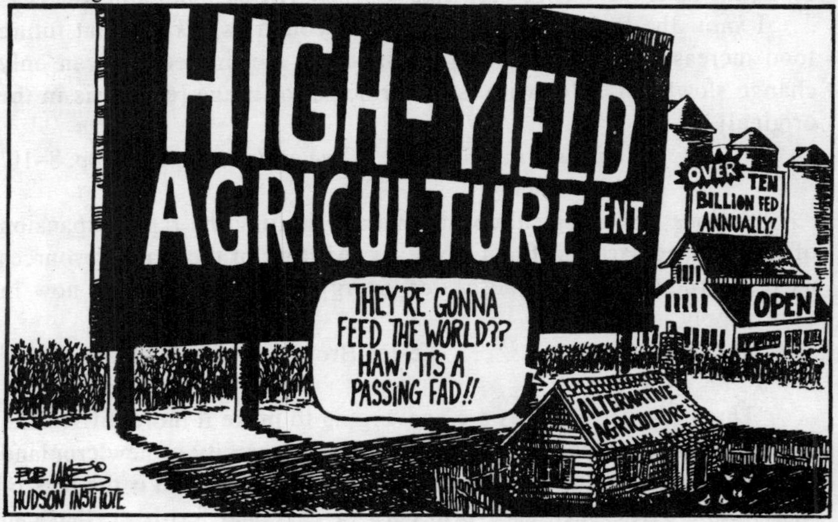

12

Seeds of Success

MYTHMAKERS SAY:

"The famines which are now approaching will not . . . be caused by weather variations and therefore will not be ended in a year or so by the return of normal rainfall. They will last for years, perhaps for several decades, and they are, for a surety, inevitable. Ten years from now parts of the undeveloped world will be suffering from famine. In fifteen years the famines will be catastrophic and revolutions and social turmoil and economic upheavals will sweep areas of Asia, Africa, and Latin America. . . . I *know* the food possibilities in these countries. I *know* that future food increases, based on today's techniques, are limited and can only change slowly. . . . Herewith is the principal of triage [emphasis in the original]."

William and Paul Paddock, *Time of Famines*, 1976, pp. 8–10[1]

"Serious technological constraints are limiting the rapid expansion of food, particularly beef and soybeans. All four of the major resources used to produce food—land, water, energy and fertilizer—are now in tight supply."

Lester Brown, *By Bread Alone*, 1974[2]

"Three historical trends are converging to make it more difficult to expand world food output. One is the growing scarcity of new cropland and fresh water that affects most of the world. The second is the lack of new technologies, such as hybrid corn or chemical fertilizer, that can dramatically boost output. And the third is the negative effects of planetary environmental degradation."

Lester Brown, *State of the World,* 1990[3]

"Rising world grain prices may be the first global economic indicator to tell us that the world is on an economic and demographic path that is environmentally unsustainable. . . . Over most of the last half-century, world grain prices have fallen, reflecting technological gains. . . . Between 1950 and 1993, world prices of wheat, corn and rice fell in real terms by 67, 83 and 88 percent. . . . Since 1993, however, the trend has been reversed. The world price of wheat climbed from a low of $3.97 per bushel in 1993 to $5.54 in 1996. While three years do not make a new long-term trend. . . ."

Lester Brown, *State of the World,* 1998

Reality Comment: Lester was wrong again, by his own guideline. Wheat prices in 1998 and 1999 averaged less than $3 per bushel. Every upward blip in grain prices is a new long-term scarcity trend in his mind—but not in reality.

REALITY SAYS:

"There are now six times as many people as there were when (Malthus') unsinkable essay was written. On average, the world's people are better fed. And clothed. And housed. And transported. What happened? Science happened. Science and technology made two blades of grass grow where one grew before. More than two blades. . . . Two hundred years ago, few people were aware that a scientific revolution had started right before their eyes. They assumed that the 'carrying capacity' of the earth . . . would not change."

Garret Hardin, *Insight*, December 20, 1993[4]

"*In World Agriculture: Toward 2000*, Nikos Alexandratos of the Food and Agriculture Organization (FAO) of the United Nations reports that only 34 percent of all seeds planted during the mid-1980s were high-yielding varieties. Statistics from the FAO show that at present only about one in five hectares of arable land is irrigated, and very little fertilizer is used. Pesticides are sparsely applied. Food output could drastically be increased simply by more widespread implementation of such technologies."

John Bongaarts, vice president, Population Council, writing in
"Can the Growing Human Population Feed Itself"
Scientific American, March 1994, p. 36–42

"Since the early 1980s . . . global cereals harvests have been rising at a rate of about 1.3 percent per year—just enough to meet the projected increase in demand. . . . But productivity increases—rises in cereal yields per hectare—have been slipping, too, from 2.2 percent per year in 1967–82 to 1.5 percent per year in 1983–94. . . . Peng, and the other agronomists who regard genetic engineering as the key to surpassing the yield barrier, have more in mind than the products of today's biotech industry . . . in which one or more genes coding for desired characteristics—such as herbicide resistance or an anti-bacterial compound—are smuggled into the organism from an outside source. . . . To break yield barriers, the plants will have to be thoroughly re-engineered."

Charles C. Mann, "Crop Scientists Seek a New Green Revolution," *Science,* vol. 283, January 1999

"As per capita availability of farmland declines and as production further intensifies, we will have to take more steps to maintain the quality of soils and waters and to optimize the efficiencies with which we use the principal inputs. We will make sure that soil bacteria have enough recycled biomass to feast on; that soil structure is conducive to soaking up rainwater as a sponge. We will get more nitrogen and more irrigation water not just from new fertilizer plants and new wells but increasingly from reducing the losses in field applications of these essential inputs . . . In most places, and not just in the poorest countries of Africa, this work is yet to begin. When largely accomplished, this change would amount to a new agricultural revolution."

Dr. Vaclav Smil, University of Manitoba agronomist, 1999

The Paddock brothers wrote a frightening book in 1967 titled *Famine 1975!* They predicted widespread famine throughout the Third World, and they urgently and sincerely recommended that the First World practice triage—selecting those countries which could be saved and letting the others suffer their self-induced famines without aid.

The Paddocks republished the same book in 1976, under a new title, *Time of Famines*—even though the famines they predicted in 1967 had not occurred. "The basic facts have not changed," they insisted in the preface for the second book. "The world is far less able to cope with famine today than in 1967, when this book first appeared." As of 1999, the major food shortfalls they have predicted have *still* not occurred.

The 1990 quotes from Lester Brown at the beginning of this chapter make the same arguments for predicting famine that Brown had made in 1974—even though the famines he had predicted in 1974 did not occur. Unfortunately, whole new generations of journalists and readers were too young to know that Brown had made the same arguments before, and been wrong for the same reasons.

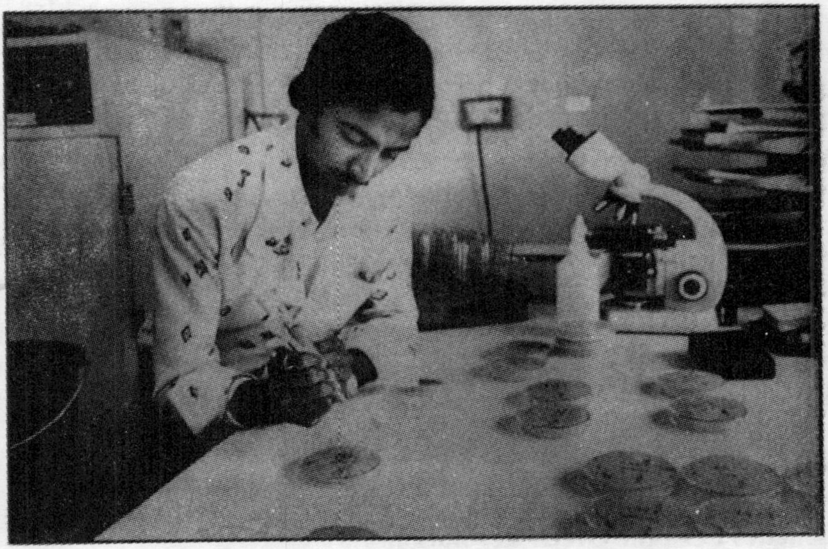

World Bank Photo by Rav Witlin

KEEPING THE GREEN REVOLUTION GOING—This research center in India is one of more than 20 internationally funded farming research units for the Third World. A leading example of the phrase "big bang for the buck," their existence is currently jeopardized by cutbacks in funding from wealthy countries.

About the only thing that Lester Brown and I have agreed on over these years is that we can't harvest much more wild seafood from the oceans. But that doesn't mean we can't have more seafood; the additional seafood we want can be raised in ponds and sea cages, and fed rations based on grains and oilseeds. Most of our growing shrimp harvest already comes from such sources.

David Pimentel of Cornell University makes a common mistake when he attempts to tie human well-being to a natural resource base. Natural resources are no longer the key to human well-being. Poverty-ridden Brazil has enormous natural resources, and wealthy Japan has virtually none. In China, since that country has begun to liberalize its centrally planned economy, agricultural output and per capita incomes have more than doubled without the help of major new discoveries of natural resources.

Virtually every country in the world has adequate farming resources to produce adequate calories for its projected population in 2050—if high-yield farming systems are pursued.

The doomsayers like Lester Brown charge that our gains in world farm output come from mining groundwater and destroying topsoil. They dismiss the power of agricultural research and technology. They continue to claim that the world is headed for famine and farming disaster.

Nothing could be further from the truth.

Even a gifted observer like Garret Hardin, who gives such a cogent explanation of why the new Malthusians were wrong that this chapter puts him in the "reality" section, then betrays his own observations and joins the doomsayers in predicting disaster. (Hardin was the *originator* of the "lifeboat ethic" of the 1970s and still claims that we will run out of farming research before population restabilizes.)

We shall now examine why the world is far more likely to eat *better* in the years of population growth that lie ahead.

Genetics Lead the Green Revolution

The Green Revolution got its primary stimulus from genetics. The simple process of crossbreeding plants, animals, poultry, and fish has driven the most dramatic changes ever seen in the world's food supply.

The genetics work has also been strongly reinforced by enormously important breakthroughs in new farming systems, fertilizer technologies, mechanization, and pest-control chemistry. Major productive investments have been made in irrigation and infrastructure (roads, silos, etc.).

But the real success story starts with genetics. The following improvements have been achieved since 1950:

- Crop yields and yield potentials have been tripled or more.

- Pests have been foiled with bred-in resistance.

- Poultry feed efficiency has been doubled.

- Extra crops have been added to the calendar with short-season seeds.

• Artificial insemination has been adding 2 percent each year to the milk yields of millions of dairy cows.

These are not isolated examples of onetime successes. They are all part of a general upward trend in food production potential.

Genetics has produced more productive strains in virtually every crop, domesticated creature, and region in the world. More new strains continue to flow from the labs and test plots with no sign of a slowdown. Such a continuing flow is particularly important; we need to continue a sizable set of breeding programs to keep from sliding backwards as pests and diseases continue to mutate and adapt.

Higher-yielding, shorter-strawed, pest-resistant, cold-tolerant, daylength-insensitive, higher-protein seeds have come from the plant breeders in profusion.

American corn breeders started the plant breeding revolution with hybrid corn in the 1920s. U.S. corn averaged about 27 bushels per acre in the 1920s[5] and more than 138 bushels per acre in 1994.[6] U.S. corn yields gained about half a bushel per acre annually in the 1930s, and are gaining 2 bushels per acre per year today.

The best farmers in most countries are still out-producing their neighbors two or three to one, indicating that there is still lots of potential for increase. The Green Revolution continues spreading to more and more countries, and to more and more commodities.

Chinese rice yields averaged 2.2 tons per hectare in the 1940s and have recently averaged 6.3 tons per hectare—an increase of nearly 300 percent.[7] In the 1980s, Chinese researchers were the first to successfully hybridize rice, achieving another 25 percent yield gain.[8] China's top rice yields have soared to more than 16 tons per hectare![9]

French wheat yields have more than tripled, from less than two tons per hectare in the 1940s to the current average of nearly 7 tons.[10] France is now planting the Green Revolution's short-strawed high-yielding wheat varieties in cold-tolerant winter strains, which get off to a stronger and earlier start in the spring than its old spring wheats. French farmers top-dress their wheat with repeated small applications of fertilizer for maximum yield stimulus, and use multiple fungicide treatments to prevent crop loss in its humid climate.

Corn yields in the Ivory Coast currently average 0.8 tons per hectare—but the potential yield is 7 tons per hectare. Consumer demand is so weak that additional corn production would drive corn prices below the cost of production. Because of the weak demand, farmers rarely use fertilizer or hybrid seed (which must be bought each year from a seed farm.)[11]

Shorter-season corn varieties have pushed the world's Corn Belt 250 miles north in recent decades, into central Canada, northward in the former Soviet Union and in China, and further south in Argentina. We have moved corn from tropic Mexico to Central Canada over the centuries, and have cut a month off the growing season in the last 40 years.[12]

Shorter-season rice varieties have cut the length of the growing season from 180 days to 110 days, permitting more double and triple cropping all over Asia. At the same time, researchers have made the rice plant more responsive to fertilizer and given it resistance to such pests as the brown planthopper, green leafhopper, gall midge, blast, blight, and grassy stunt.[13]

A redesigned rapeseed plant has produced a whole new product in addition to raising crop use value. Canadian plant breeders took out the natural chemicals which made rapeseed's oil bitter and its protein meal toxic to animals. They renamed the new plant "canola," and it now produces the most highly recommended vegetable oil for health-conscious consumers. Canola's protein meal is now a valuable addition to livestock and poultry feed supplies.

Kenya's new coffee tree, Riuru 11, has bred-in resistance to coffee berry disease and leaf rust, so growers can cut chemical sprays by one-third (and cash outlays by 60 percent).[14]

New cassava varieties for Africa are resistant to several endemic pests—and thus yield three to five times as much food.[15]

New high-protein corn has 90 percent of the food value of nonfat dry milk. The high-protein genes should help to overcome chronic protein shortages for people (and especially children) in regions where white corn dominates the diets of the poor. These include the Andean countries, remote regions of India and parts of Africa. The high-protein varieties are also being bred into yellow corn varieties, to increase their feed value by about 2 percent.[16]

Cloned cocoa trees and high-density planting will permit Malaysian cocoa growers to more than triple their production per acre. The new cloned hybrids by themselves triple yields. The new high-density planting system (3,300 trees per hectare instead of the conventional 1,000) adds further yield gains.[17]

Chickpeas resistant to root rot—and planted in a new broadbed-and-furrow farming system—produce 50 percent higher yields in Ethiopia.[18]

Wonder Wheats—In 1998, the International Maize and Wheat Improvement Center in Mexico (CIMMYT) announced a new family of "wonder wheats." This rebreeding of the wheat plant yields up to 50 percent more than any previous wheat varieties. Under ideal conditions in Chile, the new wheats have yielded up to 18 tons per hectare! The new wheats put more of their energy into larger and more numerous kernels of grain, supported by redesigned stalks and leaves.

Complaints came immediately that the new wheats would encourage the intensive use of fertilizer. However, it takes about 25 kg of nitrogen fertilizer to produce a ton of wheat, wherever and however we grow it. We can put 400 kg of nitrogen on one hectare of prime farmland and harvest 18 tons of wheat. Or we can clear 17 hectares of wildlands, put 25 kg of nitrogen on each of 18 hectares of land, and get the same 18 tons of wheat. Erosion losses will be higher with the low-yield wheat production because the additional land cleared will be of lower quality. However, the biggest difference will be the sacrifice of 17 hectares of wildlife habitat.

Disease-resistant bananas and plantains will apparently prevent the near-total loss of these key starchy foods to the dreaded Black Sigatoka disease. Black Sigatoka, first noticed in Fiji, has already spread through much of Latin America and is now established in Africa as well. Researchers at the Fundacion Hondurena de Investigacion Agricola and the Institute for Tropical Agriculture in Nigeria led the breakthrough in a scientific "first."[19]

Wait a minute! Are you telling me that all the world's bananas and plantains were at risk? My kids' bananas? Fruit salad? We were going to lose the major starchy staple for the whole Southern Hemisphere?

That's correct. A new disease, produced by natural evolution, emerged from the Fiji Islands and began attacking the banana (Musa) family worldwide.

Why haven't I heard about this?

Probably because it wasn't an immediate threat to Americans. Bananas are just a variety of fruit to us. The big food losses would have been in Latin America and Africa.

Why didn't I hear the good news when the "cure" was found?

The *Atlantic Monthly* had a small feature story. Mostly the media doesn't hype good news—though they frequently write about Lester Brown's latest famine claim, or Paul Ehrlich's complaint that we've got too many people to feed.

Did anybody get a medal?

No. And the U.S. Agency for International Development is cutting its aid for the international research center where much of the work was done.

MYTHMAKER:

"Imagine driving the plants to uniform prodigies of production with fertilizers and hormones while simultaneously eliminating their competitors with weed killers and destroying their insect and disease pests with appropriately structured biochemicals. . . . Mother Nature bound, gagged and blindfolded, helpless at last."

Dr. Joan D. Gussow, *Chicken Little, Tomato Sauce & Agriculture*, 1991[20]

Where's the Bull?

My personal experience with the power of genetics goes back to my boyhood on the dairy farm. When I was very young, we always had a young bull, penned away from the cows except when one of them was in heat.

My father was *very* careful to tell me how dangerous the bull was; that the animal was completely untrustworthy no matter how tame he might seem. It was all right to pet the cows, horses, and hogs, he said, but *never* get into the bullpen. Our bull was always young, because old bulls were even more dangerous than young ones.

Whenever a neighbor was gored by his bull—which happened fairly often—Dad would reinforce his message.

When I was about 10, we got rid of the bull. We started using artificial insemination instead. By that time, I was old enough to keep an eye on the herd, and note if any of the cows was coming into heat. (Dad was a county agent and often didn't get home during daylight.) If a cow was ready for breeding, we had to pen her up and call the artificial inseminator.

We also joined the Dairy Herd Improvement Association. To me, that meant sharing my bedroom one night a month with a young man who would weigh every cow's milk night and morning, and test it for butterfat content. I didn't realize at first that this periodic hospitality was part of a much larger concept. All the data from all the member herds' cows was being fed into a central information bank. The point was to know which bulls added more production to their offspring and which ones *didn't*.

The result of artificial insemination and record analysis has been better bulls that have added 2 percent per year to the milk of all the cows in America, each year since I was 10. This has increased average milk per cow from about 5,000 lbs. to 15,000 lbs.[21] (The world milk production record is now held by a Holstein cow in Missouri which produced 59,300 pounds of milk on twice-a-day milking in 365 days.)[22]

At the time, all I knew was that I didn't have to be afraid any more of Dad going into the bullpen.

Twinning Calves and Pest-Resistant Soybeans

In 1974, Lester Brown scoffed at the potential for higher food production in the future. He specifically noted that research had not produced higher-yielding soybeans or consistent twinning in cattle. It took a while, but research has now done both.

Mississippi State University and the USDA's Agricultural Research Service have announced a new pest-resistant soybean for America's southern states. This region has long been plagued by low yields, in large part due to heavy pest pressure. The new soybean resists soil nematodes, stem canker and several leaf-eating insects. In three years of tests, a check variety averaged 21 bushels per acre, and the new variety 45 bushels![23]

There's your higher-yielding soybean, Mr. Brown.

Genetics have also produced perhaps the biggest potential breakthrough in raising the productivity of the world's cattle herds—twin calves. An American cattle breeding service has started selling bull semen that is predicted to produce 40 percent twins, and their cow embryos 30 percent twins. Cows from the standard popular beef breeds produce about 1 percent twins.

In exchange for giving the "twinners" top-quality feed, cattle breeders can likely cut their per-pound costs of production by up to 30 percent. That's because twin calves represent almost twice the usual 70 percent of the cow's body weight "harvested" through the calf. Twins yield a far more efficient use of pasture, fencing, cow health costs, etc.

The new germplasm is from an experimental herd at the USDA's Clay Center, Nebraska, cattle research station. Breeds used in the twinning project include Hereford, Angus, Shorthorn, Simmental, Charolais, Brown Swiss, Holsteins, and a number of others.

Other Key Farming Technologies Waiting on the Shelf

- Hybrid rice, developed in China, offers potential yield gains of 25 percent for virtually all of the world's rice plantings. Currently, China is the only country using the hand labor required for pollination. However, the International Rice

Research Institute thinks it can use temperature sensitivity and apomixis to get male-line sterility and mass-produce new hybrids.[24]

- High-yielding wet rice farming systems could take advantage of 500 million acres of inland wetlands in mid-Africa. Africa could be self-sufficient in rice by planting only a small proportion of these wetlands. Most of Africa's current rice production is on low-yielding uplands better suited to trees.[25] Human diseases in the past have kept farmers out of the wetlands, but the diseases can now be controlled.

- Infrastructure also raises the effective yields of crops. In Bangladesh, a new farm-to-market road is associated with a one-third increase in crop yields. Better transport makes inputs less costly and the crop more valuable. Storage silos and grain dryers cut crop losses after harvest. Processing plants turn harvest-time gluts into year-round supplies. India's food processing industries have attracted 300 billion rupees of investment in the last two years, but the country still estimates that 20 percent of the country's farm output is wasted because of inadequate storage, refrigeration, and roads.[26] All it takes to provide more infrastructure is capital, and the world's capital resources are growing more rapidly than ever before.

Creating New Cropland

We shouldn't *need* much additional cropland if we continue to raise yields. However, it may be comforting to know that there is a great deal of good land that *could* be used for farming if more were needed—without threatening biodiversity:

- Acid-tolerant crops are now being geneticaly engineered. They will be planted successfully on the world's 1 *billion* acres of acid savannah. Before, these savannahs have been tropical wasteland. One-third of this acid savannah is in Latin America,one-third in southern Africa, and the rest in Southeast Asia.

Higher-Yielding Fish

For centuries, farmers have been breeding better cattle, hogs, chickens, grains, and oilseeds. Little has been done to improve fish genetics—until lately.

The Nile tilapia is a fast-growing, sweet-tasting fish from Africa that is renowned for its rapid growth and tolerance of tough conditions. High populations, murky water, high temperatures—the tilapia can take everything except severe winters. The tilapia converts feed efficiently. It will even grow in both fresh and salt water!

As a result, the tilapia has been fish-farmed widely for decades throughout much of Africa and Asia.

In 1988, the new International Center for Living Aquatic Resources Management (ICLARM) in the Philippines began the laborious task of breeding a still-better hybrid tilapia. It's not hard to understand why fish genetics have been slow to take off. A typical ICLARM experiment involves 500 cages, and the individual tagging of 25,000 individual little juvenile fish. (ICLARM got important help from Norway, which has already done a major breeding improvement on its high-value Atlantic salmon.)

By 1992, ICLARM's improved tilapia were growing up to 60 percent faster and with 50 percent better survival—on farms—than the commercial tilapia most commonly farmed in the Philippines.

It looks as though the increased productivity already developed in livestock and poultry can be developed for fish farming as well.

(From ICLARM Report, 1992)

- The upper Nile valleys offer major tracts of unplanted arable land. A high-yielding hybrid sorghum has already been bred for the area—but civil war in Sudan is preventing its widespread use.

- In the upper Euphrates valley, the big dams are already filling for Turkey's Greater Anatolia project. The dams will create a replica of California's famed Central Valley. Eventually, some four million hectares of low-yield dryland crops will give way to fruits, vegetables, cotton, and other high-value irrigated crops.

- Saudi Arabia test-planted 2,500 acres of a new salt-tolerant crop called salicornia in 1993. Salicornia has been bred up from a ubiquitous saltwater weed. Green-chopped before maturity, it makes a rich livestock forage like alfalfa. If allowed to mature, it yields an oilseed with a 70 percent meal proportion, like the soybean. *Most remarkably, it can be grown on the desert and irrigated with seawater*. Any coastal desert within 5–7 kilometers of the sea (so that the irrigation water can flow back into the ocean without contaminating groundwater aquifers) is now a potential site for "saltwater soybeans."[27] Saudi Arabia may have 400,000 acres of suitable desert.

High-Yield Farming and Forestry—Even for Africa

As we have seen, Africa is almost always the fearsome example held up to show the unsustainable pressure being put on natural resources by population growth. It is even offered as an example of the dangers or inadequacies of high-yield farming.

But Africa is the one continent where we have yet to *deploy* the power of high-yield farming and forestry. It is the only continent still trying to sustain rising populations on low-yield farming and wild harvest of trees nobody owns. The continent got a late start on high-yield farming. Africans didn't worry about raising farm yields during the 1960s because their land was still sparsely populated.

When agricultural researchers did begin to focus on Africa during the Sahel famines of the 1970s, they found that almost nothing developed for other continents worked on Africa's farms. The miracle rice varieties of the Green Revolution worked everywhere *except* Africa. Cassava bred for other continents did poorly in Africa. African farmers couldn't afford hybrid corn or fertilizer.

Nevertheless, strong reasons for optimism emerge from conversations with such experts on African agriculture as Nobel prizewinner Norman Borlaug; Dr. David Seckler, director of research for the Winrock Foundation; Dr. John Sanders of Purdue University; and Dr. Gebisa Ejeta (now at Purdue), who has bred some of the outstanding new crop varieties for African farmers.

Dr. Borlaug, who bred the original Green Revolution wheat varieties, is now leading an African farm productivity effort called Global 2000. The project is proving that the seeds and farming systems already available can double the yields of African grains and can repay the costs of fertilizer and pest control several times over. Global 2000 says its sorghum farmers in Ghana are tripling their yields with Africa's first short-season, short-stalk white sorghum.

Dr. Seckler returned from mid-Africa several years ago saying his biggest surprise was finding that high-yielding new seed varieties were already widely distributed on the region's farms. I asked why we weren't seeing the impact of the seeds in rising African grain production. He answered that consumers in African cities are so poor they're eating mainly root crops. Additional corn production at high cost would simply drive down the price to ruinous levels. (The best seed breeding advances *reduce* per unit production costs.)

"Africa's farmers are using their high-yielding seeds to reduce the amount of land they clear for subsistence," Seckler told me. That is a benefit, even an environmental benefit. But it is far less than the seeds could do if Africans could achieve economic growth and higher incomes.[28]

Seckler is worried about how African farmers can get and afford fertilizer, but agrees that fertilizer and soil-safe cropping systems offer substantial yield gains for that continent.

John Sanders has worked for a decade on Purdue University's animal-traction project in Burkina Faso. With animal traction, he says, farmers there can increase their grain yields fourfold, and their family farm output by sixfold by tilling "tied ridges." The key is breaking the encrusted soil surface before the rains start—and this cannot be done with feeble hand tools. But only a few of the community's farmers can become "commercial" farmers with oxen. That would leave the rest without a farming role. Sanders knows that until Africa has nonfarm jobs for more of its people, the dominant pattern of African agriculture will remain subsistence farming.

Dr. Ejeta has bred the first high-yielding hybrid sorghum for Africa (aimed at the Sudan). He is now working at Purdue to breed crops that resist Striga, a parasitic weed that is one of the key crop constraints in all Africa.

With high-yield seeds and farming systems, feeding Africa should take less cropland than the continent uses today. The elephants and gorillas need not be displaced by people.

An Ultimate "What If" Game?

About ten years ago, when I was with the State Department, I had a personal discussion with Lester Brown. He was launching yet another of his annual predictions that "this year" we would start to see the famines.

This is my recollection of the conversation:

Brown asked if I didn't think that we were running out of farm technologies. I said I saw no sign of it; in fact, I saw more productivity on the horizon.

He asked if we weren't running out of cropland. I said that we *had* more cropland, but it was less costly to raise the yields on the cropland we already farmed.

Brown: "Yes, but don't we face a limit on agricultural water supplies?"

Avery: "We could triple the efficiency of our farm water use."

Brown: "Don't we face a limit on the rate of photosynthesis?"

Avery: "Researchers have even had some success with breeding corn for a higher rate of photosynthesis."

Brown: "But we have an ultimate limit in the amount of sunlight that falls on the earth."

Avery: "We're using only 1 or 2 percent of it now."

Brown: "Then you agree that we face ultimate limits on food production."

REALITY:

"One hundred years from now the earth may have 10 billion inhabitants. . . . [T]he human population will then be approaching a stable level as

"The long answer is not quite as simple. Not only must the food supply expand, it must expand in a way that does not destroy the natural environment. For that to happen, a steady stream of new technologies that minimize erosion, desertification, salinization of the soil and other environmental damage must be introduced.

"We are confident that if the strong system of agricultural research organizations already in place is provided with enough financial support and leadership, it will develop these techniques."

Pierre Crosson and Norman J. Rosenberg, "Strategies for Agriculture," *Scientific American Special Issue on Managing Planet Earth*, September 1989, pp. 128–135.

Notes

1. William and Paul Paddock, *Famine 1975!* Boston, Little, Brown, 1967, republished in 1976 as *Time of Famines*, pp. 8–10.
2. Lester Brown, *By Bread Alone*, New York, Praeger, 1974, p. 7.
3. Lester Brown, "The Illusion of Progress," *State of the World 1990*, p. 11.
4. Garret Hardin, "Limits to Growth Are Nature's Own," *Insight*, December 20, 1993, p. 23.
5. Johnson and Gustafson, *Grain Yields and the American Food Supply*, University of Chicago Press, 1962.
6. U.S. Department of Agriculture Crop Production Estimate, November 9, 1994.
7. *FAO Production Yearbooks*, op. cit.
8. *Hybrid Rice: Proceedings of an International Symposium*, International Rice Research Institute (IRRI), Manila, Philippines, 1988, pp. 1–21.
9. Trip Report, Yunnan Province, China, by Gurdev Kush, Dennith Cassman and Shaobing Peng, IRRI, August 1993.
10. *FAO Production Yearbooks,* op. cit.
11. *Grain and Feed Annual Report*, Abidjan, Foreign Agricultural Service, U.S. Department of Agriculture, Washington, D.C., August 1992.
12. Corn breeding staff, Pioneer Hi-bred International, personal interview, 1993.
13. Huke and Huke, *Rice Then and Now*, Manila,International Rice Research Institute, 1990.
14. USDA/FAS, *Kenya: Coffee Annual Reports*, Nairobi, 1987–89.

15. IITA, *IITA Strategic Plan 1989–2000*, pp. 59–61. See also *IITA Annual Report 1989–90*, pp. 42–44; and USDA/FAS *Grain and Staple Food Outlook for Nigeria*, unclassified cable, July 19, 1990.

16. *Quality Protein Maize*, Board on Agriculture, National Research Council, Washington, D.C., 1988.

17. "World Cocoa Bean Production," *World Agricultural Production*, U.S. Department of Agriculture, WAP 10–94, Washington D.C., 1994, p.38.

18. *ICARDA Annual Report for 1993*, International Center for Agricultural Research in the Dry Areas, Alleppo, Syria.

19. Rowe and Rosales, "Diploid Breeding at FHIA and the Development of Goldfinger (FHIA-01)," *INFOMUSA*, vol. 2, no. 2, December 1993, pp. 10–11.

20. Joan Gussow, *Chicken Little, Tomato Sauce & Agriculture*, New York, The Bootstrap Press, 1991, p. vii.

21. *FAO Production Yearbook* series.

22. "New World Milk Record," *Successful Farming*, December 1993, p. 38.

23. Agricultural Research Service Quarterly Research Report, USDA, December 1993.

24. "Hybrid Rice," in *Proceedings of the International Symposium on Hybrid Rice*, Hunan, China, October 1986, by IRRI, Manila, Philippines, 1988. See also TAC Secretariat, Food and Agriculture Organization of the United Nations, *Report of the Fourth External Programme and Management Review of the International Rice Research Institute*, New York, FAO, 1992.

25. *IITA Annual Report*, 1989–90, p. 66.

26. Jimmy Burns, "Sweet Fruits of Sufficiency," *Financial Times* Special Section on India, November 8, 1994.

27. Glenn, O'Leary et al., "Salicornia Bigelovii Torr: An Oilseed Halophyte for Seawater Irrigation," *Science,* March 1991, 1065–7. See also, "A High-Grade Fodder and Seed Crop Which Thrives on Seawater," *Arab World Agribusiness*, vol. 2, no. 7–8, 1986.

28. Dr. David Seckler, Winrock Foundation, personal communication, 1989.

13

Drink Up, the Water's Fine

MYTHMAKERS SAY:

"The river systems that sustain terrestrial life and provide fresh water to 90 percent of the world human population are losing their life-giving capacity. . . . In developed countries like ours, toxic rivers illustrate dramatically the unimpeded destruction of the planet's lifeblood."

Letter to the editor, from Owen Lammers and Juliette Majot of the International Rivers Network, Berkeley, CA, printed in *E, the Environmental Magazine*, February 1994, vol. 5, no. 1

"This book has been written in response to increasing fears that nitrate from agriculture has found its way into drinking water, causing cancer, cyanosis in infants, the growth of toxic algae in rivers and seas, and untold mayhem in otherwise balanced, natural ecosystems. What is worse (so we are told) is that much nitrate has not yet had these effects but is waiting to do so, moving unseen in underground waters, sinking slowly, insidiously and inexorably towards our taps. In short, a nitrate time-bomb."

Addiscott et al., *Farming, Fertilizers and the Nitrate Problem*, preface[1]

"Agricultural groundwater pollution seriously threatens public health and welfare. Recent data from USDA indicate that 36 states have documented instances of well contamination with pesticides. Contamination with fertilizer nitrates has been detected in 30 states. These are troubling findings as most residents of rural American obtain their drinking water from underground sources."

Blueprint for the Environment: A Plan for Federal Action, report of a task force of 19 environmental organizations including

The Sierra Club, The Natural Resources Defense Council,
The Audubon Society, and Zero Population Growth, 1989[2]

"The growing use of chemical fertilizers is causing another more localized but hazardous problem: the chemical pollution of drinking water. Nitrates are the main worry, since they have risen to toxic levels in some communities in the United States. Both children and livestock have become ill, and some have died, from drinking water that contained high levels of nitrates."

Lester Brown, *By Bread Alone*, New York,
Overseas Development Council, 1974, p. 50

"Nitrate contamination of drinking water is a serious and growing problem that places thousands of infants at acute risk of contracting potentially deadly methemoglobinemia. Since 1986, over two million people drank water from municipal water systems that EPA found to be "significant non-compliers" with the Federal drinking water standard for nitrate."

Pouring It On: Nitrate Contamination of Drinking Water,
Washington, D. C., Environmental Working Group, July, 1997

"For the past 25 years, maybe longer, millions of people living in hundreds of Midwestern communities have been routinely drinking tap water contaminated with an unhealthy dose of agricultural weed killers, many of which are carcinogens. These people didn't know that pesticides were in their water, their iced tea, their orange juice, their infant formula, or in the jet of water squirting from water fountains at their schools and playgrounds."

"Tough to Swallow: How Pesticide Companies Profit from Poisoning
America's Tap Water," Washington, D. C.,
Environmental Working Group, August 1997

REALITY SAYS:

"We're safer and healthier than ever—and also more afraid of what we eat, drink and breathe. Why the reality gap? Too often, critics say, the media's coverage fuels our fears."

David Shaw, column one of "Living Scared," a *Los Angeles Times* series of stories on why Americans are safer and more frightened in their daily lives than ever before[3]

"Results from the U.S. EPA's National Pesticide Survey indicate that less than 1 percent of either rural domestic wells or community water system wells contain any pesticides in excess of lifetime [health advisories]."

U.S. EPA, 1990[4]

"Blue baby syndrome, the only health threat ever linked with high nitrate levels in drinking water, is virtually nonexistent in the modern world. Moreover, medical evidence now demonstrates that the few blue-baby cases which have occurred were not caused by nitrate in the drinking water."

Alex Avery, biologist, Center for Global Food Issues, 1999

". . . [A] continuing saga . . . began with EC Directive 778, which laid down a mandatory limit of 11 parts per million of nitrate-nitrogen in drinking water. Previously, the UK had been working with a limit of 22 parts per million with no ill effects on consumers. The reason for restrictions of any kind are based on the belief that excessive nitrates can cause blue babies and stomach cancer. The facts show, however, that the last 'blue baby' in the UK was 30 years ago. . . . And the incidence of stomach cancer in eastern England, where nitrate levels are highest—is below that of the rest of the country."

David Richardson, "The Case Against Nitrates Is Far from Watertight," *Financial Times*, June 25, 1991

"Nationwide, the total number of [pesticide] detections is a few percent of the total [water] analyses done, and in most cases the concentrations found are very small fractions of levels that are believed to be harmful to humans and aquatic life."

Pesticides in Surface and Ground Water, Council for Agricultural Science and Technology, Issue Paper no. 2, April 1994[5]

". . . [In a] detailed rebuttal to last week's report by the Environmental Working Group, dubbed 'Tap Water Blues,' which charges Midwest drinking water contains up to 30 times the acceptable Federal level of cancer-causing herbicides, . . . David Barker of the Water Quality Laboratory at Ohio's Heidelberg college . . . *using EWG's math*, calculates the additional risks pose a likelihood of one quarter of one additional cancer case per year for Illinois' 11.4 million people."

David Judson, Gannett News Service sent to Gannett, October 28, 1994

Environmental zealots—and perhaps even agencies of the United States government—want us to end the use of farm chemicals and/or spend billions of government dollars to make our drinking water "safe." They are seemingly immune to the recently rechecked reality that our drinking water is *already* safe.

The zealots believe—apparently because they want to—that fertilizer and pesticides are threatening our lives through our drinking water. That gives them another "reason" to condemn high-yield farming. In addition, "contamination" is a proven fund-raiser for an environmental organization no matter what the "contaminant" or its real danger. Even rural areas have succumbed to the tempting idea that federal money for studying their "water problems" comes free.

Health Risks from Too Much Nitrogen in the Water?

Let's deal first with nitrates.

There *are* nitrates in many of our wells. Some of them come from natural sources like legume plants and animal manure, and some from commercial fertilizer.

The EPA and the environmentalists call these nitrates "contamination." That implies danger. In reality, the small amounts of nitrate found in our wells have not been tied to *any* health threat. But nitrates are nevertheless considered contamination, and listed as such.

The current drinking water nitrate limit is not protecting the health of infants; is miscast as an environmental protection; and is forcing local governments to spend funds that could deliver more health gains in other health improvement and conservation efforts. The current standard is also being misused as a regulatory club against high-yield farmers who represent no threat to the local environment and are actually helping to save wildlands in other countries through their exports.

The only human health threat ever credibly linked to nitrates in drinking water is the infamous blue baby syndrome. (The National Academy of Sciences dismissed claims that nitrates in drinking water pose a risk of cancer or reproductive and developmental problems in 1995.)[6] However, the long-believed connection between nitrates and drinking water no longer seems plausible.

Blue baby syndrome, known to doctors as methemoglobinemia, is a form of blood poisoning. It attacks infants under six months of age because such small babies lack a key enzyme for dealing with an overload of nitrites in their systems. They literally turn blue, and in severe cases can die.

U.S. drinking water nitrate limits were established after a spate of blue baby cases in the 1940s in which the drinking water was found to contain high levels of nitrates. Doctors examined some 200 cases, and found none in which the water supply had less than 10 ppm nitrate, so that's where the limit was set.

The number of blue baby cases was relatively small even in the 1940s, and has declined radically since then—even though the world is using much more nitrogen fertilizer and has many more cattle feedlots and "hog hotels."

Today the average, properly constructed farm well contains 5–7 ppm nitrate-nitrogen. Nitrate levels in municipal water supplies have rarely exceeded 15 ppm. In contrast, an Indiana farm wife recently brought a well water sample to Purdue University technicians that tested 164 ppm nitrate. She reported that the well was 12 feet deep! It was obviously an old hand-dug well that had never made anyone noticeably sick, so it was still in use. (The water sample was not tested for bacteria.)

The last reported death from blue baby in the United States occurred in 1986, and that was the first reported death in the United States since the 1950s. The Center for Disease Control says blue baby is such a rare disease that it does not bother to collect data on cases. Great Britain has not had a blue baby death since the 1950s, nor a reported blue baby case over the last 30 years—even though British use of nitrogen fertilizer has increased radically since 1970.

This decline in blue baby cases cannot simply be attributed to better wells and water systems. The EPA estimated in 1992 that 66,000 infants per year were exposed to drinking water above 10-ppm nitrate-nitrogen.[7]

Nevertheless, the public is currently being told that high-yield farming and chemical fertilizers conflict with the safety of tiny infants. Activist organization such as the Environmental Working Group have recently demanded that the nitrogen limit in drinking water be lowered from 10 parts per million to 5 ppm.[8]

The EWG says this would provide a "greater margin of safety for infants." However, blue baby cases are now so rare that it's hard to believe the EWG is primarily concerned with infant health. It seems more likely that the EWG (already an ardent opponent of pesticides) sees the drinking water/nitrate limit as a handy way to further constrain modern agriculture.

Certainly, farming areas have recently had much more difficulty meeting the 10 ppm standard in recent decades as they have aimed for higher yields per acre to cut costs and meet the world's expanding need for more food from limited supplies of good cropland.

In such cities as Des Moines, Iowa, and Decatur, Illinois, the heavy use of fertilizer has helped drive nitrate levels in the water systems as high as 17 ppm. [9]

A number of U.S. cities are now spending millions of dollars per year in mitigating nitrate levels by treating their water with costly reverse osmosis, or by blending the water from agricultural runoff with low-nitrate water from deep wells.

Farmers are under increasing pressure from water authorities to reduce the nitrogen in the runoff from their farms. The water authorities, in turn, are under severe pressure from environmental activists.

The European Union is suing Great Britain for not enforcing the EU groundwater/nitrate limits (which is also 10 ppm nitrate). The UK government says doing so would put one-third of its cropland out of production, with no prospect of a gain in public health.

Fortunately, medical science is now telling us that the blue-baby syndrome was probably not caused by nitrate in the drinking water at all.

- When a baby suffers severe gastrointestinal inflammation, we now know that it can excrete up to 10 times as much nitrate as it consumes in its food and water.[10] This is an indication that the baby is manufacturing nitrites within its own system. That's why the nitrite levels can go high enough to turn the baby blue.

- •Severe gastroenteritis, typically accompanied by diarrhea and vomiting, can produce the blue baby syndrome even when the levels of nitrate in the drinking water are well below 10 ppm. In many cases, the drinking water contained no nitrate at all![11]

- •Healthy infants have been exposed to up to 100 ppm of nitrate in their drinking water for up to three days with no signs of methemoglobinemia.

Most of the blue baby victims have come from homes served by poor-quality wells. A poor-quality well almost always is a shallow well, and/or one with leaky casing. Often such wells are poorly sited as well—too near barnyards, septic tanks, floodplains, or other sources of bacterial contamination. Most poor-quality wells expose their users to both nitrate and bacterial contamination. After one blue-baby death, the water used in the infant's formula proved to have 40 ppm nitrate—but the well was also about six feet from a cesspool, and heavily contaminated with bacteria.

It is still possible that nitrates in water might aggravate the blue-baby attacks. But nitrates alone do not produce the condition. If nitrates in the drinking water do not cause blue baby syndrome, then we are misleading parents, communities, and public health officials by maintaining an unnecessarily strict drinking water limit of 10-ppm nitrate-nitrogen. Instead of terrorizing parents over fertilizer applications, we should be warning them about the dangers of shallow, leaky wells, too near livestock, cesspools, and floodplains.

Farmers, cities, and environmentalists should be told that nitrates can be a significant environmental problem when they cause algae blooms and oxygen starvation downstream. (Neither Des Moines nor Decatur, by the way, have problems with algae blooms.)

Raising the nitrate standard modestly, to 20 ppm, would pose no health risk and would virtually eliminate the needless expense of further modifying already-safe drinking water.

Many environmentalists say they are trying to limit farm fertilizer use "for the children." But modern medicine says their main impact would be to reduce the yields on the world's best farmland, while frightening parents of new babies about the wrong health risks.

Water and Cancer

Health professionals have looked carefully for any link between nitrate in water and stomach cancer. There *are* theoretical reasons for suspecting linkage. However, a British study of 229 urban areas found a *negative* correlation between nitrate concentrations and stomach cancer.[12] Longtime workers in fertilizer plants *do* show higher concentrations of nitrate in their saliva—but show no elevated risk of stomach cancer.[13]

Meanwhile, stomach cancer rates have been dropping rapidly in the very areas and countries where fertilizer use has increased! Stomach cancer used to be the top cancer killer among U.S. men, and third among women. The rate has dropped by roughly three-fourths with the United States now having one of the lowest rates in the world.[14]

Doctors think refrigeration and year-round consumption of fruits and vegetables are the major factors reducing stomach cancer.[15] Nitrate in the water seems to have no offsetting effect.

REALIST ON THE RIVER:

"I am an old Calvert County farmer who has lived and farmed for the last 71 years. I have loved the old river both for its bounty and for its aesthetic qualities. The Patuxent is the largest river entirely within Maryland and a major contributor of water to the Chesapeake Bay. . . . Treated [sewage] effluent now adds 55 percent of the river's fresh water. . . . During the recent ice storm, thousands of tons of urea (46 percent nitrogen) were used on airports, sidewalks, etc., all impervious surfaces. As the ice melted, I wonder where the nitrogen went?

John A. Prouty, Huntingtown, Maryland,
letter to the *Delmarva Farmer*, March 8, 1994

Pesticide Contamination—or Traces?

Trace amounts of pesticides have been found in many groundwater and surface water sources. A few pesticides are both persistent and weakly bonded to soil particles, so they can be leached out of the soil by rainfall. Examples include aldicarb on potatoes; a nematode pesticide called DBCP, which was formerly used in California; and atrazine, a corn herbicide widely used in the Midwest.

The public is easily frightened about these pesticide traces, because they have been told so often that all pesticides are "killers." What pesticides are designed to kill, however, are not humans, but weeds and insects, pest rodents, etc.

In pesticides as in virtually every "poison," danger requires a combination of (1) a substance toxic to the recipient organism and (2) enough exposure to cause damage.

Fortunately, we are long past the days when mainstream farmers tried to kill their pests with such compounds as lead arsenate and copper sulfate—which can be deadly to virtually all living organisms. We have also left behind the days of persistent pesticides like DDT—regardless of safety or efficiency. Modern pesticides are deliberately designed to have minimal impact on humans and beneficial wildlife. Most are designed for rapid breakdown, in days or weeks.

The pesticide traces found in water supplies represent even smaller risks than the pesticide residues in food—which are trivial.

It would be nice to be able to flatly say that trace contamination of water poses *no* threat, but we can't prove that negative. Any more than we can say that walking across the street (with the traffic light) poses *no* risk.

What we *can* say is that our well water supplies are safer than they've ever been. We no longer spread typhoid or cholera through our water. We are more careful now about having wells near our septic tanks or having them downhill from livestock pens.

The water experts working with rural wells say they check first for sources of bacterial contamination like the barnyard and the septic tank. They also look to make sure that the farmer isn't loading or rinsing his sprayer near the well. Those are by far the largest risks to farm water supplies.

Ohio typifies the relative freedom from pesticide "contamination" in the U.S. Corn Belt. And remember, the Corn Belt features the heaviest use of pesticides among nonirrigated U.S. farms.

A survey of Ohio water sources indicates that pesticide residues were *below the detection limits* in more than 90 percent of the private wells.[16]

Only a tiny proportion of its residents (about half a percent) are consuming waterborne pesticides in excess of their recommended lifetime risk standards. These at-risk consumers are almost all drinking from a few particularly vulnerable sources: surface springs and shallow wells.

The EPA's 1991 National Pesticide Survey indicates that less than 1 percent of either rural domestic wells or community water system wells contain *any* pesticides, even seasonally, above the lifetime health advisories.[17]

Surface water sources occasionally have peak concentrations above the level of the lifetime health advisory after springtime runoff events. However, this involves only a few surface waters (mainly smaller rivers) and short periods of time.

The major "problem" chemical is atrazine—and the EPA itself has recently decided that atrazine is about seven times safer than has been reflected in our water health advisories. If the EPA raised its Maximum Contamination Limit for atrazine to reflect its new safety rating, it would be hard for even the most dedicated fearmonger to instill much terror over American well water.

Further Cleanup

More good news: Monitoring of wells in the Midwest strongly suggests that little of the pesticide traces currently found in ground and surface waters get into the water supply from farmers' field applications. Virtually all of the contamination has come from *point sources*—where the pesticides have been accidently spilled or where the mixing and rinsing has been done carelessly in the past.[18]

In Iowa, over 80 percent of the water well systems in which pesticides were detected (other than atrazine) had known point sources—usually the local agricultural chemical dealer—near one of the town wells.

In Illinois, the state environmental agency randomly monitored more than 300 wells for pesticides and found *none*. Disbelieving, they then targeted more than 400 "high-risk" wells, testing for 34 pesticides with a *highly sensitive detection limit of 20 parts per trillion*. Only three of the wells were positive for pesticides and all three were near known point sources.[19]

These outcomes make it relatively easy to reduce further the current traces of pesticides in the water:

- •A major campaign is already warning farmers not to mix pesticides or rinse their application equipment near wells or surface waters.

- •Farm chemical dealers now virtually all have covered storages, concrete pads and holding tanks for rinsing, and their facilities are now surrounded by impermeable dikes.

- •Chemical companies are increasingly delivering their products as dry compounds (which can be cleaned up if they are accidentally spilled).

- •When the compound *has* to be delivered as a liquid, it increasingly comes in a reusable sealed container that locks onto the applicator rig to prevent spillage. The empty container is returned to the factory for reuse. Millions of plastic jugs have been eliminated as a side benefit of the reusable lock-on containers.

- •Or the companies sell their liquid products in premeasured water-soluble pouches which can be put directly into the spray tank. This means there is less spillage danger and no container to rinse.

Chemical manufacturers have also come out with new compounds that require very little active ingredient because they are so narrowly targeted at the pests' vulnerable points, such as enzymes.As a result, the amount of pesticide used by farmers has dropped significantly since 1976. Insecticide application rose from 117 million pounds (active ingredient) in 1964 to a peak of 130 million pounds in 1976—but by 1982 usage had declined to 71 million pounds.

Herbicide use, fostered mainly by the expansion of soil-conserving tillage systems, rose from 71 million pounds in 1964 to 456 million in 1982.[20]

That doesn't mean, of course, we can't further reduce pesticide traces in our groundwater. A new USDA research study combined no-till farming and ultrahigh populations of corn plants per acre. They used very narrow rows, and planted 52,000 plants per acre, compared with a normal "high" population of 26,000. *Because the high plant populations quickly shaded out weeds*, the researchers got equally high yields—150 bushels per acre—with one-fourth as much herbicide.

About 200 million pounds of herbicides are used on U.S. corn annually, especially on no-till corn. Atrazine (used to control broadleaf weeds in corn) and metolachlor (for grassy weeds) are widely used preemergence chemicals that are often found (at low levels) in ground and surface water but do not present human danger.

Eco-Economics: Billions to Fix Something Not Broken

Despite this picture of safe wells and drinking water, alarmists are sounding their cries about water contamination. The federal government is spending billions of dollars on grants, pilot projects, and demonstration efforts aimed at "making our drinking water safer."

REALIST VALUES CLEAN WATER:

"Purer drinking water, cleaner rivers and less polluted beaches have clear attractions. But what is the price of delivering them? Too high according to Mr. Ian Byatt, director-general of OFWAT, economic regulator of Britain's water industry. Meeting European Community environmental commitments agreed by the government since . . . 1989 will lead to an extra [$3 billion] a year in capital investment over the next five years on top of the [$7 billion] currently being spent.

"For the average customer, the full programme would mean an extra [$108] a year. . . . Nor is there much evidence that better standards would improve public health. Mr. Byatt argues that the new sewage measures would mean that a town with a population of 10,000 produced the pollution equivalent to 27 pigs, while the drinking water provision would result in pollution equivalent to one aspirin in an Olympic-size swimming pool."

From an editorial in the *Financial Times,* London, July 14, 1993

Notes

1. Addiscott, Whitmore, and Powlson, *Farming, Fertilizers and the Nitrate Problem*, Wallingford, UK, CAB International, 1991.

2. Alan Comp, ed., *Blueprint for the Environment: A Plan for Federal Action*, Salt Lake City, Utah, Howe Brothers, 1989, p. 13.

3. The David Shaw "Living Scared," *Los Angeles Times*, September 11, 1994.

4. U.S. Environmental Protection Agency, *National Survey of Pesticides in Drinking Water*, Washington, D.C., 1990

5. Available from CAST, Ames, Iowa.

6. National Research Council Subcommittee on Nitrate and Nitrite in Drinking Water, *Nitrate and Nitrite in Drinking Water*, Washington, D.C., National Academy Press, 1995.

7. U.S. Environmental Protection Agency, *Another Look: National Pesticide Survey Phase II Report*, Washington, D.C., 1992.

8. *Pouring It On*, Washington, D.C., Environmental Working Group, February 22, 1996.

9. M. Demissie and L. Keefer, "Watershed Monitoring and Land Use Evaluation for the Lake Decatur Watershed," *Illinois State Water Survey 1996*, ISWS Miscellaneous Publication 169, Champaign, IL.

10. E. Hegesh and J. Shiloah, "Blood Nitrates and Infantile Methemoglobinemia," *Clinica Chimica Acta* 125:107–115, 1982

11. K. F. Murray and D. L. Christie, "Dietary Protein Intolerance in Infants with Transient Methemoglobinemia and Diarrhea," *Journal of Pediatrics* 122: 90–92 (1993); J. R. Avner et al., "Acquired Methemoglobinemia: The Relationship of Cause to Course of Illness," *American Journal of Diseases of Children* 144:1229–30, 1990.

12. Addiscott et al., op. cit

13. Addiscott et al., op. cit.

14. American Cancer Society, *Cancer Rates and Risks*, 1985.

15. *Cancer Rates and Risks*, op. cit.

16. Dr. Robert M. Devlin, "Herbicide Concentrations in Ohio's Drinking Water," *Rational Readings on Environmental Concerns*, New York, Van Nostrand-Rheinhold, 1992.

17. EPA, op. cit.

18. R. S. Fawcette, "Pesticides in Ground Water—Solving the Right Problem," *Rational Readings on Environmental Concerns*, op. cit

19. Fawcette, op. cit.

20. USDA ERS Data

14

If We Stop Wasting the Water . . .

MYTHMAKERS SAY:

"At international meetings about resources we have frequently heard the statement that even in the 1990s, some countries or regions will have to stop their growth or go to war, or both, because of shortages of water."

Donella Meadows, Dennis Meadows, and Jorgen Randers, *Beyond the Limits: Confronting Global Collapse,* Post Mills, Vermont, Chelsea Green Publishing, 1992, p. 54

"'Irrigation is not likely to expand much faster than one percent per year during the 1990s, while population rises at 2 percent,' Postel said. 'This raises a red flag for the food supply.' Postel said the imposition of new population controls . . . is necessary to avoid a 'collision' between population growth and water scarcity.'"

Quote from a *Washington Post* story on *Water for Agriculture: Facing the Limits*, by Sandra Postel of Worldwatch, December 10, 1989, p. A4

"The pressure of rapid population growth, especially in the Third World, represents the fifth major strategic threat to the global water system. . . . One of the main reasons for this is the growing reliance on irrigation for agriculture. . . ."

Vice President Al Gore, *Earth in the Balance*, pp. 110–111

AND SOMETIMES THEY GET IT RIGHT:

"Farming accounts for some 70 percent of global water use. Much of the vast quantity diverted by and for farmers never benefits a crop: worldwide, the efficiency of irrigation systems averages less than 40 percent.

The technologies and know-how exist to boost that figure substantially; what is needed are policies and incentives that foster efficiency instead of discouraging it."

Sandra Postel, "Saving Water for Agriculture," *State of the World 1990*, Washington, D. C., Worldwatch Institute, pp. 39–40

One of the favorite new predictions of the eco-zealots is that the world is running out of fresh water. Those claims are false. Most of the world has ample supplies of fresh water to meet the expected human and environmental needs—if we stop wasting it.

The world's farmers use 70 percent of the water consumed, virtually all of it for irrigation. The vast majority of this irrigation water is wasted.

Worldwide, farmers' irrigation water efficiencies *in the field* average less than 40 percent, and perhaps less than 30 percent. (No one really knows accurately.) Even before the water gets to the farm field, a great deal is lost from evaporation and from seepage out of unlined canals. Once at the farm, most of the irrigation water is applied to poorly leveled fields with inherently wasteful flood irrigation systems.

USAID Photo by Chapellas

CAMEL PUMP—The typical Third World farmer wastes most of his irrigation water. The camel isn't too fond of the system, either.

(The parts of the field near the water source *have* to get too much water, so that enough will flow to the more distant parts to support a crop.)

Other factors also encourage inefficiency. Most governments offer the water to the farmers virtually free, leading them to overuse it. All too often, the irrigated fields have inadequate or nonexistent drainage systems, so that overuse of water and poor drainage lead to salinization and ruined soils.

It was probably poor drainage that ruined the Hanging Gardens of Babylon, one of the Seven Wonders of the ancient world.

Bad examples of *current* water usage:

- Saudi Arabia is using eons-old fossil water to produce millions of tons of gritty desert wheat which it dumps at giveaway prices in poverty-stricken Mideast markets under the label of "economic diversification."

- California farmers use underpriced government water to produce rice (one of the thirstiest crops) on semidesert where it is ill-suited. Some of the rice is even surplus (given the trade barriers).

- Indonesia is using heavy doses of irrigation water, pesticides, and even plant growth stimulants to boost the yields on its domestic rice, while shutting out lower-cost rain-fed rice from nearby Thailand.

Relative Water-Use Efficiency

flood–irrigation	35–60 percent
center-pivot sprinklers	70–85 percent
trailing tube pivots	85–90 percent
drip irrigation	85–90 percent

Reprinted from BridgeNews Forum, January 22, 1999

Where Will the 21st Century Get Its Water?

Will the scarcity of fresh water be the biggest constraint on food production—and human survival—in the 21st century?

Probably not, according to an in-depth new study done by Dr. David Seckler of the International Water Management Institute. But he says that avoiding water shortages will cost many billions of dollars—for more efficient irrigation technology, better water management, and new dams.

Even so, some dry regions may have to shift their current water supplies from food self-sufficiency to household and industrial uses. The dry countries will find their scarce water far more valuable for brushing kids' teeth, drip-irrigating oranges, and supporting industrial jobs than for low-value grain production. A few of the driest cities may even have to adopt the ultimate high-cost solution—desalinating salt water.

Seckler says 97 percent of the world's water is too salty for drinking or irrigation. Another 2 percent is tied up in ice caps, glaciers, swamps and deep aquifers. The world's annual rainfall totals 108,000 cubic kilometers, and 60 percent of that evaporates.

Humans currently capture about 3,400 cubic kilometers of fresh water per year. Seckler estimates that we could more than double our fresh water supply to 9,000 cubic kilometers—if we spend enough money on it. The costs may be highest, however, in the driest countries where they will need to commit their available water to its highest use even if that means radically changing traditional water use patterns.

Better irrigation is the world's strongest weapon against water shortages, says Seckler, because farming now uses about 70 percent of the water "used up" by humanity. Currently, most of the farmers' water goes into badly done flood or furrow irrigation systems with a water efficiency of perhaps 30 percent. With plastic pipe, drip tubes, center-pivot systems, land leveling and better drainage, we could raise irrigation efficiency close to 90 percent. An increasingly rich world will be increasingly able to afford such investments.

Seckler goes on to say that with "business-as-usual" in irrigation, the world would need 62 percent more fresh water than it uses today to support current standards of living in 2025. If we raise irrigation efficiency to 70 percent, we could support current living standards with only 17 percent more water.

The big problem with raising irrigation efficiency is that more than half of the water savings will be in the well-watered parts of China and India which won't help a water-short country like Pakistan. This means one-third of the world's population will need substantial investments in water availability beyond irrigation efficiency. The problems range from arid climates to well-watered places like Zaire and the Congo, which lack water purification, pumps and sewers.

If the Third World also wants to eat more meat, milk, fruits and vegetables, that will add more demand for water—or to the need to import more food and feed for arid-region cities from well-watered crop-land. Free trade in farm products will become an important strategy for minimizing the number of dams and keeping down the cost of water to poor households.

The International Water Management Institute's new "world water needs model" indicates that 12 percent of the world population lives in countries which have ample water for the next century. These water-rich countries include most of Europe, Japan, South Korea and Thailand.

A second group of countries (16 percent of world population) will need less than 25 percent more water by 2025. This also-fortunate group includes the United States, Canada, Mexico, the Philippines, Vietnam and Argentina.

The third group of countries is projected to be moderately short of water (25 to 100 percent increases). This group also has 16 percent of the world population, notably including Kenya, Tanzania, Peru, Nepal, Turkey, Brazil, Indonesia, and Australia. Most of them have water resources that could be developed. However, countries such as Kenya, Tanzania, and Myanmar may be short of investment capital.

The unlucky countries, those that will have to double their water availability include most of sub-Saharan Africa and Haiti. Most of them will be hard put to find the investment capital—even where there's plenty of water, as in Zaire and the Congo. Such poor and arid countries as Niger and Mozambique will have to harvest more of their rainfall with many small dams and an occasional big one. Port cities may find they can import food more cheaply than building water projects to grow food locally.

Eight percent of the world's people live in the water-scarcest regions, the Middle East and North Africa. They are already constrained by water shortages and some already have fairly high levels of irrigation efficiency. This means the easy gains from new irrigation technology will be limited. Seckler says some of them may have to divert the water they currently use for irrigation to household use and job-creating industries—and import a good deal more of their food.

In a very few regions, such as the Sahel, just south of Africa's Sahara Desert, water may actually put an upper limit on populations and/or standards of living. In those cases, outmigration could become another water conservation strategy; fortunately the numbers of people involved are not very large. Nor is the desert spreading; it just expands and contracts with weather cycles.

Don't look for water-shortage famines or spreading deserts to blight progress toward world affluence. Do look for new water conversion and water-conservation technologies to be big news in the 21st century—and don't be surprised if your water bill doubles.

Solutions to the Water "Shortages"

California stands as a good example of how we could resolve seemingly intractable water problems.

Californians could double the amount of water available for nonfarm use with virtually no public cost and without losing the jobs or earnings from agriculture: They would only need to give the farmers who currently own water rights *actual title to the water*. The farmers could then sell their surplus water to the cities, use the money to pay for more efficient irrigation systems, and suffer no loss of farm production. But a vocal minority of the citizens opposes giving farmers, with their current limited water rights a "windfall."

Many Californians want to shut down irrigated agriculture altogether, on the misunderstanding that it is despoiling the environment; but the irrigated land would support little wildlife if it weren't irrigated. Much of California's irrigated land would go back to virtual desert, with the rest probably in grassland lacking much biodiversity.

Dams and Fees

The land inundated by the dams is usually steep and erodible, and using it to hold water for high-yielding irrigated land protects much larger stretches of wildlife habitat downstream. The biggest real environmental problem caused by a dam is probably the stream flow for fish below the dam. For important fisheries like the salmon, this is a major difficulty. For most dams, it probably is far less important—despite the eco-activists' past solicitude for the snail darter.[1]

There are also some important dam sites that still *should* be developed for other important reasons. Noteworthy among these are sites for high dams in the mountains of Nepal:

- Dams in Nepal would not inundate much wildlife and would displace practically no people because of the awesome steepness of the valleys.

- The dams would prevent much of the downstream flooding that contributes to the desperate and precarious life in Bangladesh.

- Such dams would produce huge amounts of clean energy (without generating CO_2) that would provide non-farm jobs in both the highlands and the lowlands of Nepal and in neighboring India.

- The dams would also provide irrigation water for Nepal and India to increase crop yields and avoid expanses of low-yield agriculture.

Worldwide, water needs to be priced at its real value so farmers will use it more efficiently. Farmers probably don't pay 10 percent of the real cost of water. If they had to pay the real cost, they would irrigate more land with the same water using the best irrigation systems.

Pricing irrigation water at its real value would also encourage us to produce our farm needs on rainfed land, without the high capital costs and environmental risks related to irrigation. (Existing irrigation projects, of course, should continue to be used as efficiently as possible.)

FAO Photo by G. Tortoli

TRIPLING WATER EFFICIENCY—A pressurized sprinkler system uses its water up to three times as efficiently as the flood irrigation systems typical of poor countries. That means more crops from the same water, with less salinization and waterlogging.

The Future of Irrigation

Most of the best irrigation sites *are* already developed, at least by current standards. Most irrigation systems in the world today *are* troubled with waterlogging and salinization, which threaten their long-term viability. However, the environmental zealots are wrong to contend that we should turn our backs on irrigation and quietly accept famine or food aid for poor arid countries.

There is nothing inherently unsustainable about irrigated agriculture. The technologies and policies to support long-term productivity from irrigated fields are simple, straightforward, and well-known to water experts.

As water continues to become more valuable, it is virtually certain that urgent needs will overcome political reluctance and that the key technologies and policies will eventually prevail.

One of the immediate solutions for waterlogging and salinization is to provide better drainage for irrigated fields. That is where the farmer usually skimps because drainage is expensive.

One of the longer term solutions for salinization is to plant salt-absorbing crops, which will permit farmers to desalinate their land cost-effectively. The newest solution is genetically engineered salt-tolerant crops.

Reprinted from BridgeNews Forum, September 11, 1998

The Promise of High-Tech Irrigation

The Ogallala Aquifer, one of the premier natural water resources for the United States and the world, is being rapidly depleted. The Ogallala, which holds more than 3 billion acre-feet of water, underlies 175,000 squire miles in eight U. S. states. But its level has recently been dropping by a foot per year.

Obviously, we're taking out too much water.

We're using most of it to grow grain for the world's not-yet-stabilized population, and especially to produce meat for the world's rapidly expanding middle class. Much of the farm output goes to densely populated Asian countries—where the alternative to food imports will be clearing tropical forests to grow chickenfeed.

Fortunately, new irrigation technology can save the Ogallala—without destroying its communities, without forcing Depression-style living standards, and without shutting off the flow of farm products to land-short Asia.

I recently rode across the flat, treeless plains of eastern Colorado with two visionary irrigation experts. Irrigation has shifted the landscape from brown to green, and the cropping patterns from low-yield dryland wheat to high-yield feed crops. Cattle feedlots and hog farms turn the crops into meat and add local jobs. Rural communities which once threatened to dry up and blow away are now building new homes and modest golf courses.

I saw the graceful fountains of Ogallala water spraying from the big center-pivot irrigation sprinklers. My irrigation experts were appalled. "Half the water from those pretty fountains evaporates before it ever reaches the crop roots," they snorted.

Jack Jenkins grew up in the wheat fields of Washington State, and helped pioneer irrigated farming in Saudi Arabia. Now with the Department of Energy, Jenkins sees a way to cut the water and electricity needs of the Ogallala irrigated farmers by 40 percent—with no reduction in crop output.

"With the old sprinkler systems, we only got about 40–50 percent of the water to the root zone of the crops,"says Jenkins. "That meant we needed lots of water, and big pumps. Then we moved to drop tubes that spread the water at crop height under lower pressure. That raised our water efficiency to about 60–70 percent. But we were still getting too much runoff from the high parts of the field, and too much waterlogging in the low spots. With the latest technology, we can get over 90 percent of the water to the root zone."

Jenkins and Hal Smedley, director of the Colorado Corn Growers Association, took me to a farm that had installed much of the new technology. The high-tech pivot had nozzles two feet above the ground that put out a round sheet of water rather than droplets. The water descends straight to the ground instead of clinging to the plants and inviting evaporation. The pivot is computer controlled, utilizing buried sensors to detect when the crop roots need water. It operates at five or six pounds of pressure rather than the old 100 pounds, which means it requires a far smaller pump that uses much less electricity.

The total package will save 40 percent of the water and 40 percent of the energy on 100,000 wells, which in the driest years have drawn as much as 18 million acre-feet of Ogallala water. The next step in upgrading the system hasn't even been installed yet: variable-rate nozzles that would put exactly the right amount of water on each square yard of soil.

That could take the water savings to 50 percent or even more.The beauty of the new water-saving technology is that the savings on electricity can pay back the costs of downsizing the pumps and installing the new nozzles and controls—in just three years of operation! After that, the lower electric bills would boost the farmers' profits. The reduced electric demand would mean less coal burned in the region's power plants.

So far, few farmers have upgraded their systems. It takes $12,000–$15,000 per system and an individual farm may have from 3 to 50 pivots. Few farmers dare to walk into their local banks and ask for big "nonessential" loans.

Reequipping the farms of the Ogallala with downsized pumps, new nozzles, and computer controls would cost well over $1 billion. That's serious money, even if the upgraded systems would pay for themselves in three years and make the Great Plains' biggest water resource sustainable for the foreseeable future.

Jenkins points out that other economic sectors have set up "energy-conservation companies" which provide technical advice and financing to encourage conserving technologies. He and Smedley are trying to gin up something similar for the farms of the Ogallala Aquifer.

Perhaps the best news of all is that this high-tech irrigation can be extended for much of the world. That will save scarce water and energy, and lessen salt buildup in the soil on a huge scale—while preserving, as wildlife habitat, millions of acres of land too poor to farm.

Notes

1. The snail darter was an "endangered" small fish that was used by the eco-activists as an excuse to block the construction of the Tellicoe Dam on the Tennessee River. Because the darter was protected under the Endangered Species Act, they claimed the Tennessee was its last refuge. After the dam was built, it was amazing how many streams and rivers turned out to have snail darter populations. In environmental strategy, the snail darter was the first spotted owl. The comeback strategy that has evolved on the development side has been to offer bounties to anyone who can discover additional populations of such endangered species.

2. Maris, Apse et al., "Salt Tolerance Conferred by Overexpression of a Vacuolar Na/H Antiport in Arabidopsis," *Science,* August 20, 1999, vol. 285, pp. 1256–58.

15

The New War Against Plant Nutrients

MYTHMAKERS SAY:

"Poisoning of the soil with artificial agricultural additives began in the middle of the last century when a German chemist . . . mistakenly deduced that what nourished plants was nitrogen, phosphorus and potash . . ."

Peter Tompkins and Christopher Bird, *Secrets of the Soil*, 1989

"A 7,000-square-mile dead zone in the Gulf of Mexico, caused by the runoff of fertilizer, is getting worse and the only solution may be to change farming practices throughout the Corn Belt, according to scientists."

Paul Recer, Associated Press, January 25 1999

"Agricultural runoff that includes nutrients from animal waste is the largest contributor to the pollution in 60 percent of rivers and streams that the EPA has identified as 'impaired,' according to a report released by Senator Tom Harkin (D-IA). . . ."

Alternative Agriculture News, Wallace Institute, Greenbelt, MD, 1998

"The American Rivers [organization] says the Pocomoke River in the Eastern Shore of Maryland is threatened by factory poultry farms that 'stimulate the growth' of Pfiesteria . . . that has been killing fish and making swimmers and boaters ill."

United Press International, "Environmental Group Warns Hog, Chicken Farms Harming U.S. Rivers," April 6, 1998

"Most of the state's swine operations are located in eastern North Carolina . . . a labyrinth of rivers, streams, upland flats and wetlands. All waters wind slowly east, the majority emptying into the Albemarle and Pamlico Sounds, the nation's second largest estuary . . . habitat for waterfowl and shellfish, and spawning and nursery grounds for many fish along the Atlantic seaboard. As nutrient inputs increase upstream, they accumulate in unhealthy levels in the estuaries and sounds . . . "

Michelle Nowlin, environmental attorney, "Manure Management—No Easy Solutions," *Journal of Soil and Water Conservation*, September/ October 1997

"Nitrate contamination of drinking water is a serious and growing problem that places thousands of infants at acute risk of contracting potentially deadly methemoglobinemia. Since 1986, over two million people drank water from municipal water systems that EPA found to be 'significant non-compliers' with the Federal drinking water standard for nitrate."

Pouring It On: Nitrate Contamination of Drinking Water, Washington, D. C., Environmental Working Group, July 1997

". . . hog factories also emit into the air huge amounts of unregulated ammonia nitrogen gas. This nitrogen then is redeposited onto the landscape and waterways, choking rivers and estuaries already impaired by too much nitrogen."

Environmental Defense Fund "Hog Watch" Web site, January, 1999

"Well-managed farms play an important role in the fabric of life in the Chesapeake Bay watershed, but too many others—especially the 'factory farms' that concentrate poultry, cows and hogs at almost unimaginable densities—cause problems related to excessive manure production and fertilizer runoff."

William C. Baker, President, Chesapeake Bay Foundation, letter to the *Washington Post*, December 12, 1997

REALITY SAYS:

"Although fertilizer use can contribute to environmental contamination unless managed properly, it is often an indispensable source of the nutrients required for plant growth and food production."

Policies to Promote Environmentally Sustainable Fertilizer Use and Supply to 2020, Washington, D. C., International Food Policy Research Institute, 1996

"Nutrient replenishment to the soil is essential to sustain agricultural productivity under intensified land use, and fertilizers only replace substances naturally present in the soils. . . . At the extreme, environmentalists have been calling for chemical-free agricultural growth in Sub-Saharan Africa. . . . But such action dooms the region to stagnant crop yields and slow economic growth, in addition to ruining its wildlife habitat with slash-and-burn agriculture."

M. W. Rosegrant and C. Ringler, *Why Environmentalists Are Wrong About the Global Food Situation: Methods and Myths,* Washington, D.C., International Food Policy Research Institute, 1997

"Experience the world over also points out that no organic sources have proved sufficient to meet growing requirements of plant nutrients for sustained yield-based agricultural growth—not even in countries where they were not scarce."

Gunvant Desai, international agriculture consultant, *Agricultural Policy Notes*, Washington, D.C., John Mellor and Associates, Febuary 1995

"Few chickens are raised near the river where the most recent suspected outbreak of Pfiesteria piscicida in Maryland was reported, suggesting that poultry waste may not be the primary cause of the toxic microbe's appearance, a Maryland agriculture official said yesterday. . . . Officials do not believe that farmers around the Chicamacomico River had brought in manure to spread on their fields."

Todd Shields, *Washington Post,* September 27, 1997

"Virginia Tech biologist George M. Simmons . . . now roams the tidal creeks of his state's Eastern Shore armed with a pooper scooper. Simmons' surprising conclusion: Humans aren't always the source of the fecal coliform bacteria that contaminates some bay waters, forcing Maryland and Virginia officials to close thousands of acres of clam and oyster beds each year. . . . The most common culprits are cute, cuddly mammals: deer, otter and, most of all, raccoons."

Heather Dewar, "Other Mammals Also Pollute Bay, Researcher Says," *Baltimore Sun*, October 21, 1998

"Pork producers operate under a zero discharge rule. Section 319 of the Clean Water Act says it is against the law for any discharge from a hog operation to find its way into groundwater or a stream, river or lake. Those who do so are subject to heavy fines. Just to put things in perspective, for every pound of nitrogen that is collected on a hog farm and applied on land as a crop fertilizer, two pounds is discharged legally by municipal sewage plants into U.S. waters."

Donna Reifschneider, President, National Pork Producers Council, letter to the *New York Times,* July 13, 1998

"One of the most natural, 'organic' fertilizers you can use is manure. . . . Not only is it a good source of nitrogen, but the bulk of it breaks down and adds substance to the soil."

Bearville Organic Sheep Farm Web Site, January 1999

". . . we find little evidence that any cause . . . is multiplying the general bulk deposition [of nitrogen on the earth's land surface]."

Charles Frink, Paul Waggoner, and Jesse Ausubel, *Nitrogen Fertilizer: Restrospect and Prospect*, Proceedings of the National Academy of Science, February 1999

"If I have to haul my manure to be disposed of somewhere else, and replace that with a commercial fertilizer, that could put us out of business."

Daniel Shortall, Centerville, Maryland, poultry and grain farmer, quoted in the *Baltimore Sun*, February 15, 1999

Americans have badly abused their waters. In the 19th century, we used our creeks and rivers as sewers. When the Industrial Revolution came along, we dumped industrial effluent in the creeks and rivers too. Fortunately, as the country came out of the Great Depression and began the great economic surge that has radically increased its incomes since World War II, America discovered it had the technologies and wealth to treat its waters with far greater kindness.

The critical moment arrived in 1969, after the Cuyahoga River literally caught fire at Cleveland, Ohio. (An accidental spill of petroleum products spread a flammable film over the river's surface.) The fiery Cuyahoga remains the most infamous water pollution incident in the nation's history.

The response of the Congress was to pass the Clean Water Act in 1972, and launch the nation on a multibillion-dollar crusade to eradicate all water pollution by 1985.

The stated objective of the Federal Water Pollution Control Act (the real name of the "Clean Water Act") was to "restore the biological integrity of the Nation's waters." However, as the Senate Committee on Public Works noted at the time, we lacked any basic water quality information. No "clearly defined relationship between effluent discharge and water quality [had] been established." In other words, we didn't know the quality of our waters, the real impacts of the various pollutants, or have any benchmarks by which to measure the success of "restoring biological integrity."

Instead, Congress simply defined pollution as "the man-made or man-induced alteration of the chemical, physical, biological, or radiological integrity of water." Any product or by-product of human activity could legally be called "pollutants." Congress then focused the nation's resources on installing technologies to prevent "pollutants" from being discharged into the waters. The Administrator of the Environmental Protection Agency was given full discretion to decide who was polluting, and what must be done about it—mostly without benefit of water quality statistics or science.

Since 1972, the country has spent nearly $600 billion on the Clean Water effort. Unfortunately, we still have no data on where we started, or how much success we have achieved. We have no beginning baseline, and fewer than 20 percent of the nation's river miles are now being monitored for water quality. We have little data on pollution from agriculture—or even on the much more concentrated pollution from urban streets and sewage systems.

Undoubtedly, the quality of America's water has improved. But we mostly don't know how much, or where, or what produced the improvements. EPA's recent claim that 57 percent of the nation's water bodies are still unacceptably polluted is an unsupported assertion. It could only be true if we include soil particles as pollution—and the physics of stream flow guarantee that streams will carry sediment, even if they have to carve it from their own banks.

The truth is that 27 years after the Clean Water Act we have no scientifically valid assessment of the nation's water quality. As William Ruckelshaus, the first EPA administrator, told us, the Act applies "costly technologies in an indiscriminate and wasteful manner without regard to discernible social benefits. . . . We have set a goal of technology and not a goal of clean water."

The Clean Water Act was a child of panic. As a policy to protect the health of the environment, it was the equivalent of performing heart bypass surgery on everybody in the country because someone in Ohio died of a heart attack.

It is true that the Potomac River, "the nation's river," has radically improved in water quality since the passage of the Clean Water Act. The Potomac below Washington, D.C., once had signs warning people not to touch the water or eat fish from it because of its high levels of pollution. Today, the lower Potomac is a major water recreation area and a major bass fishery. Yet even today, following a heavy rain, a recreational canoeist is as likely to be paddling through a lake of raw human sewage as 25 years ago. The region's expanded sewage treatment plants are "legally" putting 20 million pounds of nitrogen and tens of thousands of pounds of phosphorus into the Chesapeake Bay each year.

Research biologists say that the water clarity in Lake Erie has improved dramatically in recent years due to the reduction in phosphorus pollution (along with filter-feeding by a newly invaded alien species, the zebra mussel). The populations of lake whitefish, smallmouth bass, and walleyed pike have improved importantly as a result.[1] At the same time, for exactly the same reasons, open-water fish species have faltered, and the Lake fishery is losing an important commercial species due to a shortage of phosphorus to nourish its food supply.[2] (Everything is connected to everything.)

While cities were spending the taxpayers' billions to implement clean-water technologies, America's farmers were spending billions of their own dollars on more rational approaches to water quality improvement. Examples include:

- New tillage equipment and herbicides for conservation tillage, which replaces plowing and cuts soil erosion and the runoff of water and chemicals into the streams by up to 99 percent;

- Confinement housing for poultry, hogs, and cattle, in which the wastes are saved in pits or lagoons and used as organic fertilizer on growing crops, instead of being allowed to run off into the nearest stream.

All over the country, rivers and streams have been upgraded from polluted and unusable to fishable and swimmable. Are the waters now clean enough? We don't know.

A team of water-quality specialists from the U.S. Geological Survey said recently "that perhaps in another decade, we will have the data needed to document environmentally and statistically significant trends in water quality." They note that policy makers "do not now have the information they need to make wise decisions for the future."

Despite the lack of data, the eco-drums have begun to beat again on water quality in the 1990s. The Clean Water Act was up for renewal, and the environmental movement wanted to give the EPA sweeping new water-quality powers. They also needed a new eco-publicity campaign that would engage the urban public's fears and renew their gratitude to the environmental movement. Urban water cleanup had already been "done." The new sewage treatment plants were in place, and most of the old industrial polluters had been reengineered or shut down.

Agriculture became the target again, as it had in so many previous environmental battles. Assertions began to be made about spreading algae blooms and fish kills due to agricultural runoff. (There are always at least a few algae blooms and fish kills.) The EPA and environmentalists began repeating the unsupportable claim that agriculture is "the nation's largest water polluter."

Farming uses so much of the world's land and water that it can hardly stay out of the line of fire. Moreover, today's farmers are easy to vilify because nobody in the cities understands why or how they farm—and the farmers never tell them.

The water-quality attack on farming, however, is almost entirely fraudulent:

- The environmental movement claims to be protecting U.S. infants from having too much nitrate in their water. But America has virtually no cases of blue baby syndrome and the medical evidence now indicates that blue baby is not caused by nitrate in drinking water. (See Chapter 13.)

- The claim that agriculture is "the nation's largest water polluter" is only valid if you consider soil particles as pollution. And, if you further ignore the reality that agriculture has radically cut its soil erosion since 1972.

•Eco-attackers had to pretend that indoor hogs and chickens, whose wastes were carefully saved and spread on growing crops as fertilizer, were a bigger threat to the marine environment than outdoor hogs and poultry, whose wastes wash into the nearest waters with every storm event. They did it by pretending that the wastes from confinement operations are sent directly into the nearest stream. This directly contradicts everything they know from their many hostile visits to the big hog and poultry farms, and the evidence they submitted in the many lawsuits filed against confinement meat production units. (In a big "headline" lawsuit against Premium Standard Farms in Missouri, 110 plaintiffs filed no evidence at all of air or water pollution.)

•Eco-activists have had to reverse themselves on the merits of organic fertilizer. For decades, they praised animal manure as nature's gift to humanity, the finest source of plant nutrients and soil health in the universe. Now, suddenly, this wondrous "organic fertilizer" has become "toxic waste" and a threat to our streams, rivers, and coastal waters.

•Water-quality attacks on agriculture led the environmental movement to wage war against the very nutrients that nourish the world's plants and trees, and its richest fisheries. They claim,for example, that a "dead zone" (low-oxygen zone) at the mouth of the Mississippi proves midwestern farming is a danger to the fish in the Gulf of Mexico. But such "dead zones" are characteristic of the rivers draining rich lands the world over. And the Mississippi provides most of the nutrients for the Gulf of Mexico's rich fisheries. Also, there is no evidence that the size of the low-oxygen zone is tied to the use of fertilizer on the farms in its watershed.

•An outbreak of toxic Pfiesteria on three small rivers on the Easern Shore of Maryland in late 1997 became the excuse for a new set of federal livestock and poultry regulatory programs. The environmentalists used the public hysteria as an excuse to indict the big poultry farms they already disliked—even though scientists say there's no demonstrable link between manure runoff and Pfiesteria. Now federal regulators are competing with state regulators to claim credit for shutting down the most eco-friendly livestock production in history.

The Pfiesteria Hysteria

The Pfiesteria hysteria was a heaven-sent media moment for eco-zealots. Dead fish began to turn up in three small rivers on the Eastern Shore of Maryland—right where the members of Congress and their media partners take their summer vacations! The Pfiesteria outbreak, small and natural though it was, became an immediate nationwide media event, and the launch point for the campaign to reauthorize and extend the Clean Water Act.

The Maryland governor's Blue Ribbon Pfiesteria Action Commission concluded that there was "no demonstrable cause and effect linkage" between non-point-source pollution, such as runoff from poultry manure, and toxic outbreaks of Pfiesteria. In fact, the Commission reported that high nutrient concentrations were not required for Pfiesteria to turn toxic, and that "toxic outbreaks can occur even if nutrient concentrations are relatively low."

The best science on the subject consistently links Pfiesteria's toxic blooms to concentrations of fish (it was a dry season) and some chemical agent in the fishes' own excretions. The tiny organisms have been around—and periodically killing fish—for millennia, at least. Fish kills, like forest fires, are common in nature.

The real crime of Maryland's poultry industry is that it offends the sensibilities of the environmental leadership. The industry does a number of good things; it preserves wildlife and wildlife habitat by conserving land; it produces valuable organic manure and other soil improvements; and it permits more people to live rural lifestyles on small farms. However, it does not contribute to the aesthetic enhancement of the countryside. It isn't pretty or nostalgic.

Livestock and poultry farming are not the black-hearted villains of the Chesapeake Bay drama. The Shenandoah Valley, for example, is one of the major and intensively farmed regions in the Chesapeake

Bay watershed. It has lots of outdoor dairy herds and indoor poultry flocks. Yet the Shenandoah contributes a "controllable nitrogen" loading of only 2.9 million pounds per year to the Potomac River (and thence to the Bay).

The city of Washington, D.C., has a mammoth sewage treatment plant at Blue Plains, which dumps 14 million pounds of "controllable nitrogen" per year—100 miles closer to the Bay. The estimated annual nitrogen loading for the entire Bay is 377 million pounds. What impact would it have on the Chesapeake if we shut down the whole agriculture of the Shenandoah Valley? (Our data on the Eastern Shore are less complete.)

The farmers of the Chesapeake Bay watershed need to implement "best management practices" to ensure that they are not harming the Bay. But if the overall regional goal of a 40 percent reduction in nitrogen and phosphorus loadings in the Bay is not met by 2000, it will be because population increases and nutrient-related economic activity have outpaced urban nutrient management.

Vice President Al Gore used the Pfiesteria excuse and a falsely based eco-campaign against confinement hog farms to mobilize the Environmental Protection Agency and the U.S. Department of Agriculture to draft sweeping new rules on confinement feeding of livestock and poultry.

The new EPA/USDA strategy broadens the interpretation of "point source discharge" to include water that falls on fields where manure has been applied and then runs off into tiles or ditches. Any discharge of a pollutant due to runoff from a rainfall less than the 25-year, 24-hour storm would be a violation with a potential fine. If this is strictly enforced, no runoff could leave any field by any man-made "conveyance" (ditch, grass waterway, tile drain or ditch) where manure has been applied.[3]

In Missouri, a county health official announced that no additional hog farms could be built in the county, because of undocumented "risks" to air and water quality.

In North Carolina, the Department of Environment and Natural Resources renewed a water discharge permit for Smithfield Foods' pork packing plant with a clause freezing the plant's hog kill at 144,000 per week, even though the number of hogs killed is not directly related to the amount of water discharged.[4]

The likely results of the intergovernmental competition to regulate livestock production?

- The trend toward confinement feeding in the U.S. will be slowed or reversed. Thus more outdoor hogs will produce more soil erosion by rooting on more land, and more hog wastes will be flushed into U.S. streams.

- U.S. soil quality will suffer from the failure to receive the soil quality benefits of organic fertilizer. Manure really is good for the soil (though we should use it on feed crops only, due to the new bacterial risks from E. coli O157:H7).

- More of the profits from satisfying the world's growing appetite for meat and milk will be earned by producers in other countries. America's rural landscape will become dominated by a series of huge crop farms (2000 acres each?) while rural communities lose the people and payrolls from meat production that would otherwise help keep the town library open and build the new golf course.

The Threat of Hypoxia in the Gulf of Mexico

There is a large low-oxygen zone at the mouth of the Mississippi River. The high levels of plant nutrients that the river carries down cause this "hypoxic zone" from Midwest fields, forests, and cities. In fact, there are hypoxic zones at the mouths of more than 40 of the world's major rivers, where nutrient-rich river water flows into an estuary or enclosed bay.

The nutrients in the river water trigger the growth of algae in the upper levels of fresh water—when the algae die, they sink to the bottom and use up all of the oxygen at the bottom with their decaying processes. However, the same kinds of nutrients that cause the algae blooms also nourish the fish in our bays and coastal waters. In fact, most of the nutrients for the world's big stretches of salt water come from the land, via rivers. That's why more than 90 percent of the world's fish catch comes from coastal waters.

Researchers have been unable to document any decline in the Gulf of Mexico's fisheries. In fact, the Louisiana Department of Fisheries reports increases in the important open-water fish species. The department says, "Decreasing the nutrient levels in the Mississippi River may serve to lessen the severity of hypoxia in coastal waters, but also may impact the food web of the northern Gulf and decrease fisheries production."

If Midwest farming is damaging the Gulf fishery, then aggressive steps should be taken to limit nutrient inputs, by both cities and farms. (Moreover, a pound of nitrogen reduction in Louisiana near the Gulf may be worth ten pounds of nitrogen reduction upriver, due to the river's assimilation capacity.)

But if the Gulf is getting the right amount of nutrients for the marine life, we may damage the fishery by reducing the nutrient flow. (Especially during low-flow years on the river.) This would not only hurt the fishery, but put additional pressure on Asian tropical forests by forcing Asia to meet its own food gap through its own low-yield farming.

The mouth of the Mississippi River has probably had a hypoxic zone since time immemorial. It has probably always fluctuated widely, depending on the flow of water. Dr. Nancy Rabelais, of the Louisiana State University, began measuring the size of the Mississippi's hypoxic zone in the 1980s. She found that in most years, the zone has averaged about 3,500 square miles—or 0.5 percent of the Gulf's surface area of 716,000 sqare miles. In 1988, a drought year in the Midwest, the hypoxic zone shrank to virtually nothing. After the Mississippi Valley's "flood of the century" in 1993, the hypoxic zone doubled its normal size, to 7,000 square miles.

When the "dead zone" stayed at that expanded size in 1994–97, Dr. Rabelais announced it was "growing," and that Midwest agriculture was threatening the Gulf ecology. (Most of the risk is to mollusks and mudworms, since fish and shrimp simply move to better water.) Scientific journals, eager to demonstrate the importance of ecological research, carried articles on hypoxia. Newspapers carried headlines. The White House appointed a task force. Presumably, Dr. Rabelais' research funding rose.

Dr. Rabelais is still preaching "threat," even though the size of the hypoxic zone shrank—by her own measurements—to 4,800 square miles in 1998. However, her latest measurements make the hypoxia look like a water-flow phenomenon rather than a nutrient-loading problem. (The big flood of 1993 certainly scoured up lots of extra soil and nutrients, and it should not seem surprising that there was a lag in the zone's later contraction.)

Nor is there any valid reason to assume the recent size of the big river's hypoxic zone is strongly linked with runoff from agriculture:

- U.S. fertilizer applications have varied narrowly between 43 and 55 million short tons since 1979.[5] Meanwhile, U.S. cornfields, which get about 40 percent of the country's nitrogen fertilizer, have increased their yields nearly 25 percent.[6] Obviously, higher corn yields have recently been taking up a higher percentage of the applied fertilizer.

- U.S. farmers have shifted to conservation tillage on more than 100 million acres of cropland, much of it in the states that drain into the Mississippi Basin.[7] Conservation tillage cuts water and chemical runoff by up to 90 percent by keeping cover crops or crop residue on the soil and encouraging water infiltration. There should be fewer nutrients reaching streams from Corn Belt fields today than under prior farming practices.

- U.S. livestock and poultry industries have been shifting indoors, and the modest expansion of livestock production has been largely outside the Mississippi watershed. The six biggest hog-feeding states in the Corn Belt had 35.7 million animals in 1987, and 35.1 million in 1997.[8] Also, more of our hogs and cattle are being fed in such places as the High Plains of Texas, New Mexico, and Utah, where the humidity, land costs, and human population densities are all lower.

The Ugly Secret of Modern Sewage Treatment

Though the federal government has lavished more than $100 billion on "modern sewage treatment" under the Clean Water Act, the waste-handling systems of our livestock and poultry farmers are still far ahead of our cities.

About half of the nitrogen and phosphate in our urban wastes are encouraged to volatilize into the air. The other half of the nutrients in urban wastes is dumped—as quickly and efficiently as possible—into our rivers, or spread as sludge on farmland. Nutrients have never been identified as pollutants under the Act—perhaps because the clean water effort has been using so many of our tax dollars to put more plant nutrients into the rivers.

The biggest hog waste spill on record in North Carolina was an eight-acre lagoon, which spilled—in a onetime accident—as much waste as the small city of Wilmington, North Carolina, (population 44,000) legally dumps into the Cape Fear River each year.

As Ned Meister, of the Texas Farm Bureau, pointed out to the Missippi River/Gulf of Mexico Nutrient Task Force in 1988, "Collectively, we have spent over $100 billion over 26 years in dealing primarily with urban point sources of pollution, which, by all accounts, have only achieved a 35 percent reduction in total nitrogen discharges from [such sources]."

Too Much Nitrogen in the Air?

In the June 1997, *Scientific American*, Tim Beardsley wrote that humankind has recently doubled the total amount of atmospheric nitrogen "fixed" every year by fertilizer production, legume crops, and burning fossil fuels. Beardsley raised the specter of the world being virtually drowned in nitrous oxide.

> The nitrogen glut is already causing 'serious' loss of soil nutrients, acidification of rivers and lakes, and rising atmospheric concentrations of the greenhouse gas nitrous oxide. Moreover, the oversupply [of nitrogen] probably explains decreases in the number of species in some

habitats, as well as long-term declines in marine fish catches, and in part, the algae blooms that are an unwelcome spectacle in many coastal areas.

However, Beardsley's sweeping charges in the *Scientific American* are either false or irrelevant.

There is no global evidence of serious soil nutrient losses. Acid rain was addressed by the federal government's $500 million National Atmospheric Precipitation Assessment Study, in 1991. The study found acid rain was caused mainly by the burning of fossil fuels, and represented a minor environmental problem. It affects mainly regions that lack limestone or other buffering for the natural acidity of most rainfall, including about 4 percent of eastern U.S. lakes and 3 percent of eastern U.S. forests. (Because altitude amplifies the problem, the trees affected are mainly red spruce in the Great Smoky Mountains.) [9]

The world's marine catches have been declining mainly because of overfishing. Algae blooms and fish kills are common in nature.

A comprehensive long-term study would be needed to sort out man-made impacts from natural cycles and events. Western Europe's farm price supports have been roughly twice as high as America's ever were and the EU price supports continue while America's have been largely ended. The high prices have stimulated far more intensive use of fertilizer in Europe,and much more intensive concentrations of livestock and poultry.

This has produced significant algae blooms in the coastal waters of the Netherlands, and in the nearly-stagnant Adriatic Sea. (The Netherlands is the world's most crowded nation, the United States is one of the least-crowded.)

The almost hysterical overstatement of the Beardsley article contrasts dramatically with the careful scholarship of Dr. Paul Waggoner in "Nitrogen Fertilizer: Retrospective and Prospect" in the *Proceedings of the National Academy of Science* (Feb. 1999). Waggoner says his research team can find little evidence of serious problems from nitrogen fertilizer use. [10]

He also notes that feeding 10 billion people without chemical fertilizers would take twice as much land as the world has. (Every person on the planet needs at least 3 kg of nitrogen per year, and Americans eat about 10 kg worth of nitrogen in their annual food consumption.)

In medieval Europe, the shortage of nitrogen in the soil limited wheat yields to about one ton per hectare, or less than half a ton per acre. French farmers now use nitrogen fertilizer to get seven times that much yield—thus protecting huge stretches of wildlands from conversion to farmland.

Nor does Waggoner see evidence that we're dropping too much nitrogen onto the world's land surface:

- Scientists measured about 4 to 8 kg of nitrogen deposited per hectare of land per year from 1888 to 1925 in the "high-yield farming" regions of Europe and the eastern United States. Comparisons in the Northeastern United States during the 1980s show about 6 to 7 kg per hectare deposited each year.

- The U.S. National Research Council found only questionable and contradictory evidence of an upward trend in aerial nitrogen deposition from 1965 to 1980.[11]

- The world's oldest agricultural experimental plots, at Rothamsted, England, showed a deposition increase of only 1 kg per hectare from 1888 to 1966.

We shouldn't be surprised. Waggoner says that if all of the world's commercial fertilizer—80 million metric tons of nitrogen per year—vaporized, instead of being taken up by plants, it would still mean only 1.6 kg per hectare over the earth's land surface. Since the nitrogen would also spread over water, it would average less than 1 kg per hectare for the total surface of the earth.

The current low levels of nitrogen deposit should not drive much ecological change. Waggoner says it generally takes 3–10 kg per hectare of extra nitrogen to cause ecological problems for wildlands. Of course, most of the nitrogen fertilizer isn't vaporized at all. It's turned into higher-yielding crops.

In fact, farmers are doing a better job of managing their nitrogen. In Iowa, farmers in 1964 were "mining" nitrogen from their soils; they applied only 74 percent of the nitrogen they harvested in their crops. In the export boom days of the late 1970s, Iowa farmers applied twice as much nitrogen as they harvested. In recent years, Iowa has been applying 120 to 160 percent of the nitrogen they harvest. (It's often hard for farmers to know how much their crops will use in a given year, since that depends greatly on weather and yields.)

Waggoner would still like to reduce the world's "leakage" of nitrogen from the system. He says one of the soundest ecological principles is to "worry about the leaks from the cycle, not how much is cycling." He thinks farmers are losing about 24 million metric tons of nitrogen from their fields every year, and another substantial amount of nitrogen is lost through livestock and poultry manure that isn't collected and used as fertilizer.

In the future, Waggoner thinks farmers will:

- recycle more of their animal and poultry manure onto crops through confinement livestock and poultry systems;

- permit less field runoff through conservation tillage, and waste less nitrogen on alkaline soils;

- match fertilization more closely to seasonal demand, and apply more slow-release formulations;

- use satellites and computers for precision farming that tailors fertilizer application to potential yields on each square yard of the field.

For the future, Waggoner says we need to sustain the recent 1.7 percent annual increase in world crop yields. That would quadruple crop yields by 2070 to nearly 8 tons per hectare. (In other words, we'd bring the whole world's yields up to Iowa's current levels, while taking Iowa to new heights of productivity.)

He says we also need to keep gaining 0.5 percent per year in the efficiency of our nitrogen fertilizer use. Waggoner thinks doing that would cycle slightly more than twice today's nitrogen fertilizer tonnage, with far less leakage from the system. He would also like to reduce the huge losses of nitrogen from urban sewage treatment, which involves another set of technologies.

By 2070, he sees a stable or declining human population, more wildlands than we have today, somewhat less nitrogen "leaking" into the environment from farms—and *much* less nitrogen being leaked into the ecosystems by urban sewage treatment plants.

The war against plant nutrients turns out to be as misinformed and misguided as the war against pesticides. Once again we find that modern farming is not a threat to the environment, but a major way to save it.

Notes

1. Lake Erie Committee, "Lake Erie Yellow Perch Improving," Great Lakes Fishery Commission's Home Page, November 11, 1997.

2. Lake Erie Committee, "Fisheries Productivity," Great Lakes Fishery Commission's Home Page, November 11, 1997.

3. Dr. Alan L. Sutton and Dr. Don D. Jones, Purdue University, *Comment on the EPA/USDA Unified National Strategy for Animal Feeding Operations*, Ames, Iowa, Council for Agricultural Science and Technology, January 1999

4. "Carolina Producers Appeal Hog Packing Plant Restrictions," *Feedstuffs,* November 30, 1998.

5. *Total Consumption of Fertilizers and Plant Nutrients in the U.S.*, 1960–97, The Fertilizer Institute.

6. *FAO Production Yearbooks.*

7. Conservation Technology Information Center, W. Lafayette, Indiana, 1999.

8. U.S. Department of Agriculture, *Agricultural Statistics*, Washington, D.C., 1989 and 1998 editions.

9. Greg Easterbrook, "Acid Rain," *A Moment on Earth: The Coming Age of Environmental Optimism*, Viking, 1995, pp. 161–180.

10. Frink, Waggoner. and Ausubel, "Nitrogen Fertilizer: Retrospect and Prospect," *Proc. Natl. Acad. Science USA*, vol. 96, pp. 1175–1180, February 1999.

11. Stensland, Whelpdale, and Oehlert, *Acid Deposition: Long-Term Trends*, National Academy Press, 1986.

16

More People *and* More Trees:
Saving Forests with Technology

MYTHMAKERS SAY:

"The projections indicate that by 2000, some 40 percent of the remaining forests in the [developing countries] will be gone."

"Major Findings and Conclusions," *Global 2000 Report*, Carter White House, 1980, p. 2

"Every day, some of these human beings move into places on the planet where only plants and animals used to live. Forests are cut down. Wetlands, oceans, ice caps and prairies are invaded."

Russell Train, World Wildlife Fund, quoted in *50 Simple Things You Can Do to Save the Earth*, Berkeley, California, Earthworks Press, 1989

"We must explore other building materials, such as mud bricks, that are made from soil adjacent to the building site and dried by solar power. They are widely used in Australia and New Mexican adobe houses, with beautiful results."

Helen Caldicott, *If You Love This Planet*, New York, W. W. Norton, 1992, p. 59

REALITY SAYS:

"The whole world's expected industrial wood needs for the year 2000 could be met sustainably from less than 200 million hectares of forest plantation—just 7 percent of the world's closed forest land area."

Roger Sedjo, Resources for the Future, *The Comparative Economics of Plantation Forest: GlobalAssessment.* Baltimore, Maryland John Hopkins University Press, 1983

"Annual growth in U.S. forests exceeds harvest by more than 55 percent."

U.S. Forest Service, *The Condition and Trends of U.S. Forests*, 1991

A New Era in Forestry

Most people have been taught to think that the world is being stripped of trees. They think more and more forests will inevitably be cut as the world's human population doubles again.

They are wrong.

We no longer need to sacrifice trees for people. Nor must we sacrifice the habitat of wildlife to get paper or timber. The doomsayers' link between the number of humans born and the number of trees lost is disappearing: If cities in 2050 occupy less than four percent of the earth's surface; If we use no more land for farming then than we do today; and, If the forest products for 8.5 billion people can be produced from a few acres of forest plantations with 15–20 times the yield of today's forests. Then, where is the danger to wildlife?

More than half of the world's forests are in the temperate zones. These First World forests are expanding in area. They are increasing dramatically in the number of trees per acre and speed of growth. They are no longer being cleared for crops, as they were 100 years ago, because farmers are boosting the crop yields on existing acres instead. Fewer temperate-zone trees are being cut for fuel. Far fewer trees are being lost to fire and pests. A rising percentage of the world's people are concentrating themselves in cities where their impact on trees is minimal.

In addition, First World residents now put a higher value on saving trees and especially on reforestation.

New technology and better forestry management is already achieving success in temperate zone forests. As a result of these innovations and others, U.S. forest acres logged today produce 30 percent more timber per acre than they did without management. That means more acres can be left as untouched wildlife habitat if we choose not to log them.

Tree plantations can produce far more forest products per acre than wild forests, using the best tree species for a given area. These

USDA

MANAGEMENT MAKES A DIFFERENCE—These two red pines were the same age at harvest. The bigger one was in a well-managed tree plantation.

tree monocultures are not the same as wild forest habitat. Even so, they are excellent wildlife habitat. (The lush forests shown in the frontier movie, *Last of the Mohicans*, were actually a 35-year-old tree plantation in the Carolinas.)

Tree breeding breakthroughs from biotechnology are giving us the first real high-yield forestry. Genetically enhanced trees are already yielding up to 15 times as much timber and pulpwood per acre as wild forests.[1]

New chemistry and engineering breakthroughs that produce more forest products from less wood are radically increasing the efficiency of forest production. That means more forest products from each tree, and fewer trees harvested to meet a given level of need.

Not surprisingly, the entire northern hemisphere offers dramatic examples of forestry success:

- The growing stock in Europe's forests rose 30 percent between 1971 and 1990. This information contradicts the commonly held view of European forest decline (and the myth that acid rain has devastated these forests).[2]

- Timber volumes in North America, Europe, and the former USSR together are increasing at about 700 million cubic meters per year—due to better management.[3]

REALITIES OF LOGGING:

"East Perry Lumber Company began its harvest of the Fairview timber sale in the Shawnee National Forest . . . four long years after being awarded the sale. Police arrested 17 people, four of them on Federal charges. . . . The arrests were more orderly than last year, when several protesters buried themselves in the ground and one chained himself to a logging skidder."

Greenspeak, Memphis, Tennessee, National Hardwood Lumber Association, September 1991, issue 24, p. 1

"If one were to project deforestation in the United States based on the logging rate of the late 1800s, the last tree would have fallen years ago. But the projections based on the logging and growth rates of the late 1900s would show forests covering every square inch of America in the next century. Obviously, 'if present trends continue' is misleading; the current trends never do continue."

Dixie Lee Ray, *Environmental Overkill*, Washington, D.C., Regnery Gateway, p. 111

"The fires . . . have come to northwestern Montana. Today, 150 fires are burning, from 100 acres to 5,000 each. . . . A senior fire-fighting 'hot shot' . . . says he's never seen fires behave like these conflagrations. He would have if he had been in Lincoln County in 1910, when 3 million acres went up in flames in a matter of days. It is not just coincidence that the life span of a Lodgepole Pine—80 years or more—is the amount of time that has passed since the devastating fires of 1910. For years now, the trees of the Kootenai National Forest have been ready for harvest—either by mankind, or by nature. Using appeals and litigation, environmentalists made sure that it wouldn't be mankind."

William Perry Pendley, *Summary Judgment*, Denver, Colorado, Mountain States Legal Foundation, September 9, 1994

Temperate forests *are* being logged, but logging seldom means losing the forest. Logging is the harvest of mature trees. Without logging, the trees will eventually fall down and rot—or burn. The wildlife in a logged forest may actually benefit from the logging if it is done properly. For example, warblers prefer recently logged areas above all other habitat. Virtually all of the forest's wildlife stays in the forest if the cutting is done soundly.

Mature trees have already done most of their carbon storing. Taking them away to make houses, furniture, paneling, etc. means that the forest-stored carbon will *stay* stored. If the trees are *not* harvested, they die, rot, and release their carbon quickly.

Forest regrowth can take decades, but these are the decades in which carbon storing is most rapid. That's a "global warming" advantage. The regrowth period is also when that acre of forest supports the most wildlife. Only a few temperate-zone woodland species prefer or "need" old growth forest.

LOCAL MYTHMAKER GETS LOCAL FAME:

"Some trees in the George Washington National Forest targeted for a timber harvest early next year are falling under the watchful eye of Steve Krichbaum. Krichbaum, a director for Preserve Appalachian Wilderness Network, filed an appeal protesting the 88-acre timber sale near the village of Headwaters in Deerfield Ranger District . . . because a number of the chestnut oaks in at least one stand are 183 years old. 'As far as (Virginia) goes, that's really old,' Krichbaum said."

"Preservationist Wants to Save Old Trees from Timber Sales," *Daily News-Leader*, Staunton, Virginia, October 14, 1993, p. B1

Reality Comment: The only reason Krichbaum offers for trying to prevent the logging is that the trees are "really old." How much longer are they expected to live?

Eighty-eight acres may not sound like much. But 221 million board feet of timber and 1.1 million cubic feet of pulpwood would be lost from these acres, turned back into CO_2. And we'd get nothing in return—no extra wildlife, because the trees are standing in the midst of a forest already full of wildlife; and no extra environmental quality, because the trees are in the midst of a heavily forested and sparsely populated area that already has an abundance of natural beauty.

Krichbaum got his name and picture in the local paper, and no doubt lots of "attaboys" from his colleagues in PAWN. But, we, the public, would get higher wood prices. And, the forest creatures would get a higher risk of a major fire when these aging trees die.

The trees would not be preserved. They are not antique rocking chairs to be admired in our living rooms. Nor are they old buildings that can be restored. They are living organisms with a basic life span, and they are near the end of it. The question is whether they will die uselessly or whether their timber (and stored carbon) will be harvested. We do not have the ability to maintain old trees as they are.

Why is this called "preservation"?

Postscript: Krichbaum later filed a lawsuit to block the harvest. Now we, the public, have to pay the salary of another government lawyer on top of losing the lumber, risking the fire, and gaining no wildlife. As a further irony, the Deerfield area suffered a 5,000 acre forest fire in the summer of 1998.

MYTHS OF THE NORTHWEST:

". . . the harvesting of edible mushrooms . . . could well become a major incentive for preserving rather than clearcutting mature forests in the Pacific Northwest."

Larry Evans, "Life in the Fungal Jungle," *Buzzworm, the Environmental Journal*, August 1993, p. 28

USDA

WILDLIFE HABITAT REBORN—This reforested slope in Oregon offers fine habitat to most of the forest wildlife species.

"One of North America's last great forests is being cut down and turned into telephone books. . . . Twenty years ago when directory assistance was provided free . . . phonebooks were optional. When phone companies began to charge a fee for directory assistance, however, phonebooks became a necessity."

from "Turning Forests into Phonebooks," *Sierra,* Sierra Club, January/February 1994, p. 28

Southern Softwood and Western Forests

Richard Haynes, program manager for social and economic research at the U.S. Forest Service, reports that the total timber volume from the U.S. Northwest dropped from 13 billion board feet in 1988 to 7.2 billion board feet in 1992. In the South, the total volume

American Forests Thriving

Americans seem to think that the country is losing its trees.

Weyerhauser's Charles Bingham says, "Images of disappearing rain forests are shaping the way Americans think about forestry. The public has concluded that what is happening in some Third World rain forests is what is happening here in America. For many, saving forests has become a religious experience."

The fact is, America is *not* losing its forests. The U.S. Forest Service itself says that after 500 years of harvesting, America's forest land base is still two-thirds as large as it was when European settlers arrived in the 1600s.

The big dip in North American and European forests occurred during the mid-19th century—the early days of the iron industry, when iron ore was crudely refined in small wood-fired smelters and wood was still a major fuel source.

Most of the forests on America's Eastern Seaboard were logged in the late 1800s, including the Adirondacks and the Alleghenies. After Pittsburgh began using big coal-fired Bessemer steel furnaces, the forests were largely allowed to regrow. Coal became the fuel of choice for our cities. Thus, forest area in the United States has been stable since 1920. Thanks to good forest management, we now have 30 percent more standing timber per acre in U.S. forests than we had in 1952. Moreover, timber production is already forbidden on 70 percent of national forest lands.[4]

grew from 12.7 bbf in 1988 to 14.4 bbf in 1992. Haynes says the trend is continuing.

Privately owned forests in the southeastern United States are profitably producing a rising share of U.S. forest products. Meanwhile, the huge National Forests of the western United States—which frequently produce at a net loss—are harvesting less timber.

The southern softwood expansion and new restrictions on cutting in national forests have radically shifted the timber and wood products market.

But What About the Spotted Owl?

Listen to Dr. Louis Oliver of the University of Washington. He says with just some modest changes in the way we manage our forests we can encourage more spotted owls and other old-growth species, more understory plant species and generally a higher level of biological diversity.

Oliver notes, for example, that the spotted owl likes to nest in old-growth, but likes to hunt in more open territory. (The owls eat mostly wood rats, which like more open woods better than old-growth.)

Oliver's forest management suggestions:

- Keep doing modest-sized clear-cuts, but leave a small island of old growth in the center of each section. Leave five or six trees per acre (some live, some dead) for habitat variety.

- Make small clear-cuts in dense forests to open up spaces for wildlife species that need them. Old-growth forests don't let much light into their understories. In fact most species *avoid* the dark old-growth forests.

- Thin out millions of acres of reforested trees, not only to get timber, pulp, and chips but to let the remaining trees get to their full size. Otherwise, the stands will need early harvesting to prevent most of the trees from dying young.

- Prune some of the older trees—or even girdle a few—to create dead snags for cavity-nesting birds.

- Set small, controlled fires to get rid of dead wood that might later cause a major wildfire.

Another forest economist, Dr. Peter Koch of the Wood Science Laboratory at Corvallis, Oregon, says the spotted owl reserves are resource-costly. It would take another 6 billion gallons of oil (and add another 62 million tons of CO_2 to the atmosphere) if we made up for the number of trees that would have been put off-limits by the first spotted owl forest reservation recommended by the U.S. government's interagency committee on the spotted owl, by using more steel, aluminum, brick, and concrete.[5]

Replacing timber products from the spotted owl preserves from *other* forests elsewhere in the world might have even bigger environmental costs. Bruce Lippke, Director of the University of Washington Center for International Trade in Forest Products, says that replacing the fast-growing trees on *each 100,000 acres of Pacific Northwest forest* by cutting Siberian forests would take *1.5 million acres* of slower-growing Siberian trees. (Nor are there any laws mandating reforestation in the former USSR.)

If the "spotted owl" timber products came from wild tropical forests instead, how many species would be endangered there?

MYTHMAKER REDEFINES CAPITAL:

"Old-growth forests are a form of natural capital that has taken centuries or millennia to accumulate; 'sustainable logging' of these nonrenewable resources is, strictly speaking, an oxymoron."

S. Postel and J C. Ryan, "Reforming Forestry," *State of the World 1991*

Reality Comment: What kind of "capital" do old-growth forests represent if they harbor little biodiversity—and then burn up or rot? The key to preservation is good management.

Technology and Forestry

A huge reengineering in timber products is radically increasing the efficiency of forest product conversion. We're getting more forest products from fewer trees—more product from each log.

In 1948, one acre of U.S. Douglas fir logs produced 17,900 cubic feet of timber. See Figure 16.1 for that timber's typical yield. See Figure 16.2 for what the same quantity of raw materials yielded in 1973.

Figure 16.1

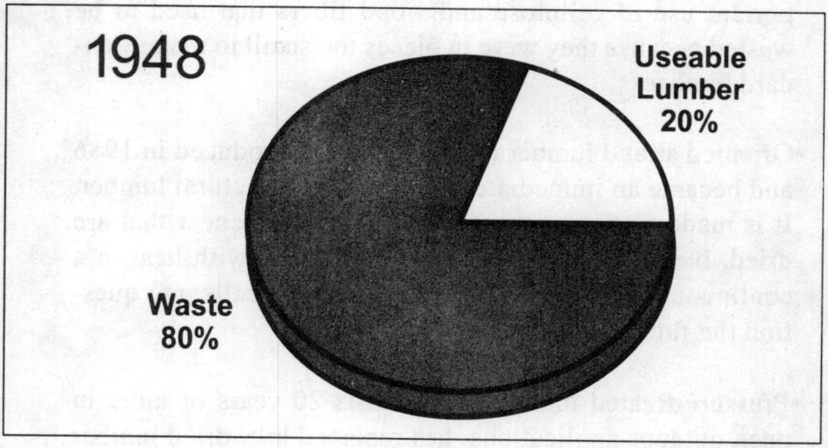

Figure 16.2

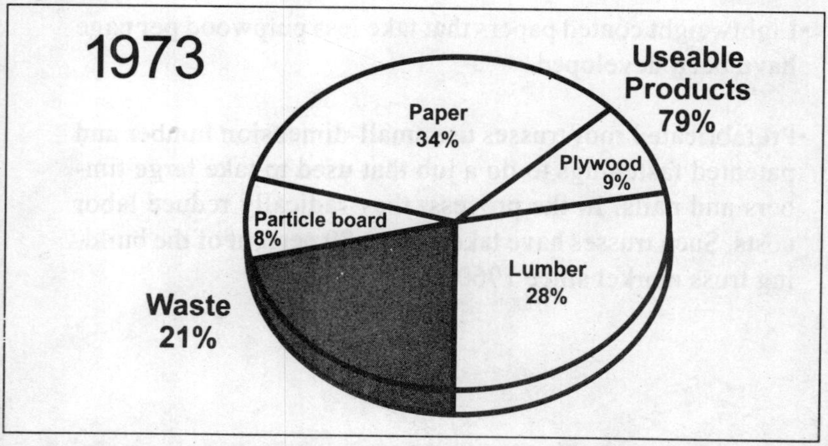

Numerous new forest-product technologies have made this higher yield possible:

- The retractable chuck lathe, in the 1960s, made it possible to economically peel small logs for veneer. This gave birth to the southern pine plywood industry.

- The chipper-canter, introduced in the 1960s, raised the efficiency of turning smaller logs into lumber and computerized log positioning, introduced in 1971, sharply improved lumber yields per log.

- Particle board, waferboard, and hardboard now make important use of cellulose and wood fibers that used to be wasted because they were in pieces too small to make standard lumber.

- Oriented strand lumber (parallam) was introduced in 1986 and became an immediate competitor in structural lumber. It is made from narrow strands of clipped veneer that are dried, blended with resin, and compressed with heat in a continuous press into solid wood billets. It calls into question the future need for large sawlogs.

- Pressure-treated lumber, which lasts 20 years or more in open outdoor applications, has replaced kiln-dried lumber which lasted only half as long.

- Lightweight coated papers that take less pulpwood per page have been developed.

- Prefabricated roof trusses use small-dimension lumber and patented fastenings to do a job that used to take large timbers and nails. In the process, they radically reduce labor costs. Such trusses have taken about 80 percent of the building truss market since 1960.

•Laminated veneer lumber is an alternative to higher-priced and less ecologically sound steel production. The major use of LVL is in flanges for wooden I-beams.

•Plywood I-beams and stronger modern adhesives are a far more efficient way to produce load-bearing joists and other structural pieces than waiting for a tree to grow big enough to produce 2 by 10 inch planks. The I-beams also produce a higher-quality floor system.

•Louisiana Pacific Corporation is producing weatherproof chipboard from fast-growing aspen trees. The products use a new adhesive system with a stronger, weather-tight bond. The aspen trees regrow rapidly from their cut stumps, taking advantage of their existing root systems.

•Plywood manufacturers in Taiwan, using more expensive equipment, are producing veneer sliced to 1/80th of an inch, compared with the longtime U.S. veneer standard of 1/26th of an inch. That produces three times as much valuable veneer from each hardwood log.[6]

Dr. Brian Greber, of Oregon State University,[7] estimates that the average yield of lumber from 1000 board feet of logs increased 14 percent just in the 1970s. Utilization rates continue to rise rapidly.

Tree Plantations to Protect Forests and Wildlife

Farmers everywhere are producing more crops for more people on fewer acres. Foresters are starting to do the same thing with trees. That leaves more land for wilderness and recreation.

Forest plantations let us produce lots of timber products from a few acres. For example, less than one percent of Latin America's forests are tree plantations—but those plantations already produce more than one-third of the region's industrial wood output. By the year 2000, the plantations are expected to produce *half* of Latin America's expanding wood requirements from less than 3 percent of its forest area![8]

Major forest plantations have been started in Brazil, Chile, Venezuela, South Africa, India, Indonesia, the Philippines, Australia, New Zealand, and a host of other tropical and southern hemisphere countries.

In the mid-1970s, developing countries had 6.7 million acres of plantation forest. Five years later, they had 9.1 million plantation acres. The projection for the year 2000 is *21 million* acres.[9]

Other major gains have come from putting existing tree species in new settings where they maximize yields.

DOESN'T THIS SEEM LIKE A FOREST?

- Eucalyptus is thriving in semiarid areas all over the world, and producing more wood per tree than native species.

- A brushy leguminous tree called Leucaena leucocephala (from Central America) is helping to fertilize crops in the new alley-cropping system of West Africa; provide dry season cattle forage in India and Ethiopia; and helping stabilize rice paddy dikes in Southeast Asia—even as it provides more firewood for the farm families in all these countries.

- A U.S. company is investing $15 million to replant Costa Rican forests cleared by homesteaders 90 years ago. It is planting 27 million gmelina arborea trees, native to India, which will soar to 90 feet in five years. (The project has already created 1,300 jobs in the small country—even before the first harvest has begun.)[10]

MYTHMAKER CAN'T SEE THE FOREST FOR THE TREES:

"By 'permanent forest,' I assume Mr. Phillips means the 30,000 acres of gmelina, a nonnative monocrop. If this is a forest, then so is an Iowa cornfield."

Andre Carruthers, "Costa Rican Rights and Wrongs," *E, The Environmental Magazine,* February 1994

Reality Comment: Apparently, if it can't be the way it was 20,000 years ago the eco-activists believe it's wrong. The gmelina trees will not only provide good wildlife habitat, but also protect the other forests in Central America from being logged or cleared. Is this not worthwhile?

MORE MYTHOLOGY:

"When diverse populations of trees are replaced with genetically uniform stands, future timber harvests are put at risk."

S. Postel and J. C. Ryan, "Reforming Forestry," *State of the World 1991*, p. 82

Reality Comment: Tree plantations *ensure* future timber harvests without risking either wild forests or wildlife. If a disease hits a forest, the trees are harvested as they stand and replanted with another variety, just as we do in other crops. The only difference is that the pace is slower.

The World's First Truly High-Yield Forestry

Now biotechnology takes a hand, producing improved tree varieties that will yield even more wood products per acre of plantation.

Cloning and tissue culture have radically shortened the tree breeding cycle. Tree breeders used to think in decades. Now, they think in months. China says it can produce a new rubber tree variety—complete with thousands of new seedlings—in four years!

An explosion in tree productivity has been touched off. The USDA's agricultural officers from Indonesia, Malaysia, Thailand, and Kenya have reported two to sixfold increases in the yields of tree crops like teak, pine, cocoa, and palm oil due to genetic engineering in recent years.

In Brazil, a genetically enhanced plantation of Georgia yellow pine (an exotic species there) is producing 50 cubic meters of pulpwood per hectare per year, compared with 15 cubic meters for the same trees in a Georgia plantation and 3 cubic meters of pulpwood for a natural forest in Sweden. Each acre of such plantations can protect as many as 14 acres of forest from being logged.[11]

Of course, this leaves each nation with the question of how best to manage its forests. In the United States, if we don't harvest the wood from a forest, we'll need to invest management dollars in burning it up—under controlled conditions so we don't get the big, too-hot forest fires from which the forests take too long to recover.

My wife remembers as a little girl driving through Oregon's huge, infamous Tillamook Burn—millions of acres of devastated woodlands still barren and untenable for wildlife more than 20 years later. (It was finally reforested.)

THE REALITY ABOUT "SUSTAINABLE" FOREST MANAGEMENT: NO PROFIT = NO JOBS:

"The Collins Pine Company has been trying to manage its 92,000 acres of California forest in the best traditions of the Sierra Club and Earthwatch since the 1870s. But the company has run into trouble.

"Leaving the healthiest and most vigorous trees and cutting only those that were diseased or fully mature turns out to be expensive. It takes a troop of foresters who analyze which trees to cut, harvesting many different size trees slows down the sawmill, and fewer trees per acre are cut. (Collins harvests less than 2 percent of its standing timber each year, against an industry average of 2.5 to 3 percent.) Collins' trees grow more slowly as well—partly because they are older.

"The result: In 1993, Collins made no profit on sales of $100 million worth of wood molding, windows, doors, and furniture. Now the company has had to fire its sawyers and lumberjacks and farm that work out to contractors."

Bill Wagner, "Not Business as Usual," *Business Ethics*, September/October 1993, p.14

Should We Use More Wood, Not Less?

"In the emerging global economy, nations should be *increasing*, not *decreasing*, their dependency on wood fiber because wood is renewable, recyclable, biodegradable, and far more energy-efficient in its manufacture and use than are products made from steel, aluminum, plastic or concrete."

So says Dr. Robert Bowyer, who is president of the Forest Products Society and a professor at the University of Minnesota. "Furthermore," he adds, "trees and the wood-turned-into-lumber they provide are both capable of storing large amounts of carbon dioxide that would otherwise escape into the atmosphere, adding to the potential for global warming."

REALITY:

"Wood, an energy form that is relatively abundant and renewable, is where we should be looking for a continued energy supply."

Mollie Beattie, later President Clinton's director of the U.S. Fish and Wildlife Service, before a Vermont natural resources conference in 1990[12]

Reality Comment: Other experts at the same conference as Beattie said even *wood-fired electrical generation* could actually help diminish the greenhouse effect and global warming trends—without forest loss. Norman Hudson of the Vermont Department of Forests, Parks and Recreation, said that the two-thirds of Vermont in forest is "vastly underused." He suggested that simple—and healthful—thinning of tree stands could produce 50 tons of wood chips per acre. The CO_2 produced by burning the wood would be offset by the trees' absorption of gases and release of oxygen during their growing years—and less fossil fuel would be released into the atmosphere.

CHAINSAWS CONFUSED WITH ASSAULT RIFLES?

". . . [A]t one of the conferences, I suggested that chainsaws be banned . . . and governments pay a bounty of $5,000 for each such machine turned in. . . . What will nations and the world do about the factories where these machines are made and about the people who work there? We cannot allow those employed in such places to remain hostage to destructive and poisonous inappropriate investment and production."

Richard Grossman, author of *Fear at Work*, in a letter to the editor of *Earth Island Journal*, Winter 1993–94

REALITY AT WORK:

"I'm CEO of a company that owns and develops timberlands and am a board member of the Alaska Forest Association. . . . You offer the opinion of some (in your January 17 article) that there has been overcutting in the Tongass (National Forest). . . . The facts are that, of the 17 million acres in the Tongass, only one-tenth of 1 percent will be harvested each year. Over the 100-year rotation cycle, 10 percent will be harvested. Forty percent of the Tongass is in land-use designations that are either wilderness or limited access. Twenty-three percent of the Tongass is designated for intensive uses (which include logging) but less than half of that area will be used for logging.

"Those of us who live and work here . . . care more than any political advocacy group that our harvest practices leave the land fertile and able to support a new crop. We obey the law, monitor the health of fish streams, build roads to specifications that control runoff, leave buffers around water bodies, avoid steep slopes, elevate logs over stream beds and generally take good care of the land. The difference between those in the timber industry and those referred to as 'environmentalists' is a matter of politics and values, not nusbandry.

"This is good timber-growing country. Rain and moderate temperature offer the opportunity. . . . The salmon streams are productive. The areas cut are in sustainable proportion to the forecast demand. The other species here have viable habitat. The world demand for fiber will be supplied from some source. . . ."

Letter to the editor of the *Wall Street Journal* from Ernesta Ballard, Ketchikan, Alaska

Notes

1. Roger Sedjo, Resources for the Future, personal interview, 1992.
2. Kaupi, Mielikanen, and Kuusela, *Science*, vol. 256, pp. 70–74.
3. UN Food and Agriculture Organization (FAO)/Economic Commission for Europe (ECE), *Forest Resources 1980*, Rome, 1985.
4. U.S. Forest Service, *1992 RPA Assessment* and *The Condition and Trends of U.S. Forest, 1991*.
5. Dr. Peter Koch, before the "Wood Product Demand and the Environment" conference, Vancouver, British Columbia, November 1991.
6. Dr. J. L. Bowyer, "Successes and Failures in Process and Product Technology," University of Minnesota Department of Forest Products; see also Bowyer, "Analysis of Growth of Competing Materials," op. cit.
7. Mark McQueen, "Pondering the Environmental Advantages of Wood," *Evergreen,* Summer 1993, p. 18.
8. InterAmerican Development Bank, *Forest Industries Development Strategy and Investment Requirements in Latin America*, Technical Report no. 1, prepared for IDB conference on Financing Forest-Based Development in Latin America, June 22–25, 1982, Washington, D.C., p. 17.
9. Lanly and Clement, "Present and Future Natural Forest and Plantation Areas in the Tropics,"*Unsylva*, 1979, vol. 31, no.123 pp.12–20
10. "Global Consequences of U.S. Environmental Policies," *Evergreen*, Summer 1993, p. 15.
11. Sedjo, personal interview, op. cit.
12. Sylvia Dodge, "Chipping Away for Alternative Energy," *Vermont Business Magazine*, 1990, vol. 18, no. 6, p. 46.

17

Can We Rescue the Rain Forests?

MYTHMAKERS SAY:

". . . [B]illowing clouds of smoke regularly blacken the sky above the immense but now threatened Amazon rain forest. Acre by acre, the rain forest is being burned to create fast pasture for fast-food beef."

Vice President Al Gore, *Earth in the Balance*, p. 23

"The world's forests declined . . . between 1980 and 1990 . . . by an area larger than Peru. Tragically, much tropical forest is being cleared in order to cultivate soils that cannot sustain crop production for more than a few years. Yet the species extinguished in the process are gone forever."

Sandra Postel, "Carrying Capacity: Earth's Bottom Line," *State of the World 1994*, p. 12

"At the present, frantic rate of deforestation, all the world's tropical forests will within twenty-five to fifty years be destroyed, along with 15 to 24 million species, and the land will be desert. . . . The Club of Earth maintains that species extinction is 'a threat to civilization second only to the threat of nuclear war.'"

Helen Caldicott, *If You Love This Planet*, p. 51

REALITY SAYS:

"Today, we know that the 1984 . . . deforestation estimate was too high. David Skole, an ecologist at the University of New Hampshire, and Compton Tucker of NASA examined in 1993 satellite photos of the Amazon rainforest to determine how much deforestation occurred between 1978 and 1988. After painstakingly examining 210 such photos, they concluded that the average rate of rainforest loss was [just 1.5 million hectares] per year. If deforestation in Brazil accounted for half of all rain forest deforestation in the world, as is generally assumed, then the new estimate means the global rate of rainforest deforestation was . . . less than a tenth of 1 percent."

Bast, Hill, and Rue, *Eco-Sanity*, Lanham, Maryland, Madison Books for the Heartland Institute, 1994, p. 84

"Emphasize this point at the beginning and throughout your chapter. Deforestation is not the same as timber extraction. . . ."

Douglas Southgate, author of *Economic Progress and the Environment: One Developing Country's Policy Crisis*, commenting on a draft of this book, November 17, 1994

"The Amazon does not serve as the lungs of the planet; the forest consumes as much oxygen as it produces. It is not being cleared to produce fast-food hamburgers; the region is a net importer of beef. Experts disagree about the extent of deforestation (reasonable estimates vary between 8 percent and 12 percent, mostly along the edges of the jungle) but it is clear that the rain forest will not 'disappear' any time soon."

Jon Christensen, a Pacific News Service correspondent, then recently returned from a year on assignment in Brazil, writing in the *Baltimore Sun*, December 10, 1989, p. 13–14

"According to satellite imagery, 532,086 sq. km of forest has been lost since 1972, around 13 percent of the entire Amazon region."

"Amazon Destruction," *Financial Times*, February 12, 1999

Reality Comment: Most of the lost forest was destroyed by low-yield slash-and-burn farmers.

Let's be grateful to the naturalists for warning about the need to save the rain forests. These forests contain a huge proportion of the world's wild species. Let's commit ourselves to saving them.

But let's *not* take the naturalists' advice on how to do it. Their advice is too often unworkable and sometimes needlessly inhumane.

The Bad News on Tropical Forests

Trees *are* being cut in the Third World. Although relatively little of the tropical wet forest has been cleared and converted to other uses, substantial parts of it are being logged. Hardly anyone lives in the wet rain forests, and there is little point in clearing them. Most of the logged forests will simply be allowed to regrow, their species largely still in place.

Some of the logged rain forests are actually cleared, mostly for slash-and-burn agriculture. This low-output farming provides only a bare subsistence living, but the people clearing the land lack better economic alternatives.

However, the world's most serious tree losses have so far been in the dry tropics. There, the trees are being cut mainly for firewood and charcoal for cooking and heating.[1] Firewood scarcity is an important cause of woodland loss and degradation in sub-Saharan Africa and India.[2]

Places like Kenya and the Sahel are facing wood crises. That is a tragedy on two counts; not only is the tree-cutting unsustainable, but the carbon is being inefficiently turned into CO_2 and released into the atmosphere. More efficient cookstoves can double firewood efficiency. But the real problem is not that trees are being cut. It is that no one is replanting any new ones.

United Nations agencies and the World Bank say tropical forests are currently being "deforested" at the rate of about 0.9 percent of total tropical area each year. This is substantially higher than the 0.6 percent estimated a decade earlier. Asia is thought to have the highest rate of cutting (1.2 percent), with Latin American cutting at 0.9 percent and Africa at 0.8 percent.[3]

Naturalists are afraid that the losses are so high and will mount so rapidly—with population and economic growth—that they will quickly doom thousands of tropic wildlife species. The picture is certainly better than that, though concern is warranted. The fact is that our data on forest losses have been poor. It's hard enough to get data on *logging* in a remote wild forest. It is harder still to keep track of land converted from forest to other uses. Even satellites have trouble distinguishing between a cut-over forest and an interplanted tract of crops and oil palms. And how do you assess the smoke from the slash-and-burn farmers, who clear another field every two or three years and let the old one regrow?

It is a vacuum that scary forecasts have come to fill. Yet the history of such unsupported forecasts should make us wary. The latest satellite data, for example, shows that the Amazon has lost about 13 percent of its original rain forest area. The rate of loss in the Amazon has now been reduced to 0.3 percent per year, with even lower rates of loss probable as Brazilians become more affluent and environmentally sensitive.[4]

And Some Good News

Dr. Ariel Lugo, one of the top U.S. experts in tropical forestry, says that only about half of the virgin tropical forests being cut annually are actually lost; the rest simply become secondary forests. Such secondary forests are still good wildlife habitat. In addition, says Dr. Lugo, replanting and natural regeneration have been producing additional secondary forest equal to another ten percent of the forest cut.[5]

Forest *loss*, then, is almost certainly lower than the environmental activists claim. More important, the current rate of rain forest loss is unlikely to be the long-term trend—any more than America's surge of cutting eastern forests to fuel little iron smelters in the 1850s was a long-term trend. We now know that there's no need to lose much more tropical forest.

- There is no global shortage of food or additional farm production potential.

- We have more cost-effective ways to increase our forest products supplies than rainforest logging.

The rain forests being lost are largely victims of lagging economic growth and institutional arrangements in the Third World. The key problem is neither "greed" nor population growth. The reality is that Third World residents are cutting trees "nobody owns" because they have no jobs and need free fuel and free cropland. Deforestation losses can be radically reduced by better institutional arrangements—along with rapid economic growth and job creation already occurring in much of the Third World thanks to liberalized nonfarm trade.

REALITY:

"Deforestation is really land use change: conversion to cropland and pasture. People do this for the reasons you identify in the chapter—weak property rights, perverse policy incentives, and (above all else) the low earning opportunities of unskilled rural labor. Clearing land for farming is basically a very unrewarding activity. Almost by definition, that means it is done by people with limited job prospects."

Douglas Southgate, author of *Economic Progress and the Environment: One Developing Country's Policy Crisis*, commenting on a draft of this book, November 17, 1994

"Investment in fuelwood production and tree farming, on a large scale, by farmers and by community groups and private enterprises will not occur unless it is profitable. The incentives are gradually emerging. . . . But fuelwood markets are developing too slowly. . . . The pace of market development will accelerate if open-access sources of fuelwood are eliminated, cutting in protected areas is restricted, farmers are not restricted in marketing wood from their own land . . . and farmers have uncontested ownership of local forests and woodlands."

Kevin Cleaver and Gotz Schreiber, World Bank staff, "The Population, Environmental and Agriculture Nexus in Sub-Saharan Africa," 1993[6]

Stealing What No One Owns

So, if population growth isn't stealing the forests, what is?

Let's look at a resource named the elephant. Africa is losing half of its elephants every eight years. Naturalists (naturally) predict the extinction of the elephant early in the next century.

Once again, the predictions will be wrong.

The reality is that elephants in some countries are doing just fine. Their offspring will undoubtedly be available to repopulate the game parks of those countries that are currently managing their game resources badly.

In most of eastern and Central Africa, governments have owned the elephants. The animals are "protected" by game wardens who are too few in number, poorly equipped, and poorly paid (so they are susceptible to bribes). In these countries, many farmers see the elephants as competitors who tear up their crops, destroy their trees, drink their water, and even trample houses and citizens.

The only way the farmers can profit from the elephant's existence is by poaching them for their ivory, hides, and meat. Or they see no reason to risk interfering with those who *do* poach. The elephants in these countries are declining rapidly. Just the opposite is happening in southern Africa, which already has 20 percent of Africa's elephants—and the elephant count is rising.

Zimbabwe is one country that is creating incentives for local people to protect the elephants:

- Hunters are charged $25,000 for shooting a trophy animal, and the money is used to pay elephant costs—including damage to local farms and homes.

- The meat is given to tribes living in the game park areas.

- Rogue elephants outside the parks are shot by the Park personnel to minimize damage, and/or villages can sell hunting permits themselves.

Zimbabwe would also like to sell its legal ivory and elephant hides to help finance elephant management. Once the elephants become an economic plus, *poachers* are shot rather than elephants. Enough people care so that the laws are fiercely enforced.[7]

The same questions of ownership, incentives, and alternatives that plague the elephant apply to the forests. Here are two good rules of thumb for evaluating "conservation" proposals anywhere in the world:

- If the proposed "solutions" depend on people acting selflessly, reject them. People always follow what they perceive to be their best interests. Get the incentives right.

- If the "solutions" involve government "saving" resources from greedy private citizens, reject them. The Third World

> cannot hire enough game wardens, or pay them enough, to defend environmental resources from the public if the incentives are wrong.

The key forest problem is the common ownership of resources. As environmental leader Garrett Hardin has noted in his famous essay on "lifeboat ethics," common ownership, almost by definition, leads to the rape of the resource.

The Third World is losing huge tracts of tropical forest, both wet and dry, because the land and the trees don't belong to anybody. The incentives lead the local people to exploit that local resource rather than sustain it.

Kenya is losing too many of its trees to charcoal.[8] Madagascar burned forest (habitat for some of the world's most unique and threatened wildlife species) because its socialist farming policies were failing. India's governmental forest authorities effectively took title to huge tracts of forest that it could not defend from the goats and cattle of the local residents. The goats and cattle prevented the growth of young trees to replace the ones poached for firewood.

We can put more and more of the Third World's natural resources in "preserves" and suffer these resource losses while we wait for the currently weak governments to become strong. Or we can change the incentives for behavior by local people.

Changing the incentives can be as simple as it has been in Zimbabwe (and perhaps will be in Kenya) with the elephants: letting the local people help manage the resource in return for sharing in the profits.

MYTHMAKERS IN HOLLYWOOD:

"Displaying a 50 by 25 foot inflatable chainsaw bearing the message: 'Hollywood: Stop the Chainsaw Massacre,' Rainforest Action Network, Greenpeace and Earth First staged a mediagenic protest at the entrance to Hollywood's Paramount Studios on September 15. Eight demonstrators, handcuffed to cement-filled barrels painted to look like tree stumps, blockaded the studio entrance as activists negotiated with Paramount over the studio's use of plywood made from lauan, a wood harvested from Southeast Asian tropical forests."

Earth Island Journal, Winter 1993–94, vol. 9, no. 1, San Francisco, Earth Island Institute, p. 19

MYTHMAKERS' JOURNEY TO HARSH REALITY:

"In November 1989, I went to . . . Venezuela, hired a dugout canoe, and set off down the Orinoco River. We stopped at one of the (Indian villages). . . . The more I saw of these wonderful indigenous people, the less I thought our culture was at all civilized. They live in harmony and peace with the forest, protecting and respecting it, while we rape and destroy it for 'economic' reasons."

Helen Caldicott, *If You Love This Planet: A Plan to Heal the Earth*, New York, W. W. Norton, 1992, p. 46–47

". . . only the left eye (of the river dolphin) is used in magic rituals and as a charm. . . . The Brazilian government has outlawed this trade, but it continues illegally."

"Rainforest Dolphin Trade," *Buzzworm's Earth Journal*, January/February 1994, p. 19

"Fund-raising concerts by Sting for his Rainforest Foundation have been successful in raising money to protect the forest home of the Kayapo people of Brazil's Amazon. But a problem has arisen. . . . Now it seems, the Kayapo have been prevented from selling off their mahogany trees . . . as a result of Sting and the others putting pressure on Brazil to stop the logging of mahogany. In case you're wondering, the Kayapo are suing for loss of income."

"The Eco-Mole," *Earth Island Journal*, Winter 1993/94, vol. 9, no. 1, Earth Island Institute, San Francisco, p. 4

The Commons Problem

Many of the world's natural resources are held "in common." That is, they have no direct owner. Air, flowing water, wild genes, and wild animals all fall into this category. Forests, fields, and pastures are often owned in common by communities or tribes. Natural resources "owned" by governments are held in common.

The difficulty is that looking after a commonly owned resource is tough and unrewarding. Someone has to cover costs for everybody else. There are too many free riders. Too often, the common resource doesn't get saved.

The grasslands of the Sahel, the rain forests of the Amazon, the air over Los Angeles and the waters of the Chesapeake Bay are all "owned" by so many people that they have been difficult to protect. Many trees in the Third World lack owners, and so risk being cut by any passerby.

Natural resources owned by weak governments are often at risk, even when they are designated as national preserves, because the governments are unable to maintain effective protection.

Garrett Hardin made a dramatic impact on America's collective mind when he published his "lifeboat ethic" in 1974. It was based on an extremely realistic (and thus dismal) examination of commons ownership and what Hardin then concluded would mean vast starvation and wildlife loss. As one commentator put it:

> Garrett Hardin's article in *Psychology Today* on 'lifeboat' ethics electrified the general public. . . . Hardin had been the first to state clearly and convincingly 'the tragedy of the commons.' His study showed how rational human self-interest within a system of common ownership or usage results, ironically but foreseeably, in a loss to everybody within it.

Unfortunately, Hardin is so committed to the myth of scarcity that he argued for limiting food aid (induced starvation) instead of changing the resource ownership structures!

Why Property Rights Were Invented

Property rights have been around so long and become so ingrained in Western society that we have evidently forgotten why they were invented.

Iceland in the 12th century was losing its eider ducks and other seabirds. Too many people were preying on the eggs. Not enough young birds were hatching to maintain the colonies.

Iceland solved the problem by declaring that the seabird eggs were henceforth the property of the landowner on whose land they were laid. The seabird colonies promptly recovered. Every landowner wanted big colonies of seabirds to lay lots of eggs on his land. So he made sure enough nests were undisturbed.

Despite the long record of resource conservation success established by ownership incentives, the tragedy of the commons is still far too typical of the Third World's forests. To date, Third World societies have lacked the institutional structures to protect their forests and wildlife. However, the answer does not lie in returning to the old tribal customs.

CONSERVATIONISTS' REALITY:

"The old paradigm of conservation held that the best way to protect biological diversity was to mark off the territory, build a wall around it, and patrol it with a machete or machine gun. . . . In the old conservation paradigm, people were frequently seen as the enemy. There were two major problems. . . . First, the preservation model was increasingly seen as unethical. Conservation workers resisted the idea of pushing people off lands they may have occupied for centuries. And they began to oppose the vision of declaring national parks while rural families suffered malnutrition around its borders. The second problem was that it didn't work. In areas of the globe with the most biological diversity—the humid tropics—rural families ignored the signs around national parks, tore down the fences, and invaded the reserves."

James Nations, vice president for Latin American Programs,
Conservation International, 1993[9]

"It can hardly be a coincidence that virtually every serious environmental problem, historically and today, occurred or occurs in those areas where well-defined systems of property rights are lacking. Indeed, upon further reflection it becomes clear that this is the essence of pollution and wildlife problems: No one owns the resources involved, and consequently no one protects them."

Bast, Hill, and Rue, *Eco-Sanity*,
Lanham, Maryland, Madison Books, 1994, p. 216

Tribalism Is Not the Answer

The American Indians are now widely praised in some circles because they did not "own" land. This is presumed to reflect their lack of greed and their respect for nature. That gentle but misguided thought betrays a misunderstanding of the American Indian culture and economy.

It was almost impossible for the Indians to assign land titles for game animals and for their typical shifting cultivation of crops. So they owned their hunting grounds as a tribe rather than as individuals. At the same time, Indian tribes reacted *violently* to any intrusion on their tribal lands. A quick death was about the best that an intruder could expect; torture was more common.[10]

When the populations of the American Indians became dense, they depopulated the forests of game, cleared more land, and put more pressure on the resources. Some of the Eastern American Indians sometimes hunted with fire, using flames to drive game to the hunters. Some early Plains Indians would drive a whole herd of bison over a cliff to get the meat they needed from a few animals. Famine is what limited the Indians' impact on nature and made the Indian culture look friendly to the environment—in distant retrospect.

Tribal and village landholding has suited the needs of indigenous peoples in Africa and Latin America in the past because they practiced the lowest-cost form of agriculture—shifting cultivation. (It is also called slash-and-burn or bush-fallow.)

Now, with rising populations and incomes, the shifting cultivation needs to give way to more intensive farming systems, such as tied ridges and alley cropping. The more intensive systems need fixed locations, and the farmers need clear title to a fixed piece of land so they can make the investments needed for the new systems. Otherwise, both hunger and severe soil erosion will keep driving the expansion of low-yield farming into more critical wildlife habitat.

Tribal landholding and/or other forms of communal ownership have often failed to defend the ecology. They are failing on a colossal scale throughout the Third World today. We probably have to get rid of either the people or the commons ownership. I vote to get rid of the commons.

MYTHBREAKERS:

". . . [T]he Kayapo, handsome warriors of the [Amazon's] Xingo basin . . . [now] have a proper reservation, and contracts for their wild oils and essences with environmentally friendly companies such as The Body Shop, a cosmetics chain. But the Kayapo's land is also rich in gold and mahogany . . . about $33 million in mahogany alone in 1988. . . . Such money as they get is often pocketed by chiefs who spend it on ranches, cars, and aeroplanes. . . . In 1988, [one] village . . . collected $1 million in timber and mineral bounty, but still had a quarter of its children die in infancy. . . .

"Under pressure to give up logging, 88 Kayapo stormed into Brasilia with an ultimatum . . . either allow the tribe to go on cutting down its forests, or pay $50,000 per village per month for timber sales forgone. The government, which is broke, would not pay; the Kayapos' trees continue to fall.

". . . The Guajajare of Northeastern Brazil seized some government agents in a bid to make the government grant them logging permits. A group of Nambikwara, from Mato Grosso, having razed their own hardwood, began to poach on their neighbors'.

"It is an awkward time for the green movement. 'We are going through a phase of disenchantment,' says one supporter of Indian rights."

"The Savage Can Also Be Ignoble," *The Economist*, June 12, 1993, p. 54

Stealing Biodiversity with Economic Stagnation

Bad economic policies in Third World countries have been another major factor in forest loss.

The environmental movements have freely criticized the governments of Third World countries—but often for the wrong reasons. Virtually never, for example, have the environmental critics complained that *most Third World governments have created too few off-farm jobs*. Yet the failure to create good urban jobs stands at the heart of key tropical resource losses in the world today.

In fact, Dr. Mostafa K. Tolba of the UN Environmental Program recently warned an international conference that there have been "loud complaints from a number of developing countries that the rich are more interested in making the Third World into a natural history museum than they are in filling the bellies of its people."[11]

A logging road doesn't bring settlers to the Amazon if they have almost any other choice. The Amazon is a steaming jungle with enough dangerous snakes, biting insects, and pervasive diseases to frighten off any rational human being who wasn't born there. Even the poorest Brazilians have mostly voted for a tin shack in a Rio slum, where the sewage runs down the unpaved streets, rather than move into the jungle.

But Brazil (and most of Latin America for the last 400 years) has been caught in an economic time warp. The Spanish and Portuguese captains who conquered the Amerindians left a tradition of repressive and pervasive government control.

Since the beginning of the 19th century, Latin governments have veered back and forth from repressive right to populist left. Both sides, however, left in place the real economic culprits—a set of indomitable government officials who tied everything up in a web of red tape and graft. Anything not specifically titled to an individual person belonged automatically to the government. Land, forests, the right to sell fruit on the street, were all part of the bureaucratic empire. For example, if you didn't pay off the officials, you couldn't run a bus line. Even if you did pay off the officials, they wouldn't let you raise bus fares to match the inflation the government created. The buses lost money. So there are few bus lines. This makes it difficult for job seekers to get to work and for customers to get to shops. And the strangling of economic opportunity continues.

Peruvian Hernando De Soto reveals in his important book, *The Other Path*, that he went through the process of getting the permits for a small clothing factory that would employ a dozen women and two sewing machines. It took 289 man-days, 19 bribe solicitations, and four bribes to get the government permissions to start the business! The same process in Miami, Florida, took De Soto four hours and the standard $25 fee for the permit.[12]

Population increase and economic growth *should* pose little danger to the rain forests. Few people want to live by subsistence agriculture in the jungle. If there are good jobs, schools, and amenities in the cities, the jungle will be left to its primitive inhabitants. (If, indeed, even they choose to stay there.)

The real reasons for tropical rain forest loss include:

- Constraints on trade in farm products due to farm subsidies in the rich countries and local farm politics in all countries.

- Ill-conceived government subsidies for forest clearing.

•Government policies that suppress economic growth, driving people to accept a short and desperate life as subsistence farmers in the rain forest; and

•Failure to invest in high-yield farming research.[13]

CASE STUDY IN THE PHILIPPINES:

"Economic stagnation . . . (in the 1980s) created such massive unemployment and poverty that internal migration patterns drastically changed. Frontier migration, already present in both countries since the 1950s, accelerated while urbanward movements, which were dominant in the 1970s, declined. Most migrants could only turn to open access forest lands. . . . As much as 25 percent of total arable lowlands in the Philippines remain underutilized. Meanwhile, cultivated area in forest has expanded from 23 to 31 percent of total cropped area between 1980 and 1987."

Maria Concepcion Cruz, Environmental Department, World Bank, "Economic Stagnation and Deforestation in Costa Rica and the Philippines,"*Proceeding of the 13th World Bank Agriculture Sector Symposium*, Washington, D.C., 1993

The most famous (or infamous) government policy mistake on tropical forests was made in the 1970s by the military junta then running Brazil. The generals decided that if they didn't populate the Amazon, "invaders" might take it away. So they built roads, and offered a government subsidy for clearing cattle pasture in the rain forest. Fortunately, the rain forest made such lousy cattle pasture that the subsidy was paid on only 1.5 percent of the Amazon rain forest area over the 16 years it was offered.[14]

Ironically, while Brazil's generals were encouraging the misuse of tropical forest, agricultural researchers in Brazil and Colombia were developing new acid-tolerant grasses and legumes that would tolerate the highly acid soils of Latin America's big acid savannas. The new acid-tolerant forage crops can *sustain over the long term* ten times as many cattle per acre as would degrade the rain forest pastures. Even better, the new grasses are planted *outside* the rain forests on the big acid savannas that have far less biodiversity. Latin America has some 300 million acres of these acid wastelands, which are covered with stunted brush and coarse grasses.

Brazil also has a successful "beef" industry in the lower Amazon, where Asian water buffalo have proven to adapt well. They thrive on jungle greenery that doesn't adequately nourish cattle. They defend their calves aggressively from snakes and jaguars. In the rainy season, when the cattle herd up on the dry ground starve, the water buffalo simply swim to food. Amazon water buffalo numbers have expanded from a few dozen at the end of World War II to more than a million.

In truth, where the rain forest has been cleared for cattle pasture, it has usually been due to government policies that devalue the trees. The rain forest land is so useless for other purposes it doesn't usually pay the cost of clearing the land. There would be more profit in maintaining the trees as the crop.

CGIAR Photo Gallery, ICRAF

AGROFOREST IN THE TROPIC—Damar trees in Sumatra. Agroforests dominated by damar and rubber trees almost simulate natural forests in their complexity and biodiversity.

Reprinted from BridgeNews Forum, June 12, 1998

Getting Serious About Saving the Rain Forests

Are we saving the tropical forests with eco-tourism, eco-harvested fruits, and eco-friendly logging? Has ice-cream maker Ben & Jerry's Ice Cream found a way to preserve the world's wildlands?

Apparently, not. At least not in Latin America, according to Dr. Douglas Southgate, an agricultural economist and author of *Tropical Forest Conservation* (Oxford Press).

The rich countries' first eco-strategy for the tropics has been the national parks for wild creatures. But it isn't working. The remote borders of the Third World wilderness are terribly difficult to police, and most of the "parks" are still inhabited by native people who routinely cut trees, burn fields, and kill game.

The national parks are supposed to be surrounded by buffer zones—where people sustainably gather forest fruits, carefully log a few trees, and tourists cheerfully wave and spend money.

Unfortunately, says Dr. Southgate, gathering nifty rain forest products for First World boutiques doesn't earn the forest folk much cash—even by their standards. "There have been some successes," he reports, "but they've been small and rare."

Ben & Jerry's Ice Cream started buying real rain forest nuts and got reams of publicity. But then it had to stop advertising its "rain forest" ice cream because they couldn't get enough jungle nuts.

The rain forests are so dense and diverse that it takes hours to find the fruits, nuts, or flowers of the desirable species. And hours more to get them back to a market town.

Local people have successfully gathered aguaje fruit and acai palms from the floodplains of the Amazon Basin. But the floodplains have bigger concentrations of fewer plants, and having a river nearby makes transport easier.

Southgate says that for forest harvesting to pay off, people need secure property rights, ready access to markets, and high concentrations of harvestable species (much like the plantations that produce our coffee, rubber, and palm oil).

The conservation job is huge. Tropical South America has some 3 million square miles of forest—nearly the land area of the United States! How many tourist visits would it take to finance the protection and management of such a huge area?

The Galapagos Islands attract a lot of tourists. They're unique, attractive, and you can travel comfortably on boats with good meals and flush toilets.

However, few tropical forests fit that description. The few tourists who do come to the rain forests use only a few square miles of it and depend very little on local products and labor.

"Environmental" logging in the rain forests is expensive, and the trees are a long way from market that cuts the profit.

Tropical loggers should go in a year ahead of the logging to cut vines. Otherwise bringing down one tree may bring down three or four. Mostly, loggers just bring the trees down and then go away for a few decades until the area regrows.

One-fourth of the trees they cut get lost in the underbrush!

There is no question that saving the tropical forests is important. If the world has 30 million wild species (a reasonable biologist's guess) then 25 million to 28 million of them are in tropical forests. Researchers have found more species in three square miles of the **Amazon** than in the whole United States.

Moreover, biotechnology will now let us use the **huge** array of genes on display in the tropical forests.

Until now, researchers could only use the one-in-a-million drug discoveries like quinine. But the potential through biotechnology to use tropical genes in other species makes the tropical forests a vast research reservoir for the 21st century

The good news on saving forests, says Southgate, is that just about any tropical country can raise its peoples' standard of living and keep its natural habitats intact.

The key is higher yielding agriculture, because slash-and-burn farmers are the big threat to the forests. They have been responsible for two-thirds of the forest that has been lost.

Farming with high-powered seeds, no-till farming systems, and chemical fertilizers will help poor countries feed more people on less land.

Then modest investments in rural education can make the local people attractive to outside employers, so they don't need to slash-and-burn the jungle to make a living.

Roads and bridges are important too—when they connect rural people and off-farm jobs.

According to Southgate, higher-yield farming is already saving millions of acres of wildlands throughout Latin America. He sees deforestation as basically the old problem of rural poverty.

Notes

1. Ariel Lugo, Institute of Tropical Forestry (Puerto Rico), U.S. Forest Service, presentation to the National Forum on Biodiversity, Washington, D.C., 1986.

2. FAO, *Forest Products Yearbook 1988*, Rome, 1990.

3. Forest Resources Assessment Project 1990, UN Food and Agriculture Agency, "Second Interim Report on the State of Tropical Forests," presented at the 10th World Forestry Congress, Paris, September 1991.

4. USDA/FAS, *Brazilian Forest Products Annual Report*, 1992, p. 3.

5. Ariel Lugo, op cit.

6. Cleaver and Schreiber, "The Population, Environment, and Agriculture Nexus in Sub-Saharan Africa," *Agriculture and Environmental Challenges, Proceedings of the 13th World Bank Agriculture Sector Symposium*, Washington, D.C., 1993, p. 208.

7. Elizabeth Larson, "Elephants and Ivory," *The Freeman*, July 1991, 261–263.

8. "Wood, Fodder and Soil-Fertility Problems in Western Kenya," *International Centre for Research in Agroforestry, Annual Report*, 1991, Nairobi, Kenya, 1992.

9. James Nations, "Does Conservation Condemn the Poor to Perpetual Poverty: A Nongovernmental Organization Perspective," *Agriculture and Environmental Challenges, Proceedings of the 13th World Bank Agriculture Sector Symposium*, Washington, D.C., 1993, p. 245.

10. Alden Vaughn, "New England Frontier," *Puritans and Indians, 1620-1675* (Vaughn and Clark, eds.), Cambridge, Massachusetts, Belknap, 1981.

11. Mostafa Tolba, before the Convention on International Trade in Endangered Species, Kyoto, Japan, 1992.

12. Hernando De Soto, *The Other Path*, New York, Harper & Row, 1989.

13. Southgate, Sanders, and Ehui, "Resource Degradation in Africa and Latin America: Population Pressure, Policies and Property Arrangements," *American Journal of Agricultural Economics*, December 1990, pp. 1259–1263. See also Southgate, *Tropical Deforestation and Agricultural Development in Latin America*, London Environmental Economics Centre, LEEC Paper DP 91-01. EC Paper DP 91–01.

14. Dennis J. Mahar, *Government Policies and Deforestation in Brazil's Amazon Region*, Washington, D. C., World Bank, 1989.

18

Saving the Planet with Plastic?

MYTHMAKERS SAY:

"During the manufacture of plastic, large quantities of different chemicals are made as by-products. They constitute toxic waste that is emitted into the air through factory chimneys, poured into sewage systems, sent to garbage dumps, drained into streams, rivers and lakes, buried in landfills, or illegally dumped by the Mafia at night when no one is looking. The food chain concentrates many of these toxic organic chemicals, so the plastic we use every day may come back to haunt us as poisonous food, water, or air, or it stays in our garbage dumps for hundreds of years."

Helen Caldicott, *If You Love This Planet*, 1992[1]

"If brewers were forbidden to put plastic nooses on six-packs of beer, if supermarkets were not allowed to wrap polyvinyl chloride film around everything in sight, if McDonald's restaurants could rediscover the paper plate, if the use of plastics was cut back to those things considered worth the social costs (say artificial hearts or video tape) then we could push back the petrochemical industry's toxic invasion of the biosphere."

Barry Commoner, "Why We Have Failed," *Greenpeace*, 1989[2]

Naturally, this chapter title is an overstatement.

We will not save the planet with plastic. However, the title makes a key point: Plastic, too, is a vital element in the high-yield farming that *can* make room on the globe for both people and wildlife.

Far from being an insult to the environment, plastic can do a great deal of environmental good through agriculture. Plastic, too, can save wildlife by raising crop yields and cropping intensity.

Plastic Sheeting

The major use of agricultural plastics is for plastic sheeting.

Mulch: Primarily, plastic sheeting is used for mulch. Laid atop the ground, the plastic raises the temperature of the soil so that tender young seedlings can be started and transplanted earlier in the spring. That promotes more rapid crop development and thus raises yields.

Plastic mulch can mean extending the ability to grow crops farther onto cool plains like those of Canada and northern China. In central China, it can mean an extra rice crop each year. In southern China, it can mean a crop of vegetables in addition to two rice crops from the same land.

Plastic sheeting has other virtues as well:

- •It prevents weeds from competing with the crop plant, without either pesticides or hand weeding.

- •It cuts water evaporation and makes the use of fertilizer safer and more effective. That can make the difference between 25 bushels of wheat per acre and 50 to 60 bushels in arid regions.

- •It helps cut soil erosion wherever it is used on slopes or in windy areas.

- •It can reduce losses to rot in crops like strawberries and tomatoes.

Soil Fumigation: Another major use of sheeting is for soil fumigation. Intensively farmed fields can build up high populations of destructive soil organisms like nematodes. To combat the nematodes and other subsoil pests, farmers sometimes have to fumigate the soils. Plastic sheeting seals in the soil fumigant, giving the soil pests a toxic dose while decreasing chemical usage.

Crop rotation is not very effective against nematodes, because they can lie dormant in the soil for up to 20 years waiting for the right crop to be planted and emit the correct feeding signal.

STRAWBERRIES IN PLASTIC—The plastic film used in this strawberry field in California helps warm the soil, conserves water use, and eliminates weeds.

Greenhouse Covering: Millions of yards of plastic sheeting are also used to cover greenhouses. Stretched over a light framework, the plastic film is far less expensive than glass greenhouses with their costly glass and glazing. (It may also take more fossil fuels to produce that glass than the plastic.)

One of the big advantages is that if the plastic greenhouse is hit by a hailstorm, the farmer can simply stretch a new roll of plastic over the frames. With a glass greenhouse, a storm means a huge reinvestment in rebuilding and reglazing.

Greenhouses play a key role in supplying good nutrition year round by growing many of our fruits and vegetables, both for harvest in cold weather and for starting seedlings early to fill gaps in the produce market.

They are especially important in cold climates and dry ones. China, again, is a major user. China's internal transportation system is still underdeveloped, and most of its population centers have to grow most of their own food. Chinese consumers still have to get their produce from their local region, simply because they can't get fruits and vegetables brought in from another climate zone.

In the Middle East, the latest greenhouse technologies are using seawater to help grow indoor crops. Forced air is blown through a mist of seawater and then into the greenhouse. The air picks up moisture but not salt. Some of the best designs get 90 percent of their crop moisture from the salt water.

Protecting Livestock Feed: Dairy cows and other ruminant livestock need huge amounts of forage year-round. In most climates this means storing food from the summer to provide for the winter. Plastic has been playing an expanding role in protecting these off-season feed supplies.

Plastic sheeting is widely used to protect silage, in both affluent countries and poor ones. Silage is simply a green forage (usually alfalfa, green corn, or high-value hay). The silage is chopped green at the peak of its total nutrient value and stored *without drying* in an airless environment where it cannot rot.

In the past, having silage meant building an expensive vertical silo, usually out of brick, cement, or cement blocks. Those silos were almost as much the hallmark of American farming as the barns they stood next to. In recent decades, big blue steel glass-lined silos have become landmarks in many farming regions. But such silos are expensive. They require natural resources to build and maintain.

With plastic, making silage can be as simple as digging a trench in the ground, lining it with plastic sheeting, and weighting a plastic cover down with old rubber tires to keep it from blowing off. It's cheap, effective, and good for the environment because it preserves productivity at low cost.

Plastic sheeting is also used increasingly to protect the big round hay bales that now dominate U.S. hay-making. Covered by plastic sheeting, dairy farmers, cattle ranchers, and sheep raisers can store these bales in the open through the winter, instead of building expensive storage barns and sheds. It is also much more convenient to have the bales in place in the fields during winter weather.

Plastic sheeting is so effective that it has become one of the major farming inputs in the world's largest agriculture. Chinese farmers, for example, currently use some 400,000 tons of plastic sheeting per year.[3]

Plastic for Irrigation

One of the most water-efficient irrigation systems is called drip irrigation. It uses permanently installed plastic tubes to feed water directly to the roots of orchard trees, grapevines, and other high-value plants. A modified drip irrigation system can be easily rigged by laying plastic tubing on the ground with holes at the appropriate places where the plants or trees are growing.

A newer irrigation system, with even higher water efficiency, is called dual-level irrigation. Developed at Iowa State University, it is the closest thing to a closed irrigation system that can be achieved outside a greenhouse.

The first level of perforated plastic pipes is laid in the root zone of the crops to be grown. This is the feed piping. The second level of perforated piping is laid perhaps two feet lower, and works as the recycling level.

No water evaporates because the water is added below ground level, and no runoff occurs because any water and nutrients not used the first time are put back through the system. Without the low cost and inert character of PVC plastic pipe, the system could not be cost-effective.[4]

Chemical Containers

Virtually all farm chemicals that are used are transported in plastic containers.

Environmentalists see this as a major complaint, because they envision millions of plastic jugs full of dangerous substances lying around farms or being dumped into landfills all over the country.

But the only other material tough and inert enough to transport the chemicals is glass. If chemicals were delivered in glass jugs, the breakage in transit—and resulting spillage—would represent a much greater environmental threat than disposing of triple-rinsed plastic containers.

Second, the plastic containers for agricultural chemicals are increasingly the big, reusable kind that go back to the factory for reuse instead of into a landfill. Handling spills have been identified as the major reason for the occasional high pesticide readings in farm groundwater sources. These big containers generally have positive lock-on features to prevent any spills in the handling and loading process.

Third, the agricultural chemicals industry is making strenuous efforts to collect old pesticides and empty pesticide jugs. The pesticides are incinerated or put into hazardous waste sites. The plastic jugs are turned over to central facilities where they can be safely rinsed and some are turned into—fence posts!

The inert nature of plastic makes for wonderful fence posts. Farmers need millions of such posts to keep livestock safely in their fields. These fence posts are a major highway safety plus for the public—as anyone who has ever driven in the Latin American countryside knows. Colliding with a horse or a cow can do enormous—even fatal—damage to highway travelers.

Making waste plastic into fence posts takes it far away from any people who might be chemically sensitive or concerned about the plastic's previous condition of servitude. Farmers love the idea of nearly permanent plastic fence posts, and the idea is spreading.

Biodegradable Plastics

Among the newer plastic developments are biodegradable plastics.

Farmers can now buy plastic sheeting that disappears from the environment. Ultraviolet light triggers a timed disintegration process in this new sheeting, which continues until the plastic has completely degraded. Once the sheeting has broken down into low-molecular-weight fragments, it is attacked by soil microbes that turn it back into carbon and water. The "timer" in the sheeting can be varied so the sheeting lasts as little as two weeks or as long as 12 months.

The biodegradable sheeting is more expensive, but eliminates the need to gather up the old sheeting and take it to a landfill.

Other types of biodegradable plastic use starch made from corn intermixed with the normal polyethylene molecules, so that the fragments break down under exposure to sunlight. Then the remaining flimsy polyethylene structure soon disintegrates.

ICI, a British-based chemical firm, is making small amounts of fully biodegradable plastic under the trade name Biopol. The firm uses bacteria that convert sugar (from corn or beets) into a natural polymer. Again, the biodegradable plastic is more expensive, and its use has been limited so far to shampoo and cosmetic bottles.

Environmentalists are not impressed. They view biodegradable plastics, including Biopol, as a red herring. They feel that they might give people the wrong message—that it's OK to throw plastics away—and distract from what they feel is the real need to recycle plastics. (So far, the biodegradables are a tiny part of the 100 million ton plastic industry.) Nor have they had much good to say about such efforts as turning chemical jugs into fence posts—a clear victory in recycling. (Approving the fence posts might seem to express approval for the pesticides.)

Meanwhile, researchers in the United States are working on plastics from lactic acid, a cheap by-product of the dairy industry.

Other Plastics for the Farm

Farmers use a wide variety of other plastics to perform necessary farm chores and to cut farming/food costs, including:

- Fiberglass shovel handles;
- Plastic seat covers for tractors and combines (machines that often sit out in the weather for extended periods);
- Plastic push broom bristles;
- Fiberglass watering tanks for livestock, that won't rust and leak water;
- Plastic insulators for the electric fences that are important for rotational grazing;
- Nylon livestock halters and lead ropes;
- Plastic non-rusting hoppers and nylon seed plates for crop planters;
- Plastic hoses and pipes for watering crops and livestock; and
- Fiberglass spray tanks for chemicals, that won't rust or corrode.

All of these products have strong, solid, functional reasons to be on our farms: to raise productivity and yield.

Growing Plastic Lumber?

An Iowa farmer and inventor named George Tyson has come up with a method of making hard, durable "wood-substitute composites" by combining agricultural biomass and discarded plastic. It's like being able to grow trees in a farm field—and reduce landfill requirements at the same time.[5]

His process uses superheated steam and chemicals to explode the cell fibers of wheat straw or other farm-produced biomass. The tiny fragments that are produced in these mini-explosions combine instantly with molecules of ground-up recycled consumer plastics. The resulting "Bondomass" is 40 percent stronger than wood, can be cut, shaped, drilled, and nailed like ordinary lumber—and is impervious to water, rot, and insects. (It can also be made flame-retardant.)

Tyson sees it being used initially for outdoor construction, highway signposts, porch furniture, insulation panels, playground equipment, and even low-cost housing. He expects the cost to be 10 or 15 percent less than pressure-treated lumber.

"The raw materials for this technology are relatively cheap and they're either already available or becoming available in abundant supply. . . . With the recycling programs that are going on throughout the country, all kinds of post-consumer plastics are looking for a home. We hope to offer that home," says Tyson.

The environmental zealots would prefer that we use virtually no plastic at all. But that is a vain hope unless you assume the world will revert to 1 billion people living in mud huts.

We need to use them carefully, but we need them.

PLASTIC BUCKETS FOR AFRICA:

"Half the population of Africa is under 15 years old. Children are everywhere: in the armies and the refugee camps, working in the fields and trading in the marketplaces. And at home it is the child who performs the most vital function of the family: fetching the water. At dawn, while so many are still asleep, small boys start up in the dark and hurry off to wells, ponds or rivers. The new technology has provided them with an essential instrument: a plastic bucket.

"The plastic bucket has revolutionized the lives of Africans. You cannot survive in the tropics without water; the shortages are always acute. And so water has to be carried long distances, frequently dozens of kilometers. Before the invention of the plastic bucket, water was carried in heavy vats made of clay or stone. The wheel was not a familiar aspect of African culture; everything was carried on the head, including the heavy vats of water. In the division of household labor, it was the women's task to fetch water. A child would have been unable to lift a vat. . . .

"The appearance of the plastic bucket was a miracle. To start with, it is relatively cheap, . . . costing around two dollars. And it is light. And it comes in different sizes: Even a small child can carry a few liters. . . .

"And now it is the child's job to fetch water. . . . What a relief for the overworked African woman! What a change in her life!. . . "

Ryszard Kapuscinski, "Plastic Buckets and Ballpoint Pens," *Granta*, reprinted in *Utne Reader,* January/February,1995, p. 39

Notes

1. Helen Caldicott, *If You Love This Planet: A Plan to Heal the Earth*, New York, W. W. Norton & Co., 1992, p. 63.

2. Barry Commoner, "Why We Have Failed," *Greenpeace,* September/October 1989, reprinted in *Learning to Listen to the Land*, Washington, D.C., Island Press, 1991, p. 165.

3. *China Light Industry Handbook*, Beijing, 1991.

4. "Subsurface Irrigation Systems: Water Management System Recycles Water," *The Grower* 24, 1991, pp. 32–34.

5. "Straw + Plastic = Wood?" *The Furrow*, January/February 1994, Deere & Co., Moline, Illinois, pp. 20–21.

19

Conserving with Cows

JEREMY RIFKIN, MYTHMAKER EXTRAORDINAIRE:

It is remarkable for one man and one book to gather so much misinformation about any single topic in one place—so we are relying on Rifkin's 1992 book, *Beyond Beef*, for most of the Mythmaker quotes in this chapter.

We will list the quotes by chapter headings, because Rifkin's chapter titles are so creative. And then we will deal one by one with the issues he raises.

"Malthus and Meat"

"It seems disingenuous for the intellectual elite of the first world to dwell on the subject of too many babies being born in the second-and-third-world nations while virtually ignoring the overpopulation of cattle and the realities of a food chain that robs the poor of sustenance to feed the rich a steady diet of grain-fed meat. The transition of world agriculture from food grain to feed grains represents a new form of human evil, whose consequences may be far greater and longer-lasting than any past examples of violence inflicted by men against their fellow human beings."[1]

One of Rifkin's major charges in *Beyond Beef* is that poor people in the Third World are being starved so their countries can export grain-fed beef to the rich countries.

This claim is utterly false.

Rifkin's "advisor" on this book was a young man who grew up on a Montana cattle ranch. He apparently believed that Third World countries produced beef the same way that North America does—

with grain and oilseeds to fatten the animals in the final stage before slaughter.

As discussed earlier, most of the world's cattle eat little in their lifetimes except grass and crop residues which humans cannot eat. About 90 percent of the feed for the world's cattle consists of things that humans cannot digest. Most of this feed is grown on land which is too dry, too steep, too rocky, or otherwise unable to support crop production. In addition, the cattle are helping to produce the high-quality food protein that the world is increasingly demanding—and doing it from safe, renewable resources.

To forgo the production of beef would be to waste the world's grass resources and crop residue. In a world that is tripling its demands on farming resources in the short space of four decades, that alternative seems ridiculous.

The countries that *do* feed grain to cattle all have their own grain surpluses. Again, none of them import fed beef from rain forest countries. (Due to their tropical climates, few rain forest countries have access to the low-cost grain needed to *fatten* cattle.) Moreover, while the affluent countries have been feeding grain to cattle, the price of grain in the world has continued to decline in real terms.

The list of grain-feeding countries is short:

- the United States
- Canada
- the European Community (now the European Union), which includes France, Germany, the Netherlands, Belgium, Luxembourg, Italy, Spain, Portugal, Denmark, Greece, Great Britain, and Ireland
- Japan

You may recognize the United States and the EC/EU as the world's two largest grain exporters. Both typically have produced far more grain than they consume. Both would prefer to be self-sufficient in meat, and they import meat only to fulfill international trade obligations.

Japan in most years has a surplus of rice, and has pushed large tracts of land not needed for rice into wheat production. Japan does import most of its feed grain, primarily from countries that have grain surpluses: the United States, Western Europe, and Thailand. Japan has recently allowed some beef to be imported from the United States and Australia.

In practice, America imports only about 0.5 percent of its beef. Most of that is grass-fed beef, used primarily in canned meat products. Much of it comes from Latin American countries and most of this is from Argentina and Uruguay, which have no rain forest. We also import some better grades of beef from Australia and New Zealand, but this beef is also primarily grass-fed.

"Malthus and Meat"

"It takes 9 pounds of feed to make 1 pound of gain in a feedlot steer. . . . Cattle have a feed protein conversion efficiency of only 6 percent.[2]

Rifkin is correct when he notes that grain use throughout the world has changed dramatically. About 40 percent of the world's grain production is currently used to feed livestock and poultry.

This use of grain for high-quality protein foods has not aggravated hunger. In fact, both beef and dairy cattle produce a net contribution to human food supplies. Cattle eat huge amounts of forage and by-products (citrus pulp, milling waste, rendered fats) that humans can't or won't eat. Humans consume less than half of the dry matter produced as "crops".

Grain Fed Per Unit of Carcass Weight

	Developed countries	*Developing countries*
Beef	2.6	0.3
Sheep and goat	0.8	0.3
Pork	3.7	1.8
Poultry	2.2	1.6

Source: *Animal Agriculture and Global Food Supply,* CAST, July 1999[3]

However, much of the field residues and processing by-products can be converted to human food by animals,substantially raising the productivity of the crops, and the acres that produced them.

Even in California, there is three times as much land that will not support crops as can be cropped. Much of this land *can* be grazed, however, and the soil erosion on most of the grazing land is as low—or lower—than on forests and "waste land."

In addition, cows eat things like citrus pulp, dried grains left over from brewing, cottonseed (after the oil is crushed out), rice straw, and corn stover. These "food sources" contribute millions of tons of cattle feed that people are not going to eat (though much of it is theoretically digestible in the human stomach).[4]

The University of California's College of Agriculture (at Davis) says even grain-fattened cattle actually consume only about two pounds of grain per pound of beef created.[5] (Most of their growth occurs before they go into the feedlot.)

"Tropical Pastures"

"Since 1960, more than 25 percent of the forests of Central America have been cleared to create pastureland for grazing cattle. . . . Each imported hamburger required the clearing of 6 square yards of jungle for pasture."[6]

Again, Rifkin has no factual basis for the forgoing claims.

I know of only one case in which tropical pasture has been deliberately cleared for beef production. That is the famous case of the Brazilian government's subsidy for clearing cattle pasture in the Amazon rain forest (discussed in Chapter 16). When trees are valued realistically, they are at least as valuable as beef pastures. Cattle *have* been pastured on former tropical forestland that was cleared for farming and proved unable to support crops. Very little of U.S. beef is produced on such land, however.

Again, no fast-food hamburger is being marketed in the United States from rain forest cattle. *Less than one-half of one percent of our beef is imported from anywhere*, because we have such a vigorous and productive animal agriculture of our own. Rifkin had no reason to target his 1993 Beyond Beef campaign against the McDonald's chain of hamburger restaurants—except that it was the biggest and most visible *media* target, and he had a book to sell.

JEREMY'S DESERTIFICATION MYTH:

"Hoofed Locusts"

"The destructive impact of cattle extends well beyond the rain forests to include vast stretches of the world's rangeland. Cattle are now a major cause of desertification. . . ."[7]

DESERTIFICATION REALITIES:

"Two-fifths of the earth's land surface . . . are drylands susceptible to desertification. However, . . . since the first desertification conference in 1977, the UN's own estimates of the areas affected have fallen dramatically, from about three-quarters of drylands to less than a quarter," says Professor David Thomas, director of the Sheffield Centre for International Drylands Research . . . in his recently published book *Desertification: Exploding the Myth*."

"The 'Grass Roots' Strategy for Holding Back the Deserts," *Financial Times*, October 14, 1994, p. 38

Grass protects fragile soils even more effectively than trees because of its dense root structure. That's why pasture is normally among the environmentally safest land uses. However, if too many animals are kept on too little grass, overgrazing can create environmental problems.

Rifkin suggests that "nearly 6 billion of the 7 billion tons of eroded soil in the United States is directly attributable to cattle and feed crop production" . . . and that America is suffering spreading desertification.

Both claims are unfounded.

Rifkin's major source of "information" here seems to be the Worldwatch Institute, an organization best known for its attempts to deplore the evils of population and prosperity.

Professor Thomas, on the other hand, says that about 20 percent of the world's soils are degraded, according to the latest global survey. However, if the "light degradation" category is omitted because it covers only minor changes that can be easily reversed, the *degraded proportion of the world's land is only 12 percent.*[8]

WE HAVEN'T LOST THE BISON—But the bison are too big and too hostile to be the basic grazing animal that turns our grasslands into meat.

(Interestingly, Professor Thomas says Europe has a high proportion of degraded soils, mostly because of the semiarid lands around the Mediterranean.)

America has no soil erosion crisis on range *or* croplands. Our soil erosion today is the lowest in modern history. This has been achieved with the help of improved pasture management, weed killers, and improved forage varieties.

NEW JERSEY MYTHMAKERS CLAIM THE GREAT PLAINS:

"Is the idea of turning 139,000 square miles of land across ten states in the Great Plains into a wildlife refuge one of the most farsighted, pragmatic and ecologically sound proposals in the history of the area? Or is it just another in the long line of misguided fantasies created by Easterners who do not appreciate or understand the geography and climate of the place?

"The idea of the Buffalo Commons occurred to (Frank and Deborah Popper) while stuck in traffic on the New Jersey Turnpike.

"Visitors to the Commons 'would see the heart of the cóntinent as Lewis and Clark first knew it,' writes Matthews.

"The Plains are emptying on their own, [the Poppers] argue.

"Matthews . . . gives a balanced though finally sympathetic portrayal of the Poppers and their idea. The truth is, they win you over themselves. Their humility and seriousness and lack of pretensions make their moments of naivete or shortsightedness acceptable."

From a book review of *Where the Buffalo Roam* by Anne Matthews, reviewed by Dallas Crow in the *Amicus Journal*, Fall 1993[9]

Reality Comment: The Poppers are correct in saying that there are fewer people in the Great Plains, but its grass will be even more important to meeting the world's protein hunger in the 21st century than it has been in the 20th century.

In the Third World, there *has* been extensive desertification, and some of it has been due to overgrazing. This is mainly so for two reasons.

First, in some tribal cultures, a family's status is based on the number of animals they own, rather than on the animals' productivity in terms of meat, milk, wool, etc. This provides an artificial incentive to keep more animals than the pastures will support.

Second, many Third World pastures are replaying the tragedy of the commons. When the pastures belong to everyone, they belong to no one. In the African Sahel, for example, too many herdsmen build up herds in the wet years which cannot be maintained in the dry years. The heavy grazing can even prevent key forage species from reproducing.

The answer to the overgrazing problem in the Third World has little to do with population numbers, and a great deal to do with ownership of the pastures.

In ancient times, the Sahel was defended by swords, and then by rifles. Herdsmen that built their herds too large suddenly found some of their animals forcibly removed. A former minister of agriculture in the Sahelian country of Mali once told me that the Sahel's real problems had begun when the colonial governments disarmed the tribes. "The Tauregs (a dominant tribe) never shot nearly so many people as die now in the droughts," he told me. (See Chapter 17 for a discussion of the commons ownership problem.)

Even in the Third World, however, neither overgrazing nor desertification are anywhere near as serious as environmental activists would have us believe. Lester Brown, for example, tried to tell us that Africa's major drought in 1983–84 was brought on by overgrazing, and that it would be followed by more and more severe drought conditions spreading across Africa. Instead, the Sahel has entered a relatively wet period, and since 1985, satellite data show clearly that the desert has been receding.[10]

Geographers say that some 2 million square miles on the edge of the Sahara have been moving back and forth from desert to grasses and back again over the centuries, depending on the rainfall cycles. Archeologists say the Sahel has been totally unpopulated for centuries at a time in the past because it was too dry to support any population. It is fortunate that only a few of the world's people and resources are affected by these marginal conditions. The entire population of the Sahelian region, for example, is probably only about 25 million people . . . 0.48 percent of the world's population.

America is unlikely to replay the tragedy of the commons on our own rangeland unless we put people like Jeremy Rifkin in charge of our grasslands.

"Marbled Specks of Death"

"Living atop the protein ladder has turned out to be very precarious. The affluent populations of the northern hemisphere are dying by the millions from grain-fed beef and other grain-fed red meat."[11]

Here Jeremy Rifkin reminds us forcefully, if not very precisely, that people should keep their fat intake moderate. However, he fails to make his case thoroughly—partly because he does not deal with the dairy products he himself consumes. I discovered his fondness for dairy products during a debate with Rifkin on Iowa Public Radio in 1992. I mentioned the potential problem of protein deficiency that goes with a vegetarian diet. He noted that he didn't worry about protein deficiencies as he and his wife ate plenty of dairy products.

But wait a minute! What is the environmental difference between a dairy cow and a beef cow? Both eat grass, trample things underfoot, drink water, and emit methane and bodily wastes.

Come to think of it, the *nutritional* impact of dairy products is essentially the same as beef. You can get just as big a dose of saturated fats from cheese and cream as from a piece of steak.

"Sacrifice to Slaughter"

"In order to obtain the optimum weight gain in the minimum time, feedlot managers administer a panoply of pharmaceuticals to the cattle, including growth-stimulating hormones and feed additives."[12]

Jeremy is right for a change. Ranchers and stockmen do use a panoply of medicines to protect their animals from sickness, and to cure them when they become sick.

Doing less would, rightly, bring a public outcry.

(It puzzles me that "organic" livestock producers claim credit for not using medicine on sick animals, or preventive medicines to prevent their becoming sick. Should we praise parents who refuse to get medicine for their children? A cattle herd is almost as likely to produce infectious diseases as a school full of kids.)

All livestock medicines have to be approved by the Food and Drug Administration, with full testing and full regard to their impact on both the animals and consumers. They must be used according to directions, or the animals may be rejected in the marketing system. Where appropriate, waiting periods are specified between the administration of the medicine and the marketing of the animal, to ensure that there is no carryover of the medicine in the meat.

Some of the feed additives help to prevent the cattle from becoming ill. Here, too, their use has to be approved by the FDA and the producers must follow directions or risk having their animals rejected at the slaughter plant.

Jeremy is also correct that some of the feed additives are hormone treatments which help the cattle gain weight faster.

The original hormone used for cattle fattening was diethylstilblestrol (DES). DES was helpful not only in fattening cattle, but also, in massive doses, in helping many women prone to miscarriage keep their babies to term. Then a few cases were discovered in which the daughters of women who had taken DES to prevent miscarriage had come down with a rare form of vaginal cancer.

To match the DES exposure of the women taking it to prevent miscarriage, a consumer would have to eat 25 tons of beef liver containing 2 parts per billion of DES—at one sitting. A woman would have to eat 1 million pounds of DES–treated beef liver to match the synthetic estrogen in one "morning-after" contraceptive pill. No matter, DES was banned.[13]

The hormone treatments that are still used must be approved by FDA and used according to directions. Key among them is withdrawing the treatment well before the marketing date, so that the medication can pass out of the animals' systems.

In the case of livestock growth stimulants, America has been a good deal wiser than Western Europe. The European Community banned all such medications completely. The result has been a rampant European black market in growth stimulants. Observers estimate that up to 90 percent of the cattle in Belgium, for instance, are illegally treated with hormones. This is done, not under the watchful care of veterinarians, but in the "dark of night" with under-the-counter drugs that may not even carry their correct names and the necessary safety precautions.[14]

The real environmental kicker on livestock pharmaceuticals is this: A recent survey of veterinarians in Western Europe indicated that to raise the current amounts of red meat, without preventive and healing medicines, farmers would have to raise nearly twice as many cows and half again as many pigs. That would be the impact of death losses and sickly animals.

Veterinarians don't usually treat poultry, but chickens, ducks, and turkeys are even more susceptible to epidemic diseases than four-footed animals. It might well be that the only way to raise poultry without medicines would be in the little movable cages that my neighbor used to put out in his alfalfa fields. Labor costs and death losses would still be much higher. Even then, the birds would have no protection against avian tuberculosis.

"Quenching Thirst"

"Now, even the freshwater reserves of the planet are threatened by . . . droughts, overcultivation and overgrazing. Nearly half the grain-fed cattle in the United States are raised . . . on a single underground aquifer."[15]

Rifkin is correct that a lot of cattle are raised over the Ogallala Aquifer which underlies parts of Nebraska, Kansas, Oklahoma, Texas, and Colorado. However, it's not clear what his ominous-sounding sentence is supposed to mean and, obviously, irrigation doesn't cause drought.

The reason there are so many cattle in the region of the Ogallala is that the rich prairie soils there produce a lot of feed grain and feeder calves. Some of the grain is irrigated, but most is not. Nor are the pastures irrigated. (See Chapter 14 for more information on the Ogallala.)

During Rifkin's media campaign against beef consumption, he kept claiming that each pound of steak required huge quantities of water to produce.

Apparently, his water consumption number was based on the idea of the steer being raised and fed from irrigated land in some arid place like Southern California or Arizona. Few beef cattle are raised in such places, in part because water and feed *do* tend to be expensive there.

Despite Rifkin's insistence to the contrary, cattle production doesn't take much water. Cows drink only a few gallons of water per day apiece, and it is returned to the environment almost immediately, enriched to encourage plant growth. The grass cattle eat is overwhelmingly rain-fed. Most of our feedstuffs are grown on rain-fed land, and *all* of it could be if we were short of water and gave up irrigating feed grain.

The true water requirements of beef and dairy cattle are tiny.

"Warming Up the Planet"[16]

"The increase in the cattle and termite populations and the burning of forests and grasslands account for much of the increase in methane . . . of the past several decades. Methane emissions are responsible for 18 percent of the global warming trend."

Once again, Rifkin's zeal has outrun reality.

Methane does contribute an estimated 18 percent of the global warming gases. But domestic livestock contribute only about 14–16 percent of the methane.[17]

Thus, all of the world's cattle apparently contribute less than 2 percent of the global warming gases. The United States has 8 percent of the world's 1.3 billion cattle and buffalo, and 10 percent of *those* are dairy cows (which Rifkin *likes*).

The world may begin to experience global warming, though we haven't had any significant rise in temperatures since 1940. If the warming does begin, remember that each cow on the planet emits about enough methane to equal the global-warming impact of the average 75-watt electric bulb.

The real question is what happens if the grasslands of the world are *not* grazed.

The answer is that the grasses would grow lush and tall—and catch fire spectacularly when the inevitable bolt of lightning hits in the dry season. Early accounts of prairie wildfires in the Great Plains say they lasted for days, and drove all kinds of frightened wildlife before their flames.

The CO_2 that would be produced by the wildfires would dwarf the global warming impact of the methane from grazing animals.

GLOBAL WARMING REALITY?

"Savanna Grasses—the Missing Carbon Sink?

"Pasture grasses planted to increase beef production in the South American savannas are countering the doomsday predictions of global warming, announced scientists at the International Center for Tropical Agriculture (CIAT) in . . . the September 15 issue of *Nature* magazine.

"'The deep-rooted grasses may remove as much as 2 billion tons of carbon dioxide—a "greenhouse gas"— from the atmosphere yearly,' says Dr. Myles Fisher, CIAT ecophysiologist.

" . . . [T]he perennial grasses *Andropogon gayanus* and *Brachiaria humidicola* convert as much as 53 tons of CO_2 per hectare yearly to organic matter, Fisher says. . . . The storage of organic matter was not noticed earlier because the extensive roots of these grasses deposit it as deep as a meter in the savanna soil. . . CIAT . . . introduced Andropogon and Brachiaria, originally from Africa, to the grassy savannas of South America in the 1970s.

"'The ocean, tropical wetlands and green plants absorb some atmospheric CO_2. But scientists cannot resolve the fate of several billion tons.

"'Improved savanna grasses must explain part of this difference,' Fisher says. Brazil alone has at least 35 million hectares of introduced pastures—enough to fix 2 billion tons of CO_2 per year . . . because

Andropogon and Brachiaria adapt well to acid soils, national programs have released one or both of the grasses to farmers in at least 12 Latin American countries."

from *CGIAR News*, vol. 1, no. 1, Consultative Group on Agricultural Research, October 1994, p. 1

"Meat and Gender Hierarchies"

"The beef mythology has been used over and over again to perpetuate male dominance, foster class divisions and promote the interests of nationalism and colonialism."[18]

No comment.

"Beyond Beef"

"Reconstructing our relationship to the bovine is a gesture of great historical significance. By making a personal and collective choice to go beyond beef, we strike at the heart of the modern notion of economics with its near-exclusive emphasis on 'industrial productivity,' a concept that has come to replace the ancient idea of generativeness. . . . Our changing relationship to the bovine, from one of revered generativeness to one of controlled productivity, mirrors the changing consciousness of Western civilization as it has struggled to define itself and its relationship to both the natural order and the cosmic scheme."[19]

This may be philosophy. Then again, it may not.

I think Rifkin is trying to say that he admires India, where they don't eat cows, more than he admires America, where we do.

But how can that be?

India has 200 million cows and 75 million water buffalo, compared to America's 100 million cattle. That means that India's ruminants are producing three times as much methane to ruin the atmosphere and bring on global warming.

India also has much more severe water constraints than the United States, so the cattle are probably ten times the water conservation problem that they represent in America. Overgrazing is another of the country's problems, as anyone who has seen a dusty Indian travelog can tell you. (Remember Rifkin's "hoofed locusts?")

And the Indians let the cows wander the streets, adding to *urban* pollution.

Could it be that Rifkin's real agenda has to do with animal rights, and not with the environment at all? The only significant difference between beef and dairy cattle is that we don't eat the dairy cattle (at least not until they're past their milking prime, and then only in processed form because the meat is quite tough). India sells its old cows to be slaughtered in Bangladesh.

Most people who are personally acquainted with cattle find them intellectually limited. But if Rifkin truly prefers cattle to people, I recommend he get his own cows and pastures. Then he can enjoy the revering generative relationship without fear of being invaded by McDonald's. He can also ensure that his ranch's neighbors are not afflicted by overgrazing and pollution of nearby streams.

Meanwhile, how do the rest of us evaluate cows on an environmental basis?

- •Cows produce small amounts of methane, but grazing prevents big wildfires that would unleash large amounts of CO_2—on virtually an annual basis.
- •Few cattle pastures could support crops. Most of the grasslands have been grasslands since time immemorial.
- •Cows and grazing even help maintain a broader variety of forage species on the grasslands. (Without grazing, the tallest grasses shade out the others.)
- •Cows help meet the world's need for high-quality protein from low-cost, renewable, and environmentally stable ecosystems.

As long as we need to graze the grasslands, we might as well make use of the meat they produce. Letting the wolves have it all doesn't seem to make sense when the world wants more protein.

And since we're going to have some grazing animals, take the advice of someone who has worked with large animals. It makes *very good* sense to pasture cattle instead of bison. Cows weigh perhaps 900 pounds each, are fairly timid, and aren't very athletic. Bison weigh up to two tons apiece, have hostile attitudes, and can jump or go through a 10-foot anchor fence at will. I would not want to give worm medicine to a bison.

Notes

1. Jeremy Rifkin, *Beyond Beef*, New York, Penguin Books,1993, p. 160.

2. Rifkin, ibid.

3. *Animal Agriculture and Global Food Supply*, Task Force Report, no. 135, Ames Iowa, Council for Agricultural Science and Technology, July, 1999, p. 42.

4. R. L. Baldwin, K .C. Donovan, and J. L. Beckett, *An Update on Returns on Human Edible Input in Animal Agriculture*, Department of Animal Science, University of California, Davis, undated paper.

5. Eric Bradford, director of the Animal Agriculture Research Center, and Edward Price, chairman of the Animal Science Department, letter to Dr. Richard Stuckey, executive vice-president, Council for Agricultural Science and Technology, April 14, 1994.

6. Rifkin, op. cit., p. 192.

7. Rifkin, op. cit., p. 200.

8. Geoff Tansey, "The 'Grass Roots' Strategy for Holding Back the Deserts," *Financial Times*, October 14, 1994, p. 38.

9. *Amicus Journal*, Fall 1993, vol. 15, no. 3, Natural Resources Defense Council, pp. 46–47.

10. Monastersky, "Satellites Expose Myth of Marching Sahara," *Science News*, July 20, 1991, p. 38. Also, Tucker and others, "Expansion and Contraction of the Sahara from 1980 to 1990," *Science*, vol. 253, July 19, 1991, pp. 299–301.

11. Rifkin, op. cit., p. 171.

12. Rifkin, op. cit., p. 12.

13. Elizabeth Whelan, *Panic in the Pantry*, New York, Atheneum, 1975, pp. 164–69.

14. Rita Boone, *An Easy Mind on Meat Again,* Research Park De Haak, Netherlands, Roularta Books, 1993, pp. 46–47.

15. Rifkin, op. cit., pp. 218–219.

16. Rifkin, op. cit., p. 225.

17. J. B. Smith and D. A. Tirpak, *Potential Effects of Global Climate Change on the United States*: Executive Summary, U.S. Environmental Protection Agency, 1989.

18. Rifkin, op. cit., p. 286.

19. Rifkin, op. cit., p. 287.

20

Biotechnology:

The Ultimate Conservation Solution?

MYTHMAKERS SAY:

"Senior Labour Member of Parliament Joan Walley claimed [genetically modified food] had been responsible for an infection which hit 5,000 people in the United States, killing 37 and leaving 1,500 disabled. The disease, called eosinophilia myalgia syndrome, or EMS, had been traced to a batch of food supplement produced by genetically engineered bacteria."

David Hughes, "Blair Refuses to Ban Frankenstein Food,"
London Daily Mail, February 4, 1999

"Monsanto trusts its genetically engineered potato to kill off the Colorado potato beetle. But whether humankind's enduring relationship to food crops is being imperiled by biotechnology is the big question."

Michael Pollan, "Playing God in the Garden,"
New York Times Magazine, October 25, 1998

"The Austrian government is under fierce pressure to stiffen its opposition to European Union guidelines on genetically modified food products, after a national petition [to ban them] attracted 1.3 million signatures, or 20 percent of the eligible voters."

Eric Frey, "Austrian Petition Urges Genetic Food Ban,"
Financial Times, April 15, 1997

"Monsanto intends to completely control crop production in Asia, Mexico and South America, areas where food shortages and growing populations represent huge potential profits. The USDA will profit too; despite being a taxpayer-funded federal agency, it plans to collect a 5 percent royalty on all terminator seed sales. Gardeners and growers everywhere should oppose this technology because it is dangerous. There is a serious risk that in the open field, terminator-grown crops could 'infect' other crops, causing crop sterilization across vast areas."

Terminator Seed Technology Threatens Farmers Worldwide,
flyer enclosed with catalog by Gardener's Supply Company,
Burlington, Vermont, January 1999

"Greenpeace, together with the International Federation of Organic Agricultural Movements and 22 U.S. farmers, filed a lawsuit today to force the U.S. Environmental Protection Agency to withdraw all approvals for genetically engineered crops containing the gene of a bacterium called Bacillus Thuringiensis (Bt). . . . According to Greenpeace, the widespread use of Bt crops will inevitably breed insect pest resistant to the Bt toxins. . . . By creating insect resistance the genetically engineered crops would render these pesticides ineffective within a few years time."

Greenpeace press release, February 18, 1999

"The [Indian] government will enact gene protection laws to ensure that farmers' rights to produce seeds are not adversely affected. 'The government will not allow the misuse of the genetic infrastructure of the country and import of terminal genes,' federal agricultural minister Sompal said on Saturday."

New Delhi, India (*Asia Pulse*), March 29, 1999

"Campaigners against genetically modified food are today celebrating a 'dramatic and unprecedented' victory after charges against two London women accused of conspiring to damage an experimental crop were unexpectedly dropped. Jacklyn Sheedy and Elizabeth Snook had faced a maximum 10 years' jail after admitting pulling up GM maize from a Devon test site. Around 250 demonstrators greeted them after the hearing."

Geoffrey Gibbs, "Crown Drops Case Against Two Who
'Wrecked GM Crop,'" *The Guardian* (London), March 29, 1999

"Genetically altered cows could produce genetically altered manure and pollute groundwater with mutant cow dung, some Craig County residents fear.

The concerns are being raised in response to Pharming Healthcare Inc's plan to locate in Craig a herd of 200 cows capable of producing human proteins in milk. The proteins are used to make medicines."

From an AP story appearing in the Staunton, Virginia, *Daily News Leader*, September 22, 1999

REALITY SAYS:

"Earlier this year, Britain was rocked by claims that genetically modified foods are dangerous. Aprad Pusztai, a biochemist who used to work at the Rowett Research Institute in Scotland, said he had shown that GM potatoes were harmful to rats because of their genetic modification alone. Were the GM potatoes toxic? On the basis of Pusztai's evidence, it's impossible to say. In fact, his results support only one obvious conclusion: rats hate potatoes. According to toxicologists who examined the data . . . starvation or known toxins in raw potatoes were the most likely culprits for any changes seen in the rats. . . . It is often difficult to feed lab animals enough GM fodder, whether or not they find it palatable . . . to reveal small differences between modified and unmodified foods."

Debra McKenzie, "Unpalatable Truths," *New Scientist,* April 17, 1999

"At the minimum, we will have three billion more people on the planet before the population stabilizes, and almost all of those will be in the developing countries. . . . There is no way of dealing with poverty reduction or environmental protection or food security without transforming agricultural production at the small holder level in the developing countries. But we are, in parallel, on the cusp of a new revolution in the biological sciences. Now that we are able to understand the structure of genes and are getting into the manipulation of genes, all sorts of possibilities are opening up . . . plants that are more drought-resistant, more salt-tolerant, less thirsty, more resistant to pests without pesticides."

Dr. Ismail Serageldin, World Bank vice president for sustainable development, "Can We Feed Our World?" *CGIAR News*, Consultative Group on International Agricultural Research, December 1998

"Which of these is definitely true? Elvis Presley is alive and living in Acapulco. Genetically modified foods are safe. Mobile phone masts are dangerous. Taking vitamin supplements makes you healthy. Genetically modified foods are dangerous. Horoscopes can tell you the future. Organic foods are better for you than non-organic. . . . None of these statements can be proven true, although it is remarkable how many people firmly believe these things. . . . Some can get positively vehement about these issues, despite having very little real evidence to prove their view."

Dick Ahlstrom, Science Editor, "Logic Goes Out the Window When Assessing Hazards," *Irish Times,* March 29, 1999

"The Webster Groves City Council has declined to pass a resolution asking the Missouri Legislature and Congress to require labeling of genetically altered food. . . . Food-labeling backers reacted angrily. 'Why can't you take a stand?' said Steve Cassilly, one of the proponents."

St. Louis Post-Dispatch, April 12, 1999

"Had the adventurous apiarists who created the honeybee been able to identify and successfully insert the gene from the African bee that encodes for high honey production, they would have been able to produce a high-output European honeybee instead of an aggressive cross-bred."

Julian Morris, *Wall Street Journal Europe,* February 19–20, 1999

"Since the early 1980s . . . global cereals harvests have been rising at a rate of about 1.3 percent per year—just enough to meet the projected increase in demand. . . . But productivity increases—rises in cereal yields per hectare—have been slipping from 2.2 percent per year in 1967–82 to 1.5 percent per year in 1983–94. . . . Peng, and the other agronomists who regard genetic engineering as the key to surpassing the yield barrier, have more in mind than the products of today's biotech industry . . . in which genes for desired characteristics—such as herbicide resistance or an anti-bacterial compound—are smuggled into the organism from an outside source. . . . To break yield barriers, the plants will have to be thoroughly re-engineered."

Charles C. Mann, "Crop Scientists Seek a New Green Revolution," *Science,* vol. 283, January 1999

We fear the unknown, and we especially fear it in that very personal area of food. Many people have said to me, "Isn't natural best? Why should we accept changes in our food?"

Natural food is hitting a rabbit with a stick, and roasting it before the meat begins to rot. Natural is gathering whatever greens and berries happen to be in season, and trying to make them palatable. Natural is hoping the grain didn't become infested with ergot mold or the flour with weevils during storage.

For eons, humanity has built storage for grain and root cellars to protect harvests; and salted, dried, smoked, pickled, and spiced their food to keep it safe for consumption. Recently, we have found that salted meat and pickles can cause tumors in rats at high doses, but they have kept us from dying of more immediate causes like botulism.

The average life span of a Roman citizen, eating all-natural foods in the 4th century B.C., was 25 years. The average life span in 18th century London was virtually identical. Today's First World life span is 77 years. The average life span in America has increased by more than 30 years since 1900 as our food has gotten less and less "natural."

Part of the increase in our life spans is due to our highly bred crops, our use of pesticides (especially fungicides), and our food preservatives and preventatives. Pasteurization has eliminated the transmission of tuberculosis and deadly bacterial listeria in milk and juices. Identifying essential micronutrients such as iron, iodine, and vitamins has allowed us to fortify foods and almost eliminate major nutritional deficiency diseases such as rickets, goiter, and pellagra.

So how many years should we lop off our lives to "eat naturally?" How much should we let biotechnology alter our foods if biotech can make the food still safer, or help us save wildlands?

Biotech or Wildlands—This Is the Choice

If it were not for conserving wildlife, we might comfortably forego biotechnology in food production. However, as we do care about wildlands and wild species and are squeamish about genocide and forced abortion, biotechnology must be an important part of the solution for 21st -century challenges.

Modern Malthusian pessimists have been saying for 25 years that the world was "running out of agricultural research." People like Lester Brown, of the Worldwatch Institute, have been carefully noting how much of the world is already using high-yield seeds, irrigation, and chemical fertilizers. "How will you keep farm yields rising," their chorus asks.

The answer is apparently biotechnology. Biotechnology is the largest piece of knowledge on raising crop and livestock yields that we have discovered and not yet broadly used. It is the most critically important piece of intellectual capital available to the global conservation effort.

How Will Biotechnology Help?

Acid-Tolerant Crops for the Tropics: Two Mexican researchers (trained in the United States) have produced acid-tolerant crops for the tropics.[1] This is a huge step forward because about 40 percent of the arable land in the tropics has high levels of toxic aluminum or manganese. The acid soils cut crop yields by up to 80 percent.

The Mexican researchers knew that some of the successful plants on the world's naturally acid savannas secreted citric acid from their roots; they took a gene that codes for citric acid from a bacterium, and inserted it in tobacco, papaya, and rice plants. It works.

The new acid tolerant plants must now complete their scientific testing, and then be incorporated in the tropical world's plant breeding and distribution programs. This will take some years.

The new varieties will work directly to protect tropical forests since the aluminum tolerance will raise crop yields in the countries that have the densest populations, the highest rates of population growth, and the highest rates of per capita income growth.

Finding Useful Genes in Wild Relatives of CropPlants: Researchers at Cornell University have used their new understanding of nature's genetic code to find promising genes among the wild relatives of our crop plants that are already stored in the world's gene banks.[2] They reasoned that our breeding lines have been getting gradually narrower over the century or so that we've been doing "scientific" plant breeding.

They tried the concept first in tomatoes, a highly bred industrial crop in which yields have been rising at only about 1 percent per year recently. Wild genes quickly produced a 50 percent increase in yields, and a 23 percent increase in solids.

They next selected two promising genes from wild relatives of the rice plant. Rice breeders had not generated any increase in peak yields since Chinese researchers had developed hybrid rice 20 years ago. Each of the two genes produced a 17 percent increase in yields, and the two are thought to complement each other. In other words, wild-relative genes may have delivered a one-third increase in global rice yields on the first try!

Animal Growth Hormones: One of the first and simplest of the genetically engineered agricultural breakthroughs was instructing laboratory bacteria to copy the natural growth hormones of cows and pigs. The hormones were too complicated for laboratory analysis, but easy enough for biotechnology to replicate.

In fact, bovine growth hormone was the very first major biotech agricultural product to get regulatory approval. BGH has been on the market for several years now, and is proving to improve the feed efficiency of dairy cows by about 10 percent. If, and when, BGH were used in all of the world's 228 million dairy cattle, they would produce an additional 450,000 tons of milk per year—with no additional land needed for pasture or feed.

Many claims have been made about the "dangers" of bovine growth hormones; but none of them have been validated by science or real-world experience. In fact, Food and Drug Administration veterinary authorities say that the growth hormones are "particularly safe" because they are proteins, and if they are ingested they are also digested like a piece of cheese or steak. Humans have been digesting such bovine growth hormone in cow's milk for thousands of years. (Milk from BGH–treated cows contains no more BGH than the milk from untreated cows.)

Fears that cows treated with BGH would suffer more mastitis problems than nontreated cows have been proven unfounded. High-production cows must be managed carefully to minimize mastitis problems whether their high output is achieved with concentrate feeds, BGH or a combination of the two. A case of mastitis in a high-yield cow causes more economic and environmental loss than a case of mastitis in a low-production cow. The correct environmental answer is not to have lots of low-yield cows grazing the wildlands, but to carefully manage the high-yield cows. That certainly means using veterinary medicines to treat them when they need treatment. (Not treating cows that need medication is hardly a humane approach in any case.)

Pork growth hormone promises to be an even better bargain for people and the environment than bovine growth hormone: It offers the potential of human health benefits (through lower-fat pork) even as it radically reduces the feed required to produce the meat. Tests have shown that PGH will produce pork with about 50 percent less body fat and 15–20 percent more lean meat—with 25 percent less feed grain. With PGH, the world of 2050 might need 60 million tons less feed grain for its hogs than would otherwise be the case. It's hard to know what corn yields will be in 2050, but currently world corn averages only 1.6 tons per acre. Saving 60 million tons of corn would mean saving 58,000 square miles of wildlands—equal to the land area of Florida.

The Politics of Biotechnology

Both the European Union and Canada have withheld regulatory approval for bovine growth hormone. This is politics, not science. Both countries are trying to defend protectionist dairy policies based on limiting their milk production. BGH would raise the government costs of dairy surpluses.

Governments on both sides of the Atlantic are under intense pressure from eco-zealots, organic farmers, and other "modern know-nothings" not to approve new food production technologies. The pressure is much harder to resist in Europe, where the big farm surpluses stimulated by high farm price supports are extremely visible and expensive to the public.

The EU is still defending a ban on meat produced with non-biotech growth hormones, even though the Codex Alimentarius (the science referee for the World Trade Organization) has ruled that the EU has no scientific basis for banning them.The EU ban on meat hormones is especially ironic since a high percentage of the beef cattle slaughtered in the EU show hormone residues. (The use of the hormones is profitable to farmers, and they are widely available from unregulated "black market" sources across Europe.)

Biotechnology Safety Questions

If we discovered that biotechnology was a huge threat to humanity, we would, of course, reject it. So far, however, that looks extremely unlikely. Even opponents of biotech agree that it's not biotech that creates risk, but specific transformations being proposed. However, each proposed biotech transformation must be examined individually for its safety and potential value to people and the environment.

The worst human health concern that's currently being presented seriously is that some of the biotech foods might turn out to cause allergies in some people. I don't mean to make light of allergies. But let's face the realities: First, humanity is surrounded by millions of allergens, including pollen, house mites, pet dander, etc. Fortunately, there are terrific new allergy treatments that allow most sufferers to cope quite comfortably with their allergies. Second, no one has yet found any of these famous allergens in biotech foods. If we found them, we'd probably withdraw the allergy-causing product, and/or modify it to eliminate the allergens. To keep things in perspective, we've lost dozens of human lives to the hybridized African bees that are making their way north from Latin America, and dozens more lives (mostly children) to injuries attributed to the inbreeding of pit bull terriers. Not a single, minor, temporary real-world health problem has been traced to biotechnology.

Many of the same people who say biotechnology is dangerous are same ones who continue to say that we shouldn't use pesticides or chemical fertilizers. Many also reject modern medicine. Most of us can't afford to reject modern medicine, and we will find that we can't reject biotechnology in either medicine or in food production.

Reprinted from BridgeNews Forum: June 4, 1999

The Moral Imperative of Biotechnology

Is Europe starting to have second thoughts about its opposition to biotechnology?

There was a minor flurry of headlines in Britain recently after the Nuffield Council on Bioethics reported that the world has a moral obligation to develop genetically modified crops to help "fight Third World hunger." An environmentalist who took part in the deliberations of the London-based bioethics group said that if biotech foods could help fight the problem, "it would be immoral of us to stop" their use.

British Prime Minister Tony Blair found it "extraordinary" that the media would neglect positive reports on biotechnology such as the Nuffield report, while giving huge coverage to reports "which fed the hysteria."

The response of European environmental groups to this "moral imperative" was woefully weak. Christian Aid, a church-based development agency headquartered in London, called the council's report "naïve" and claimed the solution to world hunger lies in organic farming. But yields from organic farming are only about half as high. Christian Aid's focus seems to be on saving the jobs of small-scale organic farmers.

Who in the affluent world of the future will be willing to leave his or her air-conditioned office cubicle to hoe corn and pull weeds in the hot sun?

Friends of the Earth, a London-based environmental group, call world hunger more of a political problem than a scientific one. Perhaps the group was talking about its own political campaign to block the use of farm technologies like hybrid seeds and genetically engineered crops?

How the world manages to triple today's farm output in order to feed the world of 2040 will be a critical decision concerning the wildlife that Friends of the Earth is pledged to protect.

But Europe no longer seems to care about either humanitarian or environmental arguments. The region is solely focused on its own vague, unproven food safety concerns.

The Financial Times recently noted a favorable report by a British House of Commons committee on the safety of genetically modified foods, saying it brought "a refreshing note of reason to a debate too long dominated in Europe by sensationalism and hysteria." *The Times* added, "Authorities have been too ready to ban products with no proven health risks, just because they arouse popular suspicion."

Europe started crying "Wolf!" about food safety 20 years ago, when it began worrying that the hormones being developed to stimulate growth in cattle might increase its expensive beef surplus.

Then, without any scientific evidence of consumer risk, the European Union banned meat growth hormones. (EU farmers didn't mind. Residues found in meat show that they use unregulated growth hormones bought on the black market.)

The EU then began offering special subsidies for organic food that gave EU bureaucrats visions of restoring the traditional small peasant farms. Organic yields were also lower, which meant less surplus food had to be subsidized for sale on the world market.

The cost of this "surplus reduction," however, was erosion of the public's confidence in modern farm inputs. Because the governments that made up the European Union said organic food was safer, European citizens came to feel the same.

Codex Alimentarius, the international food safety referee based in Rome, has blown the whistle on the EU's meat hormone ban. It formally concluded that the EU has shown no valid evidence of risk from meat hormones.

The EU's first response was to demand more time. This spring, the Codex and the World Trade Organization demanded action instead.

So the EU called a shameless press conference where a couple of captive scientists talked about findings of hormone-related cancer risk. The scientists refused to talk about the level or probability of the risk, which almost certainly means the "risk" is less likely than a person being hit by lightning twice.

Meanwhile, countries that use pesticides have average lifespans 20 years longer than those that don't and the IQ scores of American kids have continue to trend upward throughout the pesticide era.

Western Europe is essentially asking the rest of the world to reject science and accept its emotion-based regulation of food production designed to protect a set of EU farm subsidies that don't even keep Europe's small farmers on their farms.

However, the enormous human and environmental costs of banning modern farm inputs will be paid in the Third World. Michael Lipton of the Poverty Research Unit at Sussex University in Britain recently pointed out that 200 million people in the Third World suffer from vitamin A deficiency. Those people could be cured by a Rockefeller Foundation project to genetically engineer rice with enhanced vitamin A.

Fourteen million of those people are children who have severe eye damage due to the vitamin deficiency. "Many of them go blind," Lipton said. "Banning bio-engineered field trials globally would amount to blinding children."

The "harvest" of Europe's policies is easy to predict: increased hunger and malnutrition for tens of millions of people; sight impairment (and even blindness) for millions of vitamin-deficient children; and, the plowing down of at least 10 million square miles of wildlife habitat.

Western Europe should be ashamed of itself.

The Terrible Terminator Gene vs. Biocontrols

One of the big fears of eco-zealots is that a genetically modified organism will escape into the wild and alter the global ecology. It is then ironic that the same people who worry most about such biotech escapes are also generally in favor of using biocontrols instead of pesticides. (Biocontrols are living organisms, such as wasps, diseases or protozoa, which are deliberately introduced into an ecosystem to prey on a pest.)

Alien species are now widely recognized as one of the world's biggest ecological problems. Hundreds of species have become extinct because alien species took over their ecological niches. Yet people who fear pesticides and biotechnology are cheerfully urging us to release still more living organisms into the wild, to modify the ecosystems in unpredictable ways—rather than use tightly controlled sprays of safety-tested pesticides on our fields.

We have never recorded a single species extinction because of either pesticides or biotechnology. On the other hand, one biocontrol insect now threatens to cause the extinction of two rare native American thistles, and the pictured-wing fly which feeds on one of them. (The U.S. Department of Agriculture introduced the European flowerhead weevil to help control the European thistles that have proliferated in America.)

It is equally ironic that the people who say they are worried about biotech organisms escaping into the wild are also opposed to the "terminator gene." The "terminator" was conceived as a built-in safeguard against biotech escapes; it could make the offspring of the genes sterile. Consequently, they couldn't "take over the world." (Researchers at the U.S. Department of Agriculture developed the gene.)

But the people who say they want to conserve nature then told us that the "terminator" was unfair to the small farmers who traditionally save their own seeds. The "terminator" would force them to buy new seeds every year!

How does the terminator gene differ from hybrid seeds? Hybrid seeds don't breed true, so if farmers want the high yields from hybrid seeds, they have to go back to the seed company every year. Nevertheless, hybrid seeds have been one of the best investments farmers have ever made.

In the old days, every American farmer saved his own seed—but that meant yields couldn't ever get much better. Hybrid seeds have taken American corn yields from 25 bushels per acre to better than 300 bushels, and they're still trending upward.

Biotech companies also like the idea of a "terminator" gene, because it would ensure that they can get paid for their research. In 1998, Canadian farmers got upset because biotech companies were investigating whether some farmers had grown biotech seeds without paying the companies' fees. The farmers felt their privacy was being invaded. The companies thought hiring detectives was a contentious and expensive way to get paid for their products.

In India, farmers still riot over the unfairness of hybrid seeds—but that's one reason why Indian farmers average only one-third as much corn yield as U.S. farmers, and one-sixth as much sorghum yield.

The Urgency Test

Biotechnology is a tool kit. You can use the tools to cut off your hand, or to build a house. If you already have a house, you aren't that excited about the new tools. If you're hungry and cold in a cave somewhere, the tools become very important.

What we see around the world today is a wide variety of responses to biotechnology, mostly based on the urgency of needs.

- Europe sits smug amid its grain and meat surpluses, acting as though everyone in the world is eating well. Europeans see no reason why they should take any risks from genetically modified food. They essentially recommend that the whole world ban biotechnology in food.

- The United States is more ambivalent: well-fed, proud of its science, proud of its past role in helping feed the world, but wary of global overpopulation and still frightened of the idea of "unnatural" in its foods.

- China and Russia are enthusiastic about biotechnology in food production, because they still need more food.

- Most of the world's other governments are frantically trying to straddle the widening gulf between science and real needs on the one hand and urban public perception, framed by fearmongers, on the other.

- Greenpeace has an urgent need to attack biotechnology. It has recently gone through a major funding crisis and severe staff cutbacks; apparently people have decided that the whales are saved. Its campaign against biotechnology in food has put the organization back in the forefront of headlines, fundraising, and policymaking.

The Greenpeace position on banning biotech foods leaves the organization in an awkward position on conservation. The organization is basically saying its all right to go ahead with biotechnology in medicine, saving lives and lengthening life spans all over the world with gene therapy. But at the same time it is telling us not to use biotech in farming so we can feed the extra mouths from less land. That's environmentally unconscionable.

Scientists concede there is some risk potential in biotechnology. They say it's relatively small because each experiment must be approved and carefully regulated.

Without biotechnolgy, we will almost certainly see massive destruction of natural habitat in the next several decades. With biotechnolgy on the farms, the world may not need to lose any more wildlands at all.

Reprinted from BridgeNewsForum, August 27, 1999

Genetically Modified "Golden Rice" to Combat Third World Malnutrition

Two of the world's most persistent sources of human malnutrition may soon be overcome because Swiss researchers and the New York–based Rockefeller Foundation have developed new genetically modified rice plants.

The new plants have been engineered to help overcome vitamin A deficiency, which afflicts some 400 million poor rice consumers in developing nations and causes millions of children to go blind. This means biotechnology will make its first big contribution to human welfare through food production, not through medical research as most observers expected.

The same genetically engineered rice varieties will also provide hefty amounts of iron to combat the iron-deficiency anemia, a characteristic shared by many of the world's 4 billion rice consumers.

This means that the world will have to reevaluate claims by the environmental group Greenpeace that genetic engineering should be banned because of possible harm to health and ecology.

A team at the Swiss Federal Institute of Technology in Zurich led the breakthrough in rice breeding. The research network included Germany's University of Freiburg, Swiss pharmaceutical firm Hoffman-LaRoche, and the International Rice Research Institute in Manila, Philippines.

Funding for the research came primarily from the Rockefeller Foundation, with additional support from the European Commission's FAIR agricultural research program in Brussels, Belgium.

This is sadly ironic, since European governments are now backpedaling from biotechnology in the face of a media campaign by Greenpeace and the World Wildlife Fund in Washington to convince European consumers that biotechnology is producing "Frankenstein Foods."

The anti-biotech campaign has always run the risk of a biotech innovation coming along that would offer the world something obviously wonderful. With the rice success, that moment clearly has arrived. The success was greeted with full scientific honors at the 16th International Botanical Congress in St. Louis recently, and is being widely reported in the major scientific journals.

In nature, rice contains no vitamin A or beta-carotene, which the human body readily converts to vitamin A. The Swiss researchers inserted four enzymes that convert one of the rice molecules into beta-carotene.

Two of the enzymes were cloned from daffodils at the University of Freiburg, while the Swiss lab took the other from a bacterium. The researchers could tell when the transformation worked, because the beta-carotene literally turned the rice grains golden.

Rice normally contains a molecule called phytate that ties up 95 percent of the iron a person consumes. Overcoming the iron deficiency in rice diets required putting three new genes into the rice plants.

One gene codes for an enzyme called phytase, which breaks down phytate. A second gene codes for an iron-storage protein that doubles the iron level in the rice grains. The third gene provides a sulfur-rich protein to help the human digestive system absorb the newly incorporated iron.

The researchers then hybridized the vitamin A rice and the iron-enriching rice to create one nutritionally super set of plants.

The result of this elegant science is a hybrid rice that could overcome both vitamin A and iron deficiencies with just the 300-gram serving of rice per day that is already typical of Asian diets. This ultra-low cost miracle food would reach millions of poor and rural people who could not be helped through factory-fortified food additives.

The Manila-based International Rice Research Institute has taken up the task of crossbreeding the new rice strains into field-ready varieties for farmers. Gurdev Kush, the Institute's chief rice breeder, thinks this can be accomplished in two to three years, if regulators welcome the breakthrough and find no serious health or environmental dangers.

The real aim of the anti-biotech campaign is hostile regulation. If the "Frankenstein Foods" label sticks and spreads across Europe, it could delay the benefits of the science for a decade or more.

Biotechnology is surely the most powerful tool that was ever put in the hands of agricultural and medical research. Nobody can say, of course, that all biotechnology is "safe,"—that depends on what the researchers create. A sound regulatory system is thus vital to consumers and industry alike.

Will the European activists now try to demonize "golden rice"? Will the European media simply ignore the humanitarian potential of the science?

Or will the world rejoice in its new power to help poor and malnourished people?

Bob Lang, reprinted from CFGFI Quarterly, spring 1997

THE REALITY IS—Wildlife can't be saved if people aren't fed. If Borlaug's Green Revolution is losing momentum, the surest tool to acheive both is biotechnology in food production.

Notes

1. "Making Plants Aluminum Tolerant" *Science,* vol. 276, June 6, 1997.

2. Tanksley and McCouch, "Seed Banks and Molecular Maps: Unlocking Genetic Potential from the Wild," *Science*, vol. 277, August 22, 1997.

21

Subsidies or Free Farm Trade:
Which Will Win the 21st Century?

MYTHMAKERS SAY:

"[A] people unable to protect itself from starvation could not be protected from any other danger. The inescapable conclusion is that a government that makes itself the servant of international free trade is not protecting its land and people, it is protecting the supranational corporations that thrive at the expense of land and people."

Wendell Berry, Kentucky poet and philosopher, "Free Trade and the Environment," *The Amicus Journal*, Fall 1993

"The Business-as-Usual system . . . is clearly leading us rapidly to disaster. One apparent route of escape is . . . a 'soft path' system making use of . . . empathetic technologies. At the end of this path is a food supply made up of whole fresh foods produced by environmentally benign methods *somewhere in the vicinity of where they are to be eaten.*" [emphasis added]

Dr. Joan D. Gussow, *Chicken Little, Tomato Sauce & Agriculture,* New York, The Bootstrap Press, 1991, p. 34

"New international trade rules are threatening to shift food safety decisions away from Federal and state elected officials and regulators to international trade bureaucrats operating in secret in Geneva. . . . This would make it difficult . . . for citizens to have control over the safety of the foods they eat. . . ."

Public Citizen and the Environmental Working Group, *Trading Away U.S. Food Safety*, April 1999[1]

REALITY SAYS:

"Trade and environment are actually complementary . . . and could be much more so. . . . The one commodity-producing sector specifically exempted from the principles of the GATT . . . is agriculture. And, throughout this 40-year period, agriculture's environmental and economic problems have steadily worsened as the direct result . . . ."

Robert Repetto, *Trade and Environment Policies: Achieving Complementarities and Avoiding Conflicts*,"
World Resources Institute, July 1993

"Farmers in Australia and North America are not the only ones affected adversely by the protectionist policies of Western Europe and East Asia. . . . Farmers in other countries are estimated to lose $46 billion per year, while consumers in Western Europe and East Asia are worse off by more than $100 billion per year."

Kym Anderson and Rodney Tyers, *Agricultural Policies of Industrial Countries and Their Effects on Traditional Food Exporters*,
University of Adelaide (Australia), Working Paper 86–4, 1986

"Australia will tell its trading partners today that halving protection globally for agriculture, services and manufacturing would boost the world economy by more than $400 billion [per year]. Services would make the largest contribution, $250 bn, with agriculture savings projected at $90 bn. . . . The Australian study, *Global Trade Reform*, says a 50 percent cut in all barriers by 2015 would increase exports for all countries . . . the biggest winner in agricultural protection would be Japan, the country with the highest protection. Its gain would be $43 billion, while the report estimates the EU would save $13 billion."

"Trade Reform Promises $400 bn Bonus for World,"
Financial Times, May 26, 1999

The Price for Farmer Votes in the United States

Farm subsidies used be a standard policy for affluent nations. Only the countries of darkest Africa and poorest Asia didn't offer their farmers "something extra"—mainly through government-supported prices for their commodities. Farmers got subsidies because they were small, numerous, and admired.

Today, we are probably seeing the end of farm subsidies as we've known them. The simple reason is that they've become too politically expensive per vote. In 1933, an 80-acre U.S. farmer might harvest about 500 bushels of corn per year. A ten-cent increase in his corn price cost the government only $100—for a fairly large farmer—and meant quite a lot as he had few off-farm expenses. (The family raised its own food, draft animals, and fodder, and bought no fertilizer or pesticides.) These early payments were hidden in consumers' food costs, so the government didn't even have to pay the subsidy.

Today's 1600-acre corn farmer might harvest 288,000 bushels of grain. A price support which would gain the farmer comparable buying power to the 10 cents per bushel in 1933, say 50 cents per bushel, would cost the government $144,000 per farm vote.

There is no way to hide this in consumer food costs because farmers have become so price-responsive that a higher price support calls forth more production. The government must pay the cost of subsidizing the surplus into industrial uses or exports or make direct farmer payment that comes right out of the treasury's pocket.

And in Europe . . .

The EU is currently spending about $150 billion per year on its farm subsidies to about 11 million farmers, according to the Organization for Economic Cooperation and Development (OECD). That's $14,000 per farmer even though the European farms are much smaller than America's.

Critics say the EU is spending far too much on a policy that does not keep small farmers on their farms. (Twenty-five million European farms have disappeared since the policy was installed.) Worse, farm subsidies substantially raise taxes on urban workers and their employers which discourages off-farm job creation even as the ex-farmers come to town looking for work.

The European farm subsidies are already being reined in, primarily because of pressures from other countries. Both the EU's commitments in the World Trade Organization and its Blair House agreement with the United States require it to lower its farm price supports and subsidize less of its farm surplus into third-country markets. This means that farmers' benefits from the EU's traditional price supports are virtually guaranteed to decline. (The EU is shifting heavily to direct payments, mostly to smaller farmers who produce organically and/or farm difficult lands.)

The EU farm policy hasn't even preserved the traditional rural landscape. The subsidy money goes primarily to larger farmers, who often use the money to buy out their smaller neighbors. The new landowners then remove the ancient hedgerows and stone fences that used to separate the fields. As a result, the famous Loire Valley of France looks like Illinois (except that the unfenced fields that stretch to the horizon are growing wheat instead of corn).

The EU knows it is facing drastic changes in its farm subsidies because its fear of the economic and political chaos in Russia has made it a political imperative to take in East European nations as buffers between Russia and Western Europe. The EU is likely to take in 8 to 10 eastern nations in the next several years. This will almost certainly mean taking in more than 6 million additional farmers and more than 20 million hectares of farmland on which yields could be readily doubled with high-yield inputs. Extending the current Common Agricultural Policy to the additional farmers and farmland could double the already-crippling governmental costs of the EU farm policy within a decade. Neither the EU's current members nor the impoverished governments of Eastern Europe can afford the costs.

Are Asia's Barriers Crumbling?

For many years, the global pattern of high national-farm price supports and pervasive farm trade barriers seemed set in stone. Now, suddenly the stone may be crumbling.

The United States began to phase out its old farm subsidies in 1996 as part of a budget-balancing effort, and probably lacks the federal tax funding to reinstitute the system. (The unfunded, congressional obligations for Social Security and Medicare after 2010 add up to nearly 20 trillion dollars.)

The Cairns Group of nations (so-called because they first met in Cairns, Australia) includes 15 countries that are eagerly pursuing farm trade liberalization. The group includes many influential nations including: Canada, Australia, New Zealand, Argentina, Brazil, Thailand, and Hungary.

The strongest reason for believing that the subsidies will give way to farm trade is that both the Asian nations and the potential farm exporting nations need the trade urgently. In Asia, China has almost no additional land or water it can dedicate to farming. In India, the need for more livestock and poultry feed is pushing its farms closer and closer to the tiger preserves—so more farmers are being eaten, and more tigers are being shot as "man-eaters." Indonesia is clearing tropical forest from poor-quality land to grow chicken feed on its outer islands; the land has soil erosion risks at least ten times as great as the farmland in Iowa.

Subsidies have been the biggest blocking factor in farm trade negotiations. As long as American farmers had price supports and payments, they showed little interest in free trade opportunities overseas. Farmers in countries that should import feed, meat, and ice cream powder to help meet their consumers' diet aspirations in the future have used the fact that high-yield export farmers are subsidized to block farm imports.

American agriculture already earns $50 to $60 billion per year from farm exports; that should double within the first decade of free farm trade—and farmers would earn the extra income essentially with the land, water, machinery, and labor already on their farms. In nonfarm trade, the GATT and the World Trade Organization have cut the average tariff from perhaps 40 percent in 1947 to about 4 percent today. The value of nonfarm goods traded has increased more than fifteenfold as a result.

Meanwhile, the average "tariff" on farm products is probably still more than 50 percent. (It's hard to know, because much potential farm trade is simply banned, or numerical quotas keep the amount of imports tiny in comparison to consumption.) As a result, farm trade is basically stagnant. For example, the world is trading only about 200 million tons of grain per year, about what it traded 20 years ago and only about 10 percent of world consumption.

The increase in net farm income from free trade should be enormous for most farmers in most nations as each farmer focuses on the commodities in which they have the highest prices and lowest costs.

That's fortunate, because the outlook for farm subsidies is bleak. Farms in the First World are no longer small, they are no longer very numerous, and farmers are no longer much admired.

Reprinted from The BridgeNews Forum: November 13, 1998

Aid to Farmers: A Poor Substitute for Farm Trade

Just before the recent U.S. Congressional elections, American farmers got a present of about $6 billion in cash from the federal government. Congress cited this year's low world farm prices in generously deciding to give the farmers the aid.

Since U.S. farmers have about $1200 billion invested in land, livestock, machinery, and buildings, the $6 billion "gift" represents a tiny, onetime return on assets of about 0.5 percent.

What Congress didn't give farmers was free trade in farm products, which is the only real solution to the farm price problem.

Ironically, current world farm prices aren't even "low." They're at a depressed level, but one we've come to think is normal. Markets around the globe are closed to agricultural products while, at the same time, Europe subsidizes exports.

Because of these trade barriers and this export dumping, the current low world prices are exactly what farmers can expect again next year.

Will the U.S. Congress ride to another inadequate rescue then? And every year after that?

Free trade would raise the world market's farm prices by about 25 percent to 35 percent. In some heavily depressed commodities, like dairy, prices might rise 50 percent.

Free trade would allow export farmers to sell all the output they could cost-effectively produce at the higher prices. This would be worth perhaps $50 billion to $60 billion per year to U.S. farmers every year for as far as we can foresee.

Free trade would also be a major boon to Australian and Ukrainian wheat growers, Polish dairy farmers, Brazilian sugar growers, Thai poultry producers, and so on around the world.

Even subsidized farmers are losing now. They've had to buy expensive marketing quotas or land that is more expensive because of the subsidies. Farmers who were in business when the subsidies were installed got a windfall gain, but they're long retired.

Today's farmers aren't getting any gift from the politicians; they just have to support a higher cost structure.

And over coming decades, they'll be shut out of the biggest market growth in farming history.

Congress didn't even seek free trade in farm products. It failed to give the U.S. Special Trade Representative the authority to negotiate farm trade liberalization in next year's World Trade Organization talks. Congress voted its little onetime "gift" instead.

Farmers are caught in the "poor-but-grateful" political trap.

It's the same game the politicos play with old-age pensions. They've largely refused to let people invest a part of their earnings in tax-free pension funds. Thanks to the miracle of compound interest, those funds would turn into really attractive retirement incomes over the years.

Instead, the politicos have trapped people in a pension system that earns no annual return on the investment at all, because nobody invests anything.

The politicos pay today's retirees with the taxes on today's workers. Then the workers are supposed to feel grateful when the government gives them a pension that is more a token than a safety net.

Today, such pension schemes are rapidly going bankrupt because so many of us are living so long. Legislators know they can solve the problem in an instant—whenever they are willing to give up the "grateful voter" game and let people really invest in their pensions.

The politicos' problem is that voters who earn their pensions might not feel grateful to their legislators.

Farmers are trapped by the same poor-but-grateful strategy. The politicos are trying to keep farmers on the public dole with relatively small doses of federal cash rather then letting them march off into the sunset with increased profits earned from more farm exports.

The world will demand three times as much farm output over the next 40 years. Densely populated Third World countries will have trouble producing the food they'll need.

Members of Congress once stated that they were ending the price supports and cropland diversion program because the United States couldn't gain export markets so long as it was subsidizing its farmers.

Now, legislators say they can't seek free trade for farmers because U.S. unions are afraid that free trade will export nonfarm jobs. But if this is the real concern, Congress could simply give the Special Trade Representative authority to negotiate free trade in only farming.

> Not only would this make America's farms more profitable, it would add a couple of hundred thousand blue-collar nonfarm jobs as well. Those jobs couldn't be exported unless someone can find a way to export America's prime cropland, climate and big rivers.
>
> The political objections to free farm trade are fraudulent. Clearly, the politicians are more interested in buying the votes of farmers than allowing them to earn true prosperity.
>
> When will farmers stop settling for being "poor-but-grateful"?

The True Mission of Farm Trade

. . . is to help us supply the world's food needs with as little environmental and economic cost as possible. The eco-activists' confusion over what they really want and how to achieve it in agricultural trade mirrors their more basic confusion in trying to save wildlife with low-yield farming.

If the environmental movement understood agricultural production systems, they would be on the side of both high-yield farming and free trade in farm products.

The Shenandoah and Corn

Let's review the comparison between the steep, rocky land in my Shenandoah Valley and the flat rich plains of the Corn Belt:

- Corn yields in the Shenandoah are only about half the yields in central Iowa and Indiana.

- Any farmer trying to till land in the Shenandoah is going to hit lots of rocks, and even some shale ridges. He will have to travel more slowly, stop more often—and will still break more expensive parts on his machinery.

- To prevent heavy soil erosion losses, my Shenandoah farmer needs to plant on the contour. That means a lot of surveying, plus some tricky work with the tillage equipment and the planter. His labor costs are higher.

- The Shenandoah farmer should also alternate strips of clover or other "filter" crops to stop any soil that erodes from the corn rows from moving very far. The green strips lower the overall average output of the field. Again, that means less output per acre of land.

All in all, both the cash outlays and the environmental costs of corn production can be twice as high in the Shenandoah as in the Corn Belt. That's why the Shenandoah grows virtually no corn for grain. (It does grow some corn silage for its dairy cows on level pieces of ground.)

The Valley's marginal land has been returned to grass for beef and dairy cattle and trees for lumber and parks. Soil erosion has been substantially reduced, food costs have also been reduced, and the nation has more wildlife habitat.

If We Let the Tropics Grow Our Sugar

Comparative advantage is one of the basic truths of economics:

- If country A has iron ore and lots of coal, it can readily be a steel producer.

- If country B has lots of low-cost labor, it may well become a garment producer, because both capital costs and wages tend to be low in that industry.

- If A needs clothing and B needs steel, they can trade and both be better off. Plus, B doesn't have to invest resources in costly steel plants that will operate at an economic loss.

One of tropical farming's big comparative advantages is sugar. Tropical land can produce sucrose through sugarcane nearly twice as efficiently as temperate-zone farmers. Beet sugar in the United States and Europe yields only about 60 percent as much sucrose per acre, even though the beets sop up more fertilizer and need more pesticides.

Equally important, we haven't found much else that can be grown efficiently on the sugarcane land. We could grow a whole roster of grains and oilseeds on the high-quality temperate land that sugar beets need. But, because of national farm self-sufficiency, temperate-zone sugar beets provide nearly one-third of the world's sugar.

What would happen if we let the tropics grow our sugar?

- The world would be able to farm less-fragile land, suffer less soil erosion, and use fewer farm chemicals.
- Western Europe could shift nearly 9 million acres of its best farmland to better uses, probably feed crops.
- Tropical sugar-growing countries would have somewhat more foreign exchange with which to import wheat and other food needs they can't grow efficiently.

In case after case, free trade in nonfarm products has vastly increased economic growth and human well-being. Free trade is the world's most successful strategy for making the Third World rich enough to be environmentally sensitive and responsible, and to be good customers for the First World's high-tech products. By comparison, foreign aid has been a miserable, corrupting failure.

Food Self-Sufficiency: Recipe for Hunger?

In 1994, Japan needed to import more than 1 million tons of food-quality rice, due to extremely poor weather following a near-failure of its 1993 rice crop. There weren't enough high-quality rice stocks in the world to supply this demand. So, the Japanese had to make do with industrial-quality rice that rice consumers in other countries didn't want.

How did a country with the purchasing power of Japan land in such a predicament? Because of its own policies of rice self-sufficiency.

For 40 years, Japan adamantly opposed rice imports. Japan offers its rice growers prices as much as ten times higher than world market prices to ensure its self-sufficiency. It spends billions of dollars stockpiling Japanese rice. But even all this expenditure wasn't enough. When the Japanese ran into two bad rice harvests in a row, they needed imports. And because they've been so rabid against imports, nobody else had been producing or stockpiling rice for a possible Japanese sale.

Thailand could have done it. Instead, they planted more corn, because the Japanese said there would be no market.

America could have produced more rice—but instead we severely limited U.S. rice acreage, to avoid stockpiling more costly surplus rice in government storage bins.

Consequently, Japanese politicians had the "interesting" task of explaining to the Japanese consumers how their policy of self-sufficiency was "protecting" the country.

In 1987, I was in Finland, during the wettest crop year in its history. For 40 years the Finns had trumpeted the importance of growing their own grain. I watched combines trying to harvest wheat with water standing in the fields, rain still coming down, and the combines throwing rooster-tails of water from their wheels. But it was September at the Arctic Circle, and the last chance to save the Finnish wheat crop. Naturally, most of the wet grain sprouted in storage, and was ruined.

Quietly, the Finnish government went out and bought two-thirds of a year's wheat supply in the world market, for half the price it paid its own farmers, from a variety of willing exporters. Then it went back to declaiming the virtues of its high-cost food self-sufficiency.

It would have served the Finns right if there had been an international grain monopoly that charged them double the fair price. After all, they had been charging their voters about $4,000 per family per year to subsidize Finnish farmers.[2] Instead, there was a highly competitive world grain market trying to sell wheat to any and all comers at attractive prices.

All Countries Experience Crop Failures

Any country can have a crop failure, even big agricultural countries like the United States and China. When that happens, the country's food security depends on the ability to import from another country that had good weather and thus has a surplus for export.

No one country's farm output is anywhere near as stable as world-wide production measured as a whole. In almost any year, at least one country will have a crop failure and another will have a bumper crop.

Under a free-trade regimen, we wouldn't need to keep extra grain expensively sitting unused every year in every country's storage bins. Trade evens out the *inevitable unpredictableness of* weather.

What About Wars?

The most common rationale for the self-sufficiency approach has to do with historical experiences of wartime. Europe and Japan, for example, say they were hungry at the end of World War II, and must produce their own food because of a deep-down fear of being hungry again.

But would Germany and Japan have been any *less* hungry in 1945 if their agricultures had been twice as big in 1937? No. The fighting in World War II would still have devastated the farms.

In fact, there was no widespread famine in Germany or Japan at the end of the war. As quickly as the fighting ceased in a region, the Allies brought in food for the populace. The German and Japanese peoples were saved from famine in 1945–46—by imports!

FAMILY FARMERS AT WORK—America's family farms are fewer and larger but remain the backbone of our agriculture and model for the world.

What About Embargoes?

But (the argument continues) what if other countries in the world embargo our country? What if they won't ship grain to us?

History rejects that argument. When President Carter embargoed grain shipments to the USSR to protest the Soviet invasion of Afghanistan, the Soviets simply turned to Argentina and immediately replaced all the embargoed American grain contracts, without so much as a price increase. (The Argentines were hoping to build future business with the USSR.)

The Nixon embargo of soybeans backfired even more explosively. Americans were afraid of a protein meal shortage that would raise meat prices. But the Japanese were even more fearful that the U.S. embargo would create shortages of their traditional soy foods. There's no comparison in the emotional intensity of a food shortage versus a feed shortage.

Japan immediately launched a research and development partnership with the Brazilian soybean industry, which has helped Brazil raise its soybean output from around 2 million tons in 1972 to more than 20 million tons per year in the 1990s.

The entire civilized world embargoed Iraq before Desert Storm in 1990. However, Iraq's consumers suffered no major food problem until after the United States started bombing its roads and bridges. The Iraqis were very adept at smuggling food into their country. After the attacks on their transport system, however, they could no longer distribute it. The embargo, alone, had little effect.

MYTHMAKER:

"Their proposed revisions in the General Agreement on Tariffs and Trade are an attempt to place the agriculture of the world under the rule of the same economic forces that have already virtually ruined agriculture in the United States."

Wendell Berry, "Free Trade and the Environment,"
Amicus Journal, Fall 1993, p. 31

Who Stockpiles Food?

If a country is truly concerned about food security, there is a better way to achieve it than expanding the local farms. The answer is to stockpile food. Put grain in a silo. That way it's not going to be wrecked by a rainy summer, or a drought, and probably not even by a war.[3] How many countries stockpile food for security purposes?

One. India. India does it because it suffers a monsoon failure about one year in five. In 1987, it was able to replace practically all of the losses from the worst monsoon failure in a century—from its own storage stocks.

U.S. and EC grain stockpiles, by contrast, have been for politics, not food security. The stocks have been vastly larger than any food security need.

Japan, the most import-dependent feed buyer among the world's countries, keeps one month's supply of feed imports on hand. It keeps another month's supply on vessels headed toward the islands. That's their cost-effective food security policy.

Why do so many other countries "protect" their agriculture for "food security" when that lowers real food security? Because of farmer politics.

The Fallacy of Protest-Based Farm Policy

Farmers in every country hate to see food imports. The last year I was at the State Department, we saw violent demonstrations against apple imports in both Sweden and Taiwan, both "highly civilized" countries. And apples aren't even a food-security staple!

So long as farmers constitute a major proportion of a country's population, and the international farm trade rules are lax enough to let politicians buy farm votes with hidden "food taxes" on consumers, farm trade will be suppressed.

During the latter stages of the GATT Uruguay Round trade negotiations, French farmers protested the idea of freer farm trade by blocking major highways with burning tires. They also stomped

McDonald's hamburgers (full of French beef) and smashed Coke machines (full of French sugar) in their anger at America's insistence on liberalizing farm trade rules.

In a further irony, the French government announced that its farmers' incomes would be down 14 percent by 1996—primarily because of the subsidy cutbacks mandated by the EC's own budget deficit. France's commercial farmers suffered virtually the entire decline in farm income because the government made special payments to the "politically correct" small farmers. The French commercial farmers now have the worst of both worlds: Their subsidies are being cut back, and they do not have the trade access to help meet Asia's rising food needs.

Think about a world without GATT (now called WTO) with all of its trade negotiations carried on by similar protests. Detroit autoworkers might have randomly smashed foreign cars with sledgehammers—and the clean-burning Honda engines might not have made their pro-environmental impact on America's air quality. Nor would we have gotten the competitive pressure that brought us disc brakes and automatic braking systems from Europe.

This article originally appeared in *Feedstuffs,* Opinion, October 5, 1998

Would Action Take U.S. Agriculture Back to Great Depression?

North and South Dakota's farmers and congressional delegates are probably not deliberately trying to drag American agriculture back into the Great Depression. But their answers to current low farm prices might achieve exactly that.

The states are trying to unilaterally stop grain and livestock imports from Canada. Dakota's farmers and ranchers have been blockading Canadian trucks and trains at the border. North Dakota State troopers are even stopping Canadian grain trucks and demanding proof their loads are free of wild oats. (Nobody's grain is free of wild oats.)

The governor of South Dakota ordered his people to start handing Canadian truckers written warnings: go home or find an alternate route around South Dakota. North Dakota's lone congressman, Earl Pomeroy, called the harassment "a very clear sign of how disgusted we are with Canadians dumping wheat into our markets . . ."

It's great rural politics, but a ruinous farm policy. North Dakota farmers and ranchers might look around them and ask themselves how North Dakota's 650,000 residents could consume its farm output—including 900 pounds of beef and 10 bushels of grain per person per year.

Nor do the states next to the Dakotas need Dakota grain and meat. Without trade, the Dakotas would drown in their own farm surplus. If the United States can violate the North American Free Trade Agreement (NAFTA) by blocking Canadian farm products, what's to stop other States from abrogating the Interstate Commerce clause of the Constitution? What if Minnesota and Nebraska denied entry to farm products from the Dakotas?

Remember that the Smoot-Hawley Tariff in 1929 triggered the Great Depression. Smoot-Hawley was designed to "protect American jobs" by raising the duties on foreign goods to as high as 50 percent. (The current world average is 4 percent.) The U.S. tariff hike touched off a worldwide tariff war. International trade came to a shuddering halt. That collapsed the world's stock markets and the Great Depression was on.

Heaven knows, the farmers in the Dakotas have been through a lot. Many of them missed the high prices of the past two years because droughts and floods made much of the region an official disaster area. Now ample crops mean low prices with no government protection.

The question, is not whether North Dakota farmers are hurting, however. The question is what will make their situation better.

What if Mexico also abrogated NAFTA and cut a farm import deal with Argentina? Mexico will import eight million tons of grain this year, along with three million tons of soybeans and soy products, 250,000 slaughter cattle, 200,000 tons of beef, 47,000 tons of pork, and 200,000 tons of poultry. They will buy virtually all of this from the United States and Canada—because NAFTA gives us duty-free access to Mexico's 90 million consumers and its recovering economic growth. Does North Dakota plan to repay the rest of U.S. agriculture for loss of the Mexican market?

Ironically, this is the first moment the world has ever really needed Dakota farm production. The railroads didn't reach the region's rich farmlands until the 1870s. Then the only markets for Dakota grain and meat were the cities of the eastern U.S. and Europe that were already supplied from farmlands closer by.

The railroads had suddenly opened 250 million acres of cropland, plus more than 500 million acres of rangeland, west of the Mississippi. (Along with additional "new" land in Argentina and Australia.) The steam locomotive was followed by the gasoline tractor, chemical fertilizer, hybrid seeds, and modern pesticides. All raised yields. All added to the world's farm "surplus."

The world didn't need all that extra farm output—then.

Only in the 1990s have rising incomes caught up with the world's ability to produce farm products. As recently as 1980, the world had no more than 800 million consumers who could afford meat, milk, eggs, and cotton wardrobes.

Today, however, perhaps 2 billion consumers can afford high-quality diets. A high proportion of them are in China, where per-capita purchasing power has soared from less than $100 per year in 1978 to more than $3,000 today. When the rest of Asia regains its economic growth trend (it will happen in the next two years) then another billion consumers will soon join the ranks of potential grain, oilseed and meat buyers.

Many of these newly affluent consumers will be in densely populated, land-short countries that should welcome food imported from the Dakotas.

At the moment, however, the governments in India, China and Egypt are all playing the same brand of rural politics so popular in North Dakota.

They're proclaiming that "China will produce its own food." They're declaring that "foreign foods are dangerous." They're claiming that food imports would "destroy our family farms."

Dakota farmers want their Federal farm programs back—but that will mean no exports. (The world won't let us have both.)

The old subsidies gave U.S. farmers a shrinking $12 billion per year. The exports are already earning more than $50 billion per year for American agriculture—and that could readily become $100 billion with free farm trade.

In a world that will demand three times as much farm output by 2040, there will be cash customers for North Dakota's farms and ranches. But North Dakota needs to understand that those customers are not in North Dakota, or even in America. Their customers are waiting in densely populated countries where the alternatives to food and feed imports are cutting tropical forests or building huge, expensive greenhouses.

Prosperity for North Dakota's future means accepting some Canadian imports under NAFTA. (Neither U.S. nor Canadian farmers can make much selling across the border. We're too much alike, and we're doomed to get the same real prices.)

Future prosperity means supporting farm trade liberalization at the World Trade Organization negotiation next year—and fast-track for the

President to negotiate the treaty. (No other country wants to negotiate twice, with our diplomats and then with our Senate.)

Dakota farmers and ranchers aren't blockading Canada; they're blockading their own future.

Problems of Farm Self-Sufficiency

Indonesia: Food self-sufficiency means cutting down tropical forest to grow low-yield soybeans—when the United States and Argentina have millions of acres of soybean land already cleared and able to supply Indonesia's soybeans at lower cost.

Saudi Arabia: Food self-sufficiency means using petroleum-powered pumps to raise fossil water from as deep as 3,000 feet to grow gritty wheat on the desert. Yet Argentina is producing only half the wheat it could cost-effectively grow on the rich, level rain-fed Pampas.

Japan: Farmers' protests against rice imports have kept virtually all of that country's land in rice paddies, with no room for housing, parks, or recreation. (Tokyo's superhighways have been built *over its rivers* because the land area was so densely populated and legally entangled.)

China: Food self-sufficiency will mean putting enormous amounts of chemical fertilizers on China's limited supply of farmland. As recently as 1980, China was using about 14 million tons of chemical fertilizer. The government's forecast for 2000 is 150 million tons. Farm self-sufficiency for the years after 2030 would likely mean more than 300 million tons of fertilizer—on the same farmland base![4]

India: Food self-sufficiency will mean building lots of dams, irrigating lots of additional land, and putting on huge amounts of fertilizer (as in China). It may also mean pushing crops and pasture out onto the already-scarce wildlife habitat of species like the Bengal tiger and the barking deer. Milk self-sufficiency for India already means compensating for the shrinking pastures available by relying still more heavily on crop residues as virtually their sole source of feed. These crop residues should actually be going back onto the cropland to preserve its fertility and tilth. To the extent that farmers cannot compensate for the "stolen" residues with green manures and fertilizers, India risks mining some of its soil fertility. In this era when dairy products can readily be imported (as dry milk, as concentrated milk, and/or as butter and cheese) India should at least examine the trade alternative.

Brazil: Its wheat crop has dropped sharply since the government ran out of money to pay wheat subsidies. The crop is down from a peak of about 6 million tons per year in the 1980s to a little over 2 million tons at present. Now Brazil is importing the rest of the wheat it needs from next-door Argentina, where the yields are twice as high and the per-unit costs are low.

MYTHMAKER:

"The real aim of American policy is to use food as a weapon in post-Cold War geopolitical strategy. They want to control the markets on grain and oilseed exports both through their own produce and that of Third World countries, whose even cheaper exports they control in various ways; for example, the huge soybean plantations in South America, which are owned or controlled by U.S. companies."

French farmer, quoted in *Whole Earth Review*, Winter 1993[5]

Reality Comment: No one owns or controls the world's agriculture. It is driven along its erratic path by weather, technologies, and competition among millions of farmers, changing consumer demand, and the policies of 100 different governments. What the French farmer really wants is to keep his high EU price supports safely behind the EU's farm import barriers. The *Whole Earth Review* wants to preserve Asia's traditional little peasant farmers—the same peasant farmers who have been destroying the tropical forests for low-yield slash-and-burn farming.

The Imperative of Good Public Nutrition

Farm trade has been one of the important factors in improving American and European nutrition in the past two decades. It will be equally important for nutrition in Japan and other Asian countries in the two decades ahead.

USDA data indicate that farm trade has helped raise America's consumption of fruits by 25 percent and vegetables by 11 percent since 1970. These gains have reversed a long-term decline in U.S. per capita produce consumption.[6] Years ago, we didn't have many fresh fruits and vegetables in our stores and restaurants in the off-season. Instead, winter meant wilted cabbage, withering apples, and, eventually, canned vegetables. Trade with Latin America has been a major factor in the wide variety, low cost, and attractive array of fresh produce being offered to the American public year round. Lettuce and melons from Mexico, peaches and apples from Chile, tiny cocktail vegetables from Honduras, and flash-frozen asparagus from Guatemala, are but a few.

All have helped stimulate American fruit and vegetable consumption. Such imports have helped keep the salad bars and fruit bowls stocked attractively and cost-effectively, even during the winter.

MYTHMAKER:

"Great advantages will be reaped (from GATT) by agribusiness transnationals which will take over vast tracts of Third World land to grow cash crops, while peasants will suffer and forests be destroyed. Of course, they will use ample amounts of toxic pesticides and artificial fertilizers to grow this food."

Helen Caldicott, Australian antiwar and environmental activist, *If You Love This Planet*, 1992

Reality Comment: The "corporate farming" threat is a scam. Corporations don't do well at the highly intensive 24-hour-a-day management needed to contend with the biological complexities of farming. If farm trade is freed, the most vigorous and effective of the farmers already on the land will expand their production to meet the larger demand from people in other countries for the commodities that they produce most efficiently. These farmers will get higher prices for more production at lower unit costs. That means more profit for most of them. (That still won't keep a lot of small Third World farmers from taking higher-paying jobs in town, leading to larger Third World farms.)

Figure 21.1

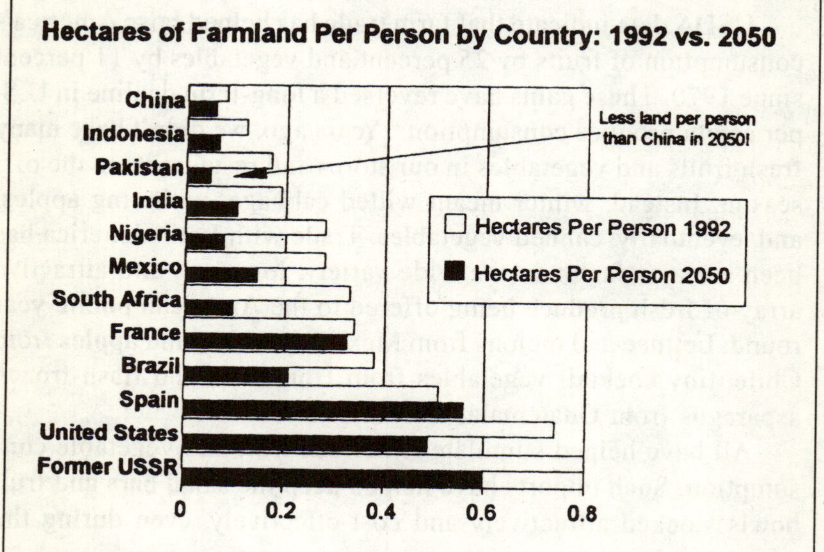

Source: Center for Global Food Issues computation

The Challenge of Population and Farmland Distribution

The biggest reality of world food production is that farming resources aren't spread equitably for the world of the 21st century.

By 2050, for example, Asia will have nine times as many people per acre of farmland as North America. Moreover, Asia is already using its farmland potential more fully than North America. Asia's development of its wet rice culture was one of the early triumphs of human knowledge. But it has left Asia with such a dense population that the 21st century food challenge can only be met by huge investments of capital and chemicals—or by trade.

A policy of nation-by-nation food self-sufficiency would foolishly leave millions of acres of the world's best cropland idle—

while encouraging Asian farmers to plow down every scrap of land they can reach. One dam now under construction, (Three Gorges) on the Yangtse River in China, will force relocation of at least 1 million people!

The slogan of the environmental movement is "Think globally, act locally." Yet eco-activists opposing farm trade are obviously thinking locally, without much regard for the world as a whole.

Anyone who advocates farm self-sufficiency for environmental reasons is either ignorant of the world's farm resource distribution—or selfishly worrying about his own neighborhood while he endangers huge tracts of wildlife and acre-feet of fossil water in other countries.

That is the antithesis of real environmentalism.

Keeping Our Food Safe—with Trade

During the Uruguay Round of trade negotiations, Public Citizen and the Environmental Working Group published a report entitled *Trading Away U.S. Food Safety*. The 86-page document did its best to frighten people and their legislators into rejecting freer trade in farm products:

"The Uruguay Round of GATT (now the WTO) Promotes Downward Harmonization of Food Safety Standards. . . . Domestic standards that do not conform to international ones must pass a battery of Uruguay Round tests in order *not* to be considered an unfair trade barrier. Specifically, food safety standards that do not conform to Codex (Alimentarius) standards:

- Must be based on scientific principles;

- Must not be maintained without sufficient scientific evidence;

- Must be based on a risk assessment;

- Must not achieve a higher level of public health protection than the Codex standard unless the regulating country . . . has scientific justification for concluding that the Codex standard will not achieve its level of protection;

•Must use the least trade-restrictive means of achieving the country's goals."

But why would an environmental group—or *any* group—oppose such a sensible set of principles?

The requirement for scientific data is basic. There's no way to assess public risk at all without scientific evidence. Without it we are forced to rely on public emotion alone, too often fanned to white heat by misinformation. *No* safety standard would be "safe" enough. Even the environmental movement, the reader will recall, has based its opposition to pesticides on the scientific data from the high-dose rat tests.

The WTO even allows member countries to have higher standards, so long as they make a scientific showing that the WTO's standards wouldn't achieve the country's level of protection.

Then the "environmental" paper gets to its real agenda:

> The Uruguay Round science and risk assessment criteria could have a severe impact on domestic food safety standards that are based on the precautionary principle, which supports the prevention of certain public health risks, or those that are based on consumer preferences in the face of such uncertain risks.

In other words, the WTO makes it harder for eco-activists to pass legislation through scare campaigns. WTO actually forces them to develop a scientific data base to support their contentions. The past campaigns against DDT and Alar, the current campaigns against atrazine, chlorine, and nitrogen—all would have flunked the WTO "science" test.

Of course, if the environmental movement comes to understand the importance of high yields and trade in saving wildlife, they will not want to ban useful pesticides or block farm trade. They will not want to fan public fears over foods that are actually safe. And, if foods or farming systems are truly dangerous, it will be a simple matter to bring forward the scientific evidence.

Nor will First World farmers want to hang onto subsidies and trade protection once they understand the combination of environmental challenge and economic opportunity that awaits them.

Once the importance of high yields and trade is clear, the only likely opponents of free farm trade will be the farmers in the newly industrializing countries, who do not yet understand the importance of preserving wildlife habitat and wild genes. Both the First World and their own political leaders will need to help them understand—for the world can reasonably take no other path.

Notes

1. Public Citizen and the Environmental Working Group, *Trading Away U.S. Food Safety*, Washington, D.C., April 1993, p. 1. Public Citizen was founded by Ralph Nader. The Environmental Working Group is relatively new and describes itself as an "environmental research" organization.

2. David Dodwell, "West's Farmers Reap $354 bn in Subsidies," reporting on an annual OECD study, *Financial Times*, June 3, 1993.

3. I used to be involved at the State Department in occasional "nuclear war" games. I could always astonish the other players by noting that most of the grain stored in the Commodity Credit Corporation's surplus silos would be usable after a nuclear exchange—just by peeling off the radioactive "hot layer" on the outside.

4. Bruce Stone, International Food Policy Research Institute, "Chinese Fertilizer Development 1990: Status and Prospects," presented at the 13th Phosphate-Sulphur Symposium, Boca Raton, Florida, January 22, 1991.

5. "Sustainability vs. Agribusiness-as-Usual in France," *Whole Earth Review*, no. 81, Winter 1993, pp. 42–47.

6. Judith Jones Putnam, *Food Consumption, Prices and Expenditures*, 1967–88, Statistical Bulletin No. 501, Washington, D.C. U.S. Department of Agriculture, Washington, D.C., May 1990. See also Stephen Hiemstra, *Food Consumption, Prices and Expenditures*, Agricultural Economic reports No. 138, USDA, July 1968 and USDA/ERS, *Vegetables and Specialties*, TVS-249, Washington D.C., November 1989.

22

New Excuses for Bad Regulations

MYTHMAKERS SAY:

"The most important thing is to reduce the overall use of pesticides. By doing that, we will automatically reduce risks and we won't have to spend all this time worrying about lots of complicated things."

EPA administrator Carol Browner, interviewed in *E, The Environmental Magazine*, December 1993

"In an interview, the [Natural Resources Defense Council's] chief lobbyist, attorney Janet Hathaway, described for me NRDC's ultimate goal. If pesticide residues can be detected on food, even in 'minute amounts,' she explained to me, and if a massive dose of that pesticide 'causes tumors in any laboratory animals, then it should be illegal.' [NRDC staffer Lawrie] Mott told me that the NRDC would ban all such chemicals 'no matter how great their benefits are.'"

Robert J. Bidinotto, *Readers Digest* staff writer, 1993

"A woman who says she is sensitive to lawn chemicals sued a homeowners' organization in her condominium complex Tuesday, charging its use of pesticides violated her fair housing rights. . . .

"The lawsuit, which seeks damages of at least $300,000, names Country Creek, four of its past and present officers and a grounds management company. . . .

"The Trial Lawyers for Public Justice, a Washington-based public-interest law firm that filed the lawsuit on Ms. Lebens' behalf, said it hopes the case will establish chemical sensitivity as a legal disability . . .

"In March 1994, [the Department of Housing and Urban Development] determined that there was reason to believe that Ms. Lebens was discriminated against. But the Justice Department, which would have had to pursue the case, declined to get involved."

From a news story in the Staunton, Virginia, *Daily News Leader*, July 20, 1994

REALITY SAYS:

"Predictions of ecological doom, including recent ones, have such a terrible track record that people should take them with pinches of salt instead of lapping them up with relish. For reasons of their own, pressure groups, journalists and fameseekers will no doubt continue to peddle ecological catastrophes at an undiminishing speed. . . . Environmentalists are quick to accuse their opponents in business of having vested interests. But their own incomes, their fame and their very existence can depend on supporting the most alarming versions of every environmental scare."

The Economist, 1997, pp. 19–21a

"Modern synthetic pesticides replaced more hazardous substances, such as lead arsenate, one of the major pesticides before the modern era. Lead and arsenic are natural, highly toxic and carcinogenic. . . . Each new generation of pesticides is more environmentally and toxicologically benign."

Drs. Bruce Ames and Lois Gold, "Environmental Pollution and Cancer: Some Misconceptions," *Rational Reading on Environmental Concerns*, 1992

It seems hard to believe that America is actually in the process of rejecting modern agriculture. There are a wide variety of groups behind this remarkable campaign, but this chapter focuses on the regulatory attacks launched against high-yield farming and the validity of the charges. These attacks are not based in science; but, if successful, they will profoundly change our lives and our environment for the worse.

The "Pesticides Cause Cancer" Scare

As a good example of the shaky science behind the activists' charges against modern farming, let's look at the history of the oldest and most successful of the scare campaigns: the "pesticides cause cancer" hoax. Because of the political process, the regulation of farm chemicals is hostage to the public's most horrifying (and often least-founded) fears. Consequently, we have spent most of 40 years worrying about pesticide residues causing cancer—without ever having found a single victim.

When pesticide regulations were initially established in the 1950s, cancer was the public's strongest fear. "Miracle drugs" like sulfa and penicillin, put to work during World War II, had begun to overcome such hitherto-deadly threats as tuberculosis and pneumonia. Better public health measures were conquering diseases like typhoid. New vaccines were eliminating epidemics of diphtheria and even polio.

In contrast, a rising percentage of the population was dying from cancer. Activists stated their belief that the rise in cancer deaths was linked with the suddenly widespread use of synthetic pesticides.

The most extreme expression of this cancer fear was written into law. The Delaney Amendment to the Federal Insecticide, Fungicide and Rodenticide Act was passed in 1958 stating that no processed food containing "any known carcinogen" could be sold. Delaney offered no tolerance whatsoever for anything that might cause cancer.

Of course, in 1958 neither the public nor their elected representatives knew that the rat tests would eventually indict as "carcinogens" about half of all the chemical compounds in the world, both natural and man-made. Or that the "cancer epidemic" was caused mainly by our lengthening average life spans, an increased number of smokers in the population, and the declining rates for most other causes of death.

Decline of the Cancer Terror

By the 1990s, the cancer scare was wearing thin. Cancer mortality (other than lung cancer) had declined 16 percent since 1950.[1] Billions of dollars in research funding had failed to link pesticides, even DDT, with any human cancers.[2/3] The only cancers with rising mortality rates were related to smoking; melanoma (probably due to sunburn); and a relatively rare cancer called non-Hodgkin's lymphoma.

Equally awkward for the Delaney Amendment, our ability to detect minute traces of chemicals had advanced a billionfold since 1950. But even parts per trillion were still illegal under Delaney.

By the 1990s, we had also learned that far more chemicals were "possible carcinogens" according to the high-dose rat tests—including virtually all of the spices used in our processed foods, and at least 19 of the 1,000 natural chemicals in coffee.

Perhaps most important in the public's mind was the new understanding of DNA and that our own genetics linked cancer to inherited genes, rather than to what we eat or breathe. National news magazines were putting "cancer genes" on their covers. Thirty years and billions of dollars late, the public was beginning to get rational about cancer.

Even our occupational exposures to carcinogens are now mild by historical standards. (Asbestos workers who smoked account for most of the occupational cancers ever documented.) Scientists, and even the EPA, now agree that less than 3 percent of our cancer deaths can be blamed on the whole roster of environmental exposures. This includes: asbestos, radon, smog, industrial chemicals, pesticide residues, and all of the other contaminants in our daily environments.

The Delaney Crisis of 1995

As the reality of cancer became clearer, even the EPA became convinced that the Delaney Amendment should be changed. Even though it had never been truly enforced, it was lurking uncomfortably in the regulatory closet. In 1995, the EPA published a regulation that would have substituted an EPA–set standard of "reasonable risk" for the unyielding intolerance of Delaney. Environmental groups, however, sued EPA contending that the law was the law. A federal court agreed that the Delaney language was clear, that only Congress could change it, and that Delaney must be enforced.

That court order caused panic in the food industry and, to some degree, in Congress. Strict enforcement of Delaney would cause chaos in the food system. Just as an example, strict Delaney enforcement would have forced the withdrawal of almost all processed foods that contain spices. (Check the labels at your supermarket to find something that would have been left on the shelves.)

The food industry feared a virtual shutdown of food processing, along with a huge public outcry about food safety. Farmers feared further loss of pesticide registrations, with resulting higher costs for crop and livestock protection.

Eventually, the anti-pesticide activists became resigned to the loss of Delaney, but they were determined to hold out for as high as price as they could. As it happened, they got quite a high price indeed.

The Endocrine Disruption Story

In the early 1990s, the W. Alton Jones Foundation hired Dr. Theo Colburn, a newly minted Ph.D. in zoology, to write a book on her theory that synthetic chemicals were wreaking havoc on the hormonal systems of people and wildlife. The book, *Our Stolen Future: Are We Threatening Our Fertility, Intelligence and Survival?—A Scientific Detective Story,* was published in 1996.

Colburn says she got her Ph.D. in order to prove Rachel Carson was right about the cancer risks of pesticides. In the end, she could not prove any cancer risks. In fact, she did not even have enough evidence to get her work on endocrine disruption published in a scientific journal.

She had only loose associations, anecdotes, and questions: Occasional homosexual behavior had been observed among some seagulls near waste dumps. Smaller-than-normal alligator penises had been found in a Florida lake next to a defunct pesticide plant. (The lake was also receiving a heavy load of human sewage, including large doses of estrogen from birth control pills.)

Colburn theorized that hormonal disruption was related to a 1992 report by a Danish researcher, Niels Skakkebaek, who claimed in a British medical journal that the sperm counts of human males had declined by almost 50 percent between 1940 and 1990. Skakkebaek later speculated the drop in sperm counts might be associated with pesticides.

Shaky it might be, but the W. Alton Jones Foundation marshaled all of the public relations power it could behind the new theory and the new book. It hired a public relations firm named Fenton Communications, which had gotten CBS-TV's *Sixty Minutes* to air the now-discredited Alar scare. The foundation also made a $50,000 donation to the Society of Environmental Journalists and donated another $50,000 to the Radio and TV News Directors Foundation.

The vice president of the United States, Al Gore, was persuaded to write the foreword to the book. He pronounced it "a legitimate sequel to *Silent Spring*."

Colborn's book ran into some problems, however. A study published about the time *Our Stolen Future* was released struck a major blow at the Skakkebaek claims of a drop in male sperm counts. Dr. Harry Fisch of Columbia University published a paper in the May 1996 issue of *Fertility and Sterility* warning that male sperm counts vary widely between geographic regions.

Skakkebaek had assumed they were constant around the world, and had compared different populations of men in different time periods. Most of the men in his early studies were from New York City where sperm counts are almost 50 percent higher than in Los Angeles, for reasons not yet understood. When the New York data were removed, the supposed drop in sperm counts disappeared. (Skakkebaek's peers would later criticize him for careless science and overstating his results.

Other scientists pointed out that pesticide residues are only a tiny fraction of our total exposure to endocrine disruptions. Dr. Stephen Safe of Texas A&M estimated that we get 40 million times as much exposure to endocrine disruption in our food (mainly from fruits and vegetables) as from pesticide residues.

In mid-1996, the debate among scientists on endocrine disruption was producing smoke, but not much fire. Within the Congress, the pressure for a new pesticide regulatory act was mounting because of the court order to enforce the unenforceable Delaney Amendment. Then a study from Tulane University exploded like a bombshell. The Tulane Center for Bioenvironmental Research announced in June 1996 that it had found a huge 1,600-fold increase in the endocrine disruption effects when two particular pesticides were tested together compared with when they were tested alone!

The scientific and political worlds were turned upside down overnight. No reputable science organization had previously confirmed strong endocrine disrupting effects or synergy in the impact of multiple pesticides. This report confirmed both, and at a very high level.

EPA administrator Carol Browner told the *Washington Post*, "The new study is the strongest evidence to date that combinations of estrogenic chemicals may be potent enough to significantly increase the risk of breast cancer, birth defects and other major health concerns." So encouraged, she promised to revise EPA research priorities accordingly.

The new Food Quality Protection Act was quickly pushed through Congress, with virtually unanimous approval. The Delaney Amendment disappeared. A standard of "reasonable safety" was established for cancer risks.

But, largely because of the Tulane Report, the Environmental Protection Agency was given new authority to tighten pesticide tolerances by a full order of magnitude "for the sake of the children," and to test for endocrine disruption and synergy between chemicals with similar modes of action.

"I never saw a paper have such impact," said a science policy observer.[4]

It was a huge victory for the environmental movement. Especially for the W. Alton Jones Foundation, which hired the author, published the book, coordinated the public affairs campaign, handed new powers to the EPA, and preserved the "fear factor" in pesticide regulation. It was nearly a full year before the rest of the world understood exactly how big a coup the Jones Foundation had pulled off.

During that year, teams of researchers at four other major laboratories tried to replicate the Tulane research—and failed. None of them found synergy or major endocrine disruption effects. Eventually, Tulane itself had to admit it couldn't replicate its own results. In the July 25, 1997, issue of *Science*, Dr. John A. McLachlan of Tulane publicly withdrew the famous study. A later investigation by the university found that "the original data failed to support the major conclusions of the paper"! (The school cleared McLachlan, who had put his name on the paper as lead author, and blamed a staff researcher who had already resigned.)[5]

Of course, withdrawing the study didn't change the wording of the new Food Quality Protection Act. Armed with its new congressional authority, the EPA quickly published a *Special Report on Environmental Endocrine Disrupters*. It concluded: "The findings contained in our assessment send a strong signal for more research on the effects of endocrine disrupting chemicals, particularly into the possible effects on humans, where we currently do not have enough information to conclusively determine the potential risks of existing exposures."

Only later did Diane Katz of the *Detroit News* report that the W. Alton Jones Foundation had given an $80,000 grant to the Tulane Center for Bioenvironmental Research. It was a small part of the $113 million in grants given by that foundation in the 1990s, but it was more than enough to have focused the McLachlan's laboratory on endocrine disruption and synergy.

The grant does not mean that Tulane faked its research. Mistakes are frequent on the cutting edge of science. But seldom are they as sensational as in the case of the Tulane report. Worse, most of the journalists who wrote big stories on the Tulane study when it was released failed to cover its retraction at all. Certainly not reported was the information that part of the work was funded by a foundation with a strong bias on endocrine disruption.

What would the media have done if a major agricultural college study confirming the safety of pesticides had been published—and then withdrawn because the results couldn't be replicated? What if reporters had only then learned the study was partly financed by a pesticide manufacturer? There would have been a journalistic firestorm.

DANGEROUS REALITY:

". . . [Banning] fungicides could lead to food scarcities. An increase in the contamination of foods by fungal products that include carcinogens and nerve, liver and kidney poisons would also follow. . . . An immediate cause for concern is . . . a Consent Decree dated 20 September 1994, in which the Environmental Protection Agency is a participant. . . . Many of the most effective fungicides would ultimately be banned . . . because they can be shown to induce cancer in one strain and one sex of a rodent when huge, nearly lethal doses are administered. An example is the important, widely used fungicide, captan. Captan is relatively nontoxic. It is rarely detected on produce or in ground water. It is readily decomposed. . . . The benefits of fungicides in the production and distribution of health-enhancing fruits and vegetables should not be jeopardized by the folly of the Delaney Clause and the actions of a regulation-proliferating agency."

Dr. Philip Abelson, editorial, *Science*, vol. 266, November 25, 1994, p. 1303

WEED-FREE WITH HERBICIDES—Such fields produce higher yields because they don't have to share the moisture and plant nutrients with competing weeds. Nor do the herbicides pose a threat to people or wildlife.

Target: Atrazine!

The eco-activist drums have now started rolling for another "vicious" pesticide called atrazine. It's a big target—and a highly valuable one, both to farmers and to eco-activists. Atrazine is a weed control herbicide used annually on 50 million acres of U.S. corn production. There is no cost-effective replacement for atrazine; it can be sprayed *once*, at a pound or so per acre, and give full-season weed control.

Wisconsin, which banned atrazine for 1992, found that its corn farmers subsequently had to spend an extra $11 per acre on weed control. Extended nationwide, *a ban on atrazine would apparently cost American farmers more than $1 billion extra per year*—or force lower corn yields and eventually larger plantings to meet the world's corn demand.

Atrazine is a valuable target for the activists for several reasons:

- It is the pesticide that turns up most commonly in our drinking water. It often shows up seasonally at measurable levels in reservoirs in the spring and early summer (depending on rainfall patterns).

- Because atrazine has been associated with mammary tumors in one strain of laboratory rat, the activists can also raise the dreaded spectre of "breast cancer."

- Recently, we have been finding three metabolites (breakdown products) of atrazine in water. These breakdown products, which are also formed in laboratory rats and as such have been studied in animal cancer tests, have the same health safety profile as atrazine itself. Hence, when activists say that there are more "triazines" (atrazine is just one) in our water than we thought, they are technically correct.

- The EPA has warned many cities and community water systems that they are facing multimillion-dollar investments for atrazine testing and remediation, because of seasonal peaks that exceed the current safety levels.

- All that's lacking for a first-rate scare story is some threat to New York and Los Angeles. (No corn grows in their watersheds, so there's no atrazine in the drinking water of those media meccas.)

How risky *is* atrazine?

- It has been broadly used for 30 years and has not been linked with any demonstrated human cancer risk.

- It is not a reproductive toxin, a teratogen, or a mutagen.

- It does not bioconcentrate, or "build up" in the food chain.

- Studies of workers in the plants where it is made show no elevated risks, even over periods of 30 years. That's true even though its production dates back to times when industrial plants were far less careful than they are today.

- A definitive study on atrazine and farmers was done recently to examine the potential for increases in non-Hodgkin's lymphoma. The study found "little or no increase in the risks . . . attributable to the agricultural use of atrazine."[6]

Atrazine does have one "safety" problem.

The potential for atrazine to produce cancer has been evaluated in at least seven lifetime studies in rats and four studies in mice. No potential to induce cancer was seen in the studies using the U.S. Government's choice of rats or in three different strains of mice. It does seem to hasten the onset of mammary tumors in the females of one variety of laboratory rat. However, scientists say this is probably not applicable to humans for several reasons:

- These rats are especially susceptible to mammary tumors; 40 to 70 percent of the females get such tumors with a normal diet and in the absence of any chemical exposure.

- A woman would have to drink 22,000 gallons of water per day at the current water-safety level of 3 parts per billion just to reach the *no-effect threshold* in rats.

- Atrazine has been widely used in both farming and forestry for 30 years. If atrazine caused breast cancer, women in farming and forestry should have more breast cancer than the national average. Yet these women have *less* breast cancer—only 84 percent as much as the average American woman.[7]

- The tumors in this strain of rat apparently result from prolonged periods of estrogen secretion; these occur more often in the older female rats. (Atrazine seems to accelerate this process.)
- By contrast, human females have *less* estrogen as they get older. It is, therefore, unlikely that atrazine affects human breast cancer rates.

The biggest reason of all for confidence in atrazine's safety?

The EPA has recently reevaluated statistical "false positives" in the atrazine rat reproduction tests and raised the "no-effect" level by *tenfold*. In effect, that says atrazine is *far safer than originally rated*.

The maker of atrazine agrees with the environmentalists that the metabolites of atrazine should be lumped with the atrazine itself. The net effect of the latest findings, however, is that atrazine's reference dose (used to assess safety) is seven times higher than originally determined (seven times safer than originally assumed). That includes both the increase in residues (atrazine *plus* the metabolites) and the reevaluated rat reproduction test results.

The manufacturer has accordingly asked the EPA to raise the Maximum Contaminant Level (MCL) from the current 3 ppb to 20 ppb. This would make the MCL consistent with the EPA's own new Reference Dose—and with the scientific evidence on atrazine's health effects.

If the EPA calculates the MCL using the new reference dose, it will eliminate virtually all of the multimillion-dollar testing and remediation requirements under the Safe Drinking Water Act. No expensive carbon filters would be needed by city and community water systems.

Well water, even with the metabolites counted, would be rated about three times as safe as before. Surface water, where the metabolites seldom appear, would be about *seven times as safe* as currently rated. (It is the surface water which is involved with most of the seasonal peaks in American drinking-water exposures.)

To put atrazine into proper perspective, let's look at what the new atrazine safety rating means to the average urban female whose water comes from a riverine reservoir. Since most of the urban "threat" from atrazine was in surface waters (small rivers), she would now have to drink 154,000 gallons of water per day to reach the no effect level. Even then, she'd have to take atrazine doses directly during the 9 months of the year when there are virtually no traces of atrazine in the urban water supplies. (Presumably she could buy atrazine directly from some agricultural supply house.)

An atrazine ban in the United States would do far more than raise U.S. corn production costs.

It would probably produce a ripple effect in other countries, as was the case with DDT. (The United States is, after all, looked to as a leader in environmental science.) In the long run, it would likely mean lower corn yields throughout large parts of the world.

Absent a shift to vegetarian diets (none is apparent yet) this lowering of yields would mean more world acres required for corn production—and less wildlife habitat.

The most likely consequences of an atrazine ban, then, would be less wildlife, more cancer and more soil erosion—all because Washington, D.C. is now populated with the first generation that grew up reading

Bob Lang, reprinted from CFGFI Quarterly, spring 1999

Reprinted from BridgeNews Forum, May 7, 1999

Deformed Frogs: Another Environmental False Alarm

Five years ago, a group of Minnesota school kids hunting frogs in a wetland discovered that nearly half of the frogs they found had deformities of the hind legs.

Their findings spread over the Internet, feeding into other reports of deformed frogs, declining frog populations, and speculation that pesticides were causing deep-seated damage to the global ecology.

Eventually, misshapen frogs were found in over 40 U.S. states. The Minnesota kids were told by environmental groups that they had gathered the final proof that man-made pesticides should be banned.

The Sierra Club's magazine trumpeted, "How Pesticides Are Creating Deformities in Frogs." A headline in Iowa's *The Des Moines Register* newspaper read, "Deformed Frogs Stun Scientists." A Reuter story warned, "U.S. Frog Deformities Could Be Linked to Pesticide."

Now *Science* magazine has published two papers that demonstrate a natural cause for the deformed frogs: natural parasites. It seems that tiny flatworms burrow into tadpoles and cause frog abnormalities ranging from no hind legs to six extra legs.

One research project examined five species of frogs in 12 U.S. locations and concluded that the deformities are characteristic of parasitic attacks, not chemicals.

A second group of researchers exposed tadpoles to the flatworms in controlled laboratory experiments and found the frogs got the same kinds of deformities as were found in the wild. A control group of tadpoles, protected from the parasites, developed normally.

Undaunted, Dr. Judy Helgen of the Minnesota Pollution Control Agency insists, "for us, chemicals are [still] the leading hypothesis."

Of course, the state's environmental protection agency is not totally disinterested in the question. If pesticides are causing deformed frogs, then the agency will get more budget to deal with the problem. It will write important new regulations to control what farmers do. Its press conferences will draw lots of reporters.

If deformed frogs are simply another harsh fact of nature, then it will get no extra money and might even have to answer embarrassing questions about why it was conducting large-scale scaremongering based on a school nature trip.

Other researchers across the country still loudly announce that "we don't have all the answers yet," and claim that their lines of inquiry should continue to be funded.

But if the problem turns out to be parasites, not pesticides, they realize their funding will dry up. Almost nobody cares about deformed frogs in the wilds of Minnesota unless they can be used for a political statement.

The Natural Resources Defense Council, World Wildlife Fund, Consumers Union (publisher of *Consumer Reports*) and other eco-groups have resigned from President Clinton's food safety advisory panel because they can't produce enough proof of risk for even this green-oriented administration to ban big groupings of man-made pesticides.

The same newspapers that ran scare headlines on deformed frogs probably also ran *Consumer Reports* magazine's self-deluded "toxic index" story on the supposed dangers of pesticide residues on fruit. After all, pesticides have been one of the most reliable and longest-running scare stories in the history of journalism. No doubt these papers will simply await the next terror opportunity.

Once again, eco-suspicion has run far ahead of science and once again embarrassed itself. A few years ago, "pesticides and pollution" were accused of killing dolphins; the killer turned out to be a virus.

Now the National Cancer Institute says that our non-smoking cancer rates began to decline 30 years ago, even though pesticide use has continued unabated. Medical researchers say our diets and our genes are the major non-smoking cancer sources. (As a country, we eat too many fatty foods and too few fruits and vegetables.)

Someday, when advances in biotechnology allow us to rely less heavily on pesticides, historians will look back and snicker at the hysteria over those chemicals that helped cut cancer risks and saved room on the planet for wildlife.

Meanwhile, the frog scare is gone, but stay braced for the next scare headlines on pesticides.

The Erratic and Irrational Pattern of Agricultural Regulation

Agriculture is being regulated more and more heavily, in ways that make little sense either for protecting the public or the environment. The current generation of federal, state, and local regulators is the first to have grown up with the unquestioning repetition of the messages condemning pesticides and modern farming—in their books, within their families, on their TV programs, and in their peer-group discussions. They have fully digested the concept that "pesticides are dangerous to people because they kill pests." Some examples of the strange spectrum of regulatory activity:

- America spent billions of research dollars trying to prove that pesticides cause cancer, with no evidence that they have caused even one case. The result is higher priced fruits and vegetables and a public that has been made fearful of its food supply.

- Herbicides are crucial in conservation tillage farming systems, the most sustainable farming systems ever devised. Yet the United States government is attempting to ban some of the most cost-effective herbicides for conservation tillage, with no evidence of harm.

- If it is ratified and enforced, The Kyoto Treaty on global warming will radically increase First World farmers' costs for diesel fuel and nitrogen fertilizer. Primarily, this will push farming out of the First World and into the Third World, where yields are low and most of the world's biodiversity is harbored in the tropical forests.

- The biggest dangers in our food system are bacterial. E. coli O157:H7, in particular, is a deadly new strain of bacteria, which can not only kill healthy people but cause permanent kidney and liver damage in its survivors. In America, E. coli O157 kills about 250 persons per year, and salmonella kills another 250. Irradiation (bombarding the food with low doses of gamma rays) kills E. coli, salmonella, and the microorganisms that cause our food to rot. Irradiation would sharply reduce deaths from bacterial contamination and keep our food fresher. There is no federal effort to encourage the use of irradiation in protecting the public food supply or study the implication that organically grown food (because of its reliance on manure) is responsible for far more than its share of the bacterial infections.

- Within the past few years, regulators from the federal, state, and even county governments seem to have entered a competition to see who could get credit for driving confinement meat and egg producers out of the country. This even though it is the safest meat production system ever devised for people, for the environment, and for the birds and animals being raised.

- The U.S. and European regulations on nitrate in drinking water are unnecessarily restrictive, forcing cities to spend millions of dollars to lower the nitrate levels in their drinking water though this does virtually nothing to improve public health.

- The federal government is about to de-register whole categories of pesticides that have been safely and economically guarding our food supply for decades. This new regulatory strangle will be based, in part, on "endocrine disruption"—even though science has thrown out the studies that attempted to demonstrate the dangers of endocrine disruption.

- U.S. labor laws have been used to virtually eliminate the seasonal jobs in fruit and vegetable production and harvesting. One ruling said apple growers could not ask a job applicant—for apple picking—to hoist a ladder. It was "not related" to the job! Growers go out of business and low-skill workers have fewer jobs so urbanites can feel that they've eliminated the "horror" of migrant labor.

- The federal government will use water as a major regulatory device. In the West, dams will be removed in a supposed effort to have more fish, no new ones will be authorized, and more of the reduced water supply will go to the cities, recreation, and "the environment." In the East, "water pollution from farming" will excuse the federal regulation of farming systems.

I met a federal regulator a few years ago who had grown up on a farm. His father had contracted malaria while working the cotton fields. That man really understood the shortcomings of traditional farming, and carried that understanding into his work. There's a strong chance that his replacement grew up on a small lot in the suburbs listening to unproven claims and fears of the media and environmental activists.

We Had Global Warming—and We Liked It
by Dennis T. Avery

"Climate extremes would trigger meteorological chaos—raging hurricanes such as we have never seen, capable of killing millions of people; uncommonly long, record-breaking heat waves; and profound drought that could drive Africa and the entire Indian subcontinent over the edge into mass starvation."

U.S. Senate Majority Leader George Mitchell,
World on Fire, (1991)

"The whole aim of practical politics is to keep the populace alarmed—and hence clamorous to be led to safety—by menacing it with an endless series of hobgoblins, all of them imaginary."

H. L. Mencken, newspaper columnist,
In Defense of Women, (1920)

We've all read the global warming scare stories lately. Be prepared for more of the same as the Kyoto Treaty sits stalled in the Senate and Al Gore ties his presidential run to environmental issues of the day.

When Senator Mitchell presented his doomsday scenario, crude computer climate models were predicting two to three times as much warming as they currently do. Now researchers say that the earth is likely to warm by about three degrees Fahrenheit during the next century.

That may sound like a lot, but it isn't. The world has experienced that much warming fairly recently in history. And we loved it! Between A.D. 900 and 1300, the earth warmed by some three degrees, according to the Oregon Institute of Science and Medicine. Scholars refer to that period—one of the most favorable periods in human history—as the Medieval Climate Optimum.

Written and oral history tells us that the warming created one of the most favorable periods in human history. Crops were plentiful, death rates diminished, and trade and industry expanded—while art and architecture flourished.

Soon after the year 1400, however, the good weather ended. The world dropped into the Little Ice Age, with harsher cold, fiercer storms, severe droughts, more crop failures, and more famines. According to climate historian H. H. Lamb, speaking of this period, "for much of the [European] continent, the poor were reduced to eating dogs, cats, and even children." The cold persisted until the 18th century.

The Medieval Experience with Global Warming

The Medieval Climate Optimum was a boon for mankind and the environment alike. The Vikings discovered and settled Greenland around 950. Greenland was then so warm that thousands of colonists supported themselves by pasturing cattle on what is now frozen tundra. During this great global warming, Europe built the looming castles and soaring cathedrals that even today stun tourists with their size, beauty, and engineering excellence. These colossal buildings required the investment of millions of man-hours—which could be spared from farming because of the higher crop yields.

Europe's population expanded from approximately forty million to sixty million. The increase was due almost entirely to lower death rates. Trade flourished, in part because there were fewer storms at sea and fewer muddy roads on land. (There was more rainfall, but it evaporated more quickly.)

England was warm enough to support a wine industry. The Mediterranean Basin was wetter than today. Farming moved further north in Scandinavia, Russia, Manchuria, northern Japan, and North America. Farmers in Iceland grew oats and barley.

At the same time, technology flourished. The water mill, the windmill, coal, the spinning wheel, and soap entered daily life. Sailors developed the lateen sail, the rudder, and the compass. New iron-casting techniques led to better tools and weapons.

Real earnings in China reached their highest point in 3,000 years, thanks largely to the more-plentiful crops. There were half as many floods and one-fourth as many big droughts as in the Little Ice Age that followed. The increase in wealth produced a great flowering of art, literature, and invention, the products of which we still enjoy and appreciate.

The Indian subcontinent prospered as well, producing colossal temples, beautiful sculptures, and elaborate art. The Khmer people built the huge temple complex at Angkor Wat. The Burmese built 13,000 temples at their capital, Pagan.

We know less about what went on in North America. We do know that the Great Plains, the upper Mississippi Valley, and the Southwest apparently received more rainfall than they do now. The Anasazi civilization of the Southwest grew abundant irrigated crops—and then vanished when the Little Optimum ended and the rainfall declined. North Africa received more rain than today. The Sahara—and presumably many other desert regions—shrank in response to the increase in rainfall.

There were some negatives, of course. The steppes of Asia and parts of California apparently suffered dry periods. But, over all, the medieval experience with global warming should reassure us greatly.

What Would a New Climate Optimum Bring?

If moderate warming recurs, it is expected to moderate nighttime and winter low temperatures more than increase daytime and summertime highs. Consequently, it will produce relatively little added stress on crop plants and trees, or on people.

The expected increase in CO_2 will be an additional blessing. Carbon dioxide acts like fertilizer for plants. Extra CO_2 also helps plants use their water more efficiently. More than a thousand experiments with 475 crop plant varieties in 29 separate countries show that doubling the world's carbon dioxide would *raise* crop yields an average of 52 percent.

The amount of carbon dioxide in the atmosphere does seem to be rising. In fact, we are nearly halfway to the expected CO_2 peak of 550 parts per million. The current levels of CO_2 in the earth's atmosphere are very low, however, compared to past periods. In fact, most of the earth's species of plants and animals evolved in much-higher levels of carbon dioxide than we have today, up to twenty times the recent preindustrial level of 280 ppm.

The increase in CO_2 will make forests all over the world healthier and more robust and allow them to support more wildlife. Canadian forestry researchers estimate that in a new warming their forest growth would increase by 20 percent.

In *Global Warming and Biodiversity,* Dr. Gary S. Hartshorn notes that the tropical forests already undergo enormous variability in rainfall. He writes, "It is unlikely that higher temperature per se will be directly deleterious to tropical forest [wildlife] communities."

In the same book, Dr. Vera Alexander notes that Arctic marine systems would be seriously threatened if the sea ice melted. The Arctic, however, has already survived major temperature changes, including the Medieval Climate Optimum, without shrinking appreciably. Even with average worldwide temperatures six to nine degrees Centigrade warmer than today's, Alexander notes, the sea ice would re-form in the winter, and the polar ice caps would stay intact.

Assessing an Arctic tundra ecosystem, Dwight Billings and Kim Moreau Peterson predict that such a warming would have no major species impact. They expect more snow-free days in the summer, more photosynthesis, and somewhat more peat decomposition, but these factors would mainly benefit the primary food chain.

Decrease in Disasters

Most of the trillion-dollar estimates of global warming "costs" headlined in the 1980s were based on forecasts that cities such as New York City and Bangladesh would be drowned under rising seas. Those are frightening scenarios, but completely untrue. It may seem paradoxical, but a modest warming in the polar regions will actually mean *more* arctic ice, not less. The polar ice caps depend on snowfall, and polar air is normally very cold and dry. If polar temperatures warm a few degrees, there will be more moisture in the air—more snowfall, and more polar ice.

The world's ocean levels have been rising at approximately the same rate, 7 inches per century, for at least a thousand years. In 1992, *Science* magazine published a paper based on ice core studies suggesting that the projected warming would reduce the sea level by one foot.

Global warming scaremongers have also claimed that a warmer world would suffer more extreme weather events. This too is unlikely. History records that the Medieval Optimum brought fewer floods and droughts. There is good reason to believe that this pattern would repeat in a new

Little Optimum. Dr. Fred Singer, professor emeritus of Environmental Sciences at the University of Virginia, says, "One would expect severe weather to be less frequent because of reduced equator-to-pole temperature gradients."

In other words, the smaller the temperature differences between the North Pole and the equator, the milder the weather. Most of the warming, if it occurs, will be toward the poles, with very little increase near the equator. This means there would be less of the temperature difference that drives big storms.

Some alarmists claim that a warmer world would suffer huge increases in deaths from horrible plagues of malaria, yellow fever, and other warm-climate diseases. Fortunately, these claims are unlikely to come true, because they ignore some important, fundamental realities. As mentioned, global warming would be very slight near the equator and thus would only slightly expand the range of the malarial mosquitoes. There is little reason to expect tropical plagues to increase.

Moreover, these diseases are nowhere near as relentless as the scare scenarios assume. In the United States, for example, malaria and yellow fever once ranged from New Orleans to Chicago. We conquered those diseases, but not by changing the climate. We did it by suppressing mosquitoes, creating vaccines, and putting screens on doors, windows, and porches. Other countries can do the same. Third World countries have had high disease rates because they were poor, not because warm climates cannot be made safe.

Why Be Wary?

The original global warming scare-stories were authored by eco-activists, who subsequently admitted that they were looking for ways to persuade people to live leaner lifestyles. There is no reason to believe that the authors of the global warming scares have any special knowledge about the future climate. In fact, their leading scientist, Dr. Stephen Schneider, was predicting global cooling just a few years ago, and he candidly states that he is willing to misrepresent the facts if it will stir up the public over the "correct" causes.

"But what if we're right?" the activists respond. History says that they are not. And the problem is that the "solutions" these activists recommend, however well intended, would leave much of the world without an energy system—and that would be deadly for both people and animals. If we were to triple the cost of coal, double the cost of oil, ban nuclear power, and tear out hydroelectric dams, which would be the result of the activists' approach, humanity would essentially be left without energy. And in a world of expensive energy, people would not be able to afford the window screens, toilets, clean water, and refrigeration that prevent millions of deaths per year.

The widespread poverty caused by expensive energy would reverse the current worldwide trend toward greater affluence, decreasing birth rates, and better health. The low-energy option would destroy millions of square miles of wildlife habitat. High energy taxes would all but destroy modern agriculture, with its tractors and nitrogen fertilizer (with the nitrogen taken from an atmosphere that is 78% nitrogen). Shifting back to draft animals would mean clearing millions of additional acres of forest to feed the beasts of burden.

Giving up nitrogen fertilizer would mean clearing five to six million square miles of forest to grow clover and other nitrogen-fixing "green manure" crops. The losses of wilderness would nearly equal the combined land area of the United States and Brazil.

History and the emerging science of climatology tell us that we need not fear a return of the Little Climate Optimum. If there is any global warming in the 21st century, it will produce the kind of milder, more-pleasant weather that marked the Medieval Little Optimum—with the added benefit of more carbon dioxide in the atmosphere and therefore a more luxuriant natural environment.

The modest global warming now predicted should bring back one of the most pleasant and productive environments humans, and wildlife, have ever enjoyed. We have nothing to fear but the fear-mongers themselves.

This article is adapted from the original publication in American Outlook, *Spring 1998. A condensed version was published in* The Readers Digest, *August 1999.*

Notes

1. Bruce Ames and Lois Gold, *Misconceptions About Environmental Pollution, Pesticides, and the Causes of Cancer*, Dallas, Texas, National Center for Policy Analysis, 1998.

2. *Carcinogens and Anticarcinogens in the Human Diet*, Washington, D.C., National Research Council, National Academy Press, 1996.

3. "Report of a Panel on the Relationship Between Public Exposure to Pesticides and Cancer," Canadian Network of Toxicology Centers, *Cancer*, vol. 80, pp. 2019–33, 1997 (Copyright American Cancer Institute).

4. *Science*, vol. 225, March 28, 1997, p. 1879.

5. "Tulane Inquiry Clears Lead Researcher," *Science*, vol. 284, June 18, 1999, p. 1905.

23

Who Are These People?

MYTHMAKERS SAY:

"Just now one of the significant historical roles of the primal people of the world is . . . to call the entire civilized world back to a more authentic mode of being."

Thomas Berry, *The Dream of the Earth*, Sierra Club Books, 1988

"On December 28, 1954, the American Association for the Advancement of Science held a symposium on 'Population Problems,' at which Dr. Alan Gregg, vice-president of the Rockefeller Foundation (1951–56) came up with a startling idea: the thought that the human species is to the planet Earth what a cancer is to an individual human being."

Van Rensselaer Potter, *Global Bioethics: Building on the Leopold Legacy*, Michigan State University Press, 1988

"We, the human species, have become a viral epidemic to the earth, . . [an] AIDS of the earth . . ."

Paul Watson, a founder of Greenpeace and Sea Shepard, speaking at the University of Oregon law school. Quoted in the Minneapolis *Star Tribune*, October 30, 1999, p. 25A

"But the assumption of continued affluence at today's level is unfounded. If our numbers continue to rise, our standard of living will fall so sharply that by the year 2000, any surviving Americans might consider today's average Asian to be well off."

Wayne H. Davis, "Overpopulated America," *The New Republic*, January 10, 1970

(*Author's note:* Far from being "dated" and unrepresentative of current environmentalism, this 1970 essay continues to be cited as a classic at meetings such as the 1994 Cairo Population Conference. It was recently reprinted in *Learning to Listen to the Land,* a collection of important environmental essays published by Island Press in 1992.)

REALITY SAYS:

"Until very recently, ordinary people spent most of their time outdoors—farming, hunting, gathering nuts and berries, pillaging the countryside in armed bands. The more contact people actually have with nature the less likely they are to 'appreciate' it in a big, mushy ecumenical way. And the more likely they are to get chiggers. . . . For most of history, mankind has managed to keep a reasonable balance between thinking nature's adorable and thinking it wants to kill us."

P. J. O'Rourke, *All the Troubles in the World: The Lighter Side of Overpopulation, Famine, Ecological Disaster, Ethnic Hatred, Plague, and Poverty,* New York, Atlantic Monthly Press, 1994, pp. 122–24

"We cannot produce harmony simply by setting up sacred wilderness temples, while downgrading, excluding and eventually learning to despise human beings. . . . Humanity has spent most of its history trying to 'tame' wilderness . . . because people found it hostile and constricting. . . . Only after we have provided ourselves with the basic necessities is it fun to go back and see what untamed nature is really like."

William Tucker, *Progress and Privilege: America in the Age of Environmentalism*[1]

We've never before seen the huge, powerful, narrowly fixated environmental movement known today. Most of us are startled by it and the ways in which it seems to be changing our lives.

So, let's examine this new phenomenon.

First, we must realize that the environmental movement has welded together several major groups of people. Most of these people are sincere in their belief that they are doing good. However, they all come to the environmental movement with values and beliefs that make them willing to risk major changes in our social and economic systems without having much solid evidence that those changes will produce a better society—or even a more environmentally responsible one.

Environmental Purists

The environmental movement is led primarily by a group of purists. These are people who value "nature" over humanity.

The intensity of the eco-activists has been captured by Philip Shabecoff, until recently the chief environmental reporter for the *New York Times*, in his book, *A Fierce Green Fire*: "Our negligent use of the Promethean forces of science and technology has brought us to the verge of disaster."[2]

The beginnings of the ecology movement can be traced to the success of the Industrial Revolution—and the reactions against it—during the early part of the 20th century. People welcomed the new productivity, but many hated the ugly urbanization, the pollution, and the exploitation of human and natural resources. Many affluent idealists then looked back fondly on the pastoral past, even though for most rural people that had also been a harsh world of unremitting toil and poverty.

USAID

TRADITIONAL FARMING—It may have allowed husband and wife to work together, but you would have a hard time convincing them that it represents the farming ideal for the future.

One of the early pivotal figures in the creation of environmentalism was Gifford Pinchot, the first chief of the U.S. Forest Service under President Teddy Roosevelt. Another was John Muir, a lover of raw nature who founded the Sierra Club.

Both men believed fervently in public ownership of natural resources. Such collectivism, they believed, was the only way to prevent the exploitation and ultimate destruction of the wild lands. Pinchot, in fact, was instrumental in expanding American public ownership to one-third of the land area of the United States.

The next step in deepening the ecology movement is credited to Aldo Leopold, the cofounder of the Wilderness Club, who wrote his widely quoted *Sand County Almanac* in 1949.

Leopold's concept was a "pyramid of life," which required preserving the diversity of all species. He wanted to "enlarge the boundaries of the community to include soils, waters, plants and animals," thus justifying "changes in the role of Homo Sapiens from conqueror of the land-community to plain member and citizen of it."[3]

It is no accident that several of the founders of the environmental movement are among the few people who *have* lived alone in the wilderness but without having to wrest their livings directly from natural resources:

- Muir lived in the California mountains for some years as a trail guide. (He met a number of important philanthropists and political figures that way.)

- Brad and Vena Angier, Bostonians who fled to the Canadian forest, became famous writing on survival skills and wilderness living. Yet they were ready to quit and return to Boston—rather than, say, take jobs in the nearby logging town—when they sold their first wildlife story.

- Aldo Leopold was a park ranger in New Mexico's Gila National Forest.

Each of these people lived *in* the wilderness without having to live *from* it. The pioneers of the deep ecology movement nevertheless acquired a proprietary feeling about natural resources. No matter that the very idea of a "friendly wilderness" is totally unrealistic unless the ecologists are backed by the products and safety factors of modern civilization, such as down sleeping bags, freeze-dried foods, and butane stoves.

Their solution was creative, and for the most part successful: They got the government to reserve much of the wilderness for the things they liked—hiking, backpacking, photographing birds, and the few other activities that can be done in the wilderness without leaving a "human imprint."

Then came the next step for the purists. By the middle of the antiestablishment 1960s, UCLA historian Lynn White, Jr., was calling for a "new religion" based on "the spiritual autonomy of all parts of nature," and "the equality of all creatures, including man."[4]

Today, the basic tenet of the environmental movement is that "all living things are created equal" and are valuable in and of themselves, regardless of their relationship to man. Some environmental purists go even further. William McKibben, in *The End of Nature*, quotes a biologist for the National Park Service, David Graber:

> Human happiness and certainly human fecundity, are not as important as a wild and healthy planet. . . . We have become a plague upon ourselves and upon the Earth. . . Until such time as Homo Sapiens should decide to rejoin nature, some of us can only hope for the right virus to come along.[5]

This is not man and nature, it is nature *instead* of man.

Such radical groups as Greenpeace and Earth First! are led by angry "deep" ecologists who are inclined to strike out violently against the perceived threats to "their" wilderness. These angry and violent purists, though few in number, have nonetheless managed to set the stage for the rest of the movement.

THEY ADMIT IT:

"The fact is we wish to preserve because we wish to preserve. If that's not a valid concept, then we haven't got one. To make believe we hold a different view is pure hypocrisy."

Environmental consultant Ian Parker, interviewed in the *New York Times Magazine* in 1982[6]

"I founded Friends of the Earth to make the Sierra Club look reasonable. Then I founded the Earth Island Institute to make Friends of the Earth look reasonable. Earth First! now makes us look reasonable."
David Brower, author of *Confessions of an Eco-Warrior*[7]

The Greens

The Greens, in contrast to the purists, are the ecology movement's more worldly politicians and pragmatists. They profess at least a nominal concern for human values and modern culture, but their goal seems to be a highly regulated, socialist society that will impose and enforce a simpler, more austere lifestyle on the rest of us.

There are quite a few Greens, and they have proven very adept at politics.

Working together, the deep ecologists, or purists, and the broader group of Greens have achieved some major gains for society. They were the groups that set aside the national parks years ago. These parks now represent an enormously valuable set of national assets for recreation, timber, grazing, fishing—and even that cherished wilderness contemplation.

They also led the cleanup of rivers, smog, and industrial effluents that has given the Western world major improvements in its environmental quality since 1970.

However, the deep ecologists and the Greens may now have gotten *too* effective at preservation:

> They are virtually withdrawing the national forests from the timber business, are halfway to eliminating grazing on public lands—and are casting covetous eyes on the rest of the nation's land uses. This could represent major environmental problems in time, such as fire damage from overaged tree stands and the destruction of communities, industries, and agriculture.

- They have not only banned DDT, but are now attempting to ban chlorine from our lives, even though without it we risk sudden death from cholera (water treatment) and going without paper and other modern necessities. The eco-activists have not proven any threat from chlorine, but then they didn't prove a threat from DDT either.

The general public may not realize that preservation and anti-technology policies have gone too far until they are already locked in place as the law of the land. By then, such policies will be enormously difficult to change.

THEY BELIEVE:

"We projected our own vicious qualities onto such animals as the wolf, the rat, the snake, the worm and the insects."

Thomas Berry, *The Dream of the Earth*, Sierra Club Books, 1988[8]

"A civilization is comparable to a living organism. Its longevity is a function of its metabolism. The higher the metabolism (affluence) the shorter the life. . . . We have now run our course."

Wayne H. Davis, "Overpopulated America," *The New Republic*, 1970

Fans of Big Government

There are also millions of people in the world who honestly trust government more than private enterprise, and/or think we're too rich for our own good. Most of these people are now also collected under the Green banner—because they have virtually nowhere else to go.

After all, most other adversarial political movements have lost out to democratic capitalism. The Soviet Union has collapsed. The Communist myth that human dictatorship and central planning are good for us has been completely discredited. Socialism, especially as demonstrated in Western Europe, looks moribund. Utopian communities are out of fashion. Nevertheless, many of the people in the environmental movement still urgently believe in the efficacy of nondemocratic collectivist "solutions."

Jane Fonda, who used to recommend Communist big government, now recommends environmental big government instead. What hasn't changed is her belief in big government (and, of course, the importance of Jane Fonda in a highly visible role as a key advisor to that government).

Power Seekers

Many members of the environmental movement have not chosen to pursue power through the traditional methods, but that doesn't mean they don't want power. They pursue it for what they feel are valid reasons, following in the footsteps of other special interest groups throughout American history.

Jeremy Rifkin, for example, spent his student years learning to lead antiwar protests that attracted TV cameras. He liked the work and the attention. Later in life, he turned his talent for protests into a lucrative career, writing fear books, making speeches, and even setting up a 900 number so anxious housewives can pay for his advice on what foods to buy and where to buy them. Fortune 500 food companies tremble at his works.

Many 1960s-style law students didn't want to go into corporate law—and found they could carve out lucrative niches with environmental lawsuits that not only got them fees but praise from their peers.

Whole troops of environmentalists are generating millions of dollars per year from memberships, grants, book sales, and speaking fees—and getting the psychological rewards of "saving the world" in the bargain.

Bureaucrats, especially, get power from the environmental movement. The EPA, state regulatory agencies, local trash recycling agencies, water authorities, and Soil Conservation Service employees get higher salaries, bigger empires, and added job security if they can make themselves seem more important to an environmentally sensitive public.

THE REALITY:

"John Stossel, ABC reporter extraordinaire and host of the cliché-busting special, 'Are We Scaring Ourselves to Death?' confirmed the bias that allowed such hysteria to go from activist group to news story unchallenged. . . . 'We approached it from the bias that on the one hand is business, which is greedy and has an ulterior motive, and will distort the data, and on the other hand is the noble environmental group, which has no motive other than to help the public. I'm embarrassed to say that it took me years to realize that their data were often soft, if not absurd, and that they had their own venal motives . . . to get on TV, to get famous, to get more grant money.'"

Brent Bozell III, "When the Media Looks at Risk," *Washington Times*, October 17, 1994, p. A17

Vice President Al Gore is another example of how well-meaning activists can obtain power through environmentalism. His best-selling book, *Earth in the Balance*, helped to strengthen his national political ambitions. The case of Gore is particularly illuminating, in fact, as the following box shows.

The Rich and the Near-Rich

William Tucker's *Progress and Privilege: America in the Age of Environmentalism* is one of the most perceptive and powerful books on environmentalism. Tucker says:

> Every survey that has ever been taken (including the Sierra Club's extensive polling of its own membership) has shown that support for environmentalism has been concentrated in the upper-middle-class, professional segment of society. Academics, attorneys, doctors, dentists, journalists and upper-income suburbanites have been, without question, the backbone of the movement. One extensive polling showed that support for environmental causes picks up strongly when income levels reach about \$30,000. . . . The "plain old rich people" have brought the ideas and attitudes. . . . The idea of looking on material progress and economic security as an irrelevant and

The following is a *Washington Times* editorial, March 2, 1994, reprinted with permission.

Mr. Gore in the Balance

In the summer of 1992, the *New Republic* published an article reproving then-Senator Albert Gore for inviting journalists to ignore scientific findings that undermined his warnings of impending environmental doom. Encouraging that kind of "self-censorship," wrote Gregg Easterbrook, is dangerous ground for liberals who are supposed to be champions of skeptical debate. . . .

Apparently, Mr. Gore's fears have worsened. Now vice president, he is . . . personally attempting to put scientists skeptical of the sort of apocalyptic outlook one finds in Mr. Gore's *Earth in the Balance* on a media blacklist. . . .

Mr. Gore urged [*Nightline*] to examine the connections between scientific skeptics and assorted politically incorrect business, religious and other groups. . . . *Nightline* examined the material and found that "in a manner of speaking" there were links between the scientists and the groups. For example, Fred Singer, oft-published in peer-reviewed scientific journals . . . "is on the executive advisory board of . . . *The World and I*, which is funded by the Unification Church International, . . . however . . . Mr. Singer has other noteworthy credentials as a former Environmental Protection Agency official, University of Virginia professor of environmental sciences and Department of Transportation scientist.

Patrick Michaels, an associate professor of Environmental Sciences at University of Virginia . . . receives funding from a consortium of coal companies to publish his *World Climate Review*. But every major environmental group in the country receives industry funding of one sort of another.

Does it taint the findings of climate change skeptics? Apparently not. Three years ago on *Nightline* Mr. Singer predicted, correctly as it turns out, that the oil fires in Kuwait would have only limited environmental effects. Mr. Gore's fellow apocalyptic, Carl Sagan, wrongly predicted disaster.

Computer models apparently led Mr. Sagan astray, and Mr. Koppel pointed out computer models are the basis for Mr. Gore's predictions on so-called global warming. With that in mind, Mr. Koppel allowed

Continued on next page

scientists to debate the relative merits of climate models and proposed government remedies. And there was plenty of debate. If the program showed anything, it's that there clearly is no scientific consensus on the matter.

Showing that debate was itself a rebuke to Mr. Gore, who wants to stifle one side of it. But Mr. Koppel . . . concluded, "There is some irony in the fact that Vice President Gore, one of the most scientifically literate men to sit in the White House in this century . . . is resorting to political means to achieve what should ultimately be resolved on a purely scientific basis. . . . The measure of good science is neither the politics of the scientists nor the people with whom the scientist associates. It is the immersion of hypotheses into the acid of truth. That's the hard way to do it, but it's the only way that works."

To date, Mr. Gore has taken the easy way out. He's tried guilt by association. He's tried blacklisting scientific critics—using tax-paid staff, by the way. He's tried journalistic self-censorship. Such tactics prove nothing about the Earth's environment. They only put his integrity in the balance.

> vulgar nuisance cannot be picked up overnight. . . . It is usually the sons and daughters of people who have achieved complete material security who make the most strident environmentalists.

Those Who Fear Chemicals

Another group of environmental leaders are the people with an intense dread of man-made chemicals. All of us, of course, must cope today with far more technology, a much more rapid pace of technological change, and more highly specialized socioeconomic systems than any people in the previous history of the world.

It's tough to be constantly adapting to change. Still, some people retreat from modern living's intimidating challenges by focusing instead on the "solvable" challenge of making their lives "toxin-free."

They often carry the concept to amazing lengths.

Some are afraid of aluminum cookware. They fear that the traces of aluminum that can be picked up from the food cooked in such pans can cause severe reactions in the nervous system—or even be the source of Alzheimer's disease. (It was never very likely that such tiny amounts of aluminum were causing problems, and it now appears that the whole scare may have been due to dust-contaminated samples.)

Others are afraid of pressure-treated lumber. One writer noted that "tests had proved" that carrots grown under a deck made from pressure-treated lumber had a higher level of arsenic. The arsenic level was not high in any absolute sense of representing a health hazard—but that fact didn't stop pressure-treated lumber from going on the danger list.

They hate and fear plastics of virtually all kinds.

Again, there are not many chemophobes. But they are vocal and persistent.

Here are a few quotes from just one issue of one magazine—*Green Alternatives for Health and the Environment*, vol. 3, no. 4, October/November 1993:

"Admittedly, the many factors involved in causing such complex diseases as Alzheimer's and osteoporosis are not known. However, aluminum has no known use in the human body. . . . While no definitive conclusions have been reached (about aluminum cookware) there are enough results to warrant caution about ingesting too much aluminum."

Kathy Gibbons, "Aluminum in My What?" p. 14

"Our Certified Kitchen Designer staff is well versed in alternative products. . . . [S]teel cabinets are inert, meaning virtually no outgassing of toxic chemicals. . . . [We can provide] optional non-tox doors and sides with special non-tox stain finishes to create the 'wood look' over the steel cabinets. We can supply whatever wood species would be chemically tolerant to the user."

kitchen-design advertisement, p. 35.

"Rachel Perry believes that while natural makeup is good for the surface layer of the skin, its ingredients do not actually penetrate into the bloodstream. . . . On the other side . . . Logona stated that one's skin is a membrane that absorbs materials and that whatever you want to avoid eating you should also avoid applying to your skin."

"Natural Cosmetics," pp. 38–41

"The chemicals in bedding most often cited as potential sources of concern are pesticides, herbicides, fire retardants, the various substances in synthetic fibers and the formaldehyde sometimes used to wrinkle-proof sheets. Even minute amounts of any of these chemicals could cause immediate health problems in a chemically sensitive individual. . . . Michael Dimock of Jantz Design, another natural bedding manufacturer, says 'There is no hard data, but people's (health) problems clear up' when they switch to organic bedding."

"Bedding," pp. 44–45

Organic Farmers

Organic farmers are obviously part of the environmental movement, but a small part of it because there are very few of them. Moreover, organic farming doesn't leave much time for organizing the general public.

The major significance of organic farmers is to serve as visible examples.

I attended the 1993 organic farming association's conference in Amherst, Massachusetts. It was charming. Kids ran sack races. People tasted organic wine. Gardeners were taught how to attract more butterflies. Members sold one another handmade brooms and llamas (which produce a "really fine fiber for hand-woven garments"). There was a wood-chopping contest.

A local grower taught us how to prune raspberry bushes so that they would live twice as long (but produce only half as many berries per year).

There was only one problem. Nobody said anything about productivity. There was none of the usual "farmer talk" about yields per acre or milk per cow. Nobody mentioned the need to raise the world's farm output by threefold to supply an adequate diet to a larger, more affluent world population in 2050.

I spoke to the conference on biotechnology's potential to produce more food from fewer acres. I pointed out that shifting to organic farming at its current low yields would mean plowing down wildlife habitat equal to the land area of North America. I noted that the world had less than 20 percent of the organic nitrogen needed to *support* global organic farming.

In response, one hot-eyed organic grower likened biotechnology and farm chemicals to nuclear radiation. Another wanted a "philosophical" decision on biotechnology. I admitted I am no philosopher; I admire biotech purely for its practical ability to save people and wild creatures from famine-related destruction.

The organic community felt *no* urgency about the world food problem. They mainly wanted enough cash from the produce of their little farms to maintain their hand-crafted communities.

I came away wondering, *who are the press agents for organic farming?* Who's touting them as the world food solution? Certainly not these sandaled folks!

Organic farmers believe in what they're doing, but few of them have demonstrated any media skills or effective lobbying capacity. The people pushing organic farming are apparently the eco-activists, most of them wearing business clothes and roaming the governmental and media capitals of the country.

Population-Phobes

There seem to be millions of people who are afraid of more people. In fact, there may be more "environmentalists" in this category than any other. Because of their numbers and secretly radical beliefs, they are politically quite dangerous.

They don't *want* high-yield farming to feed more people. Naturally, they have no open intent to cause the deaths of billions of humans. However, they are extremely uncomfortable with the idea of living in a more crowded world.

They want to solve the population equation *only* by suppressing births, even though this is not realistic. Hence Senator Bumpers's proposal to gut the funding for the Green Revolution agricultural research effort, even though it has been the salvation of humanity and wildlife over the last 30 years and is even more urgently needed for the next 30. Instead, the politicians and the environmentalists are quite comfortable putting that money into "population management" programs, that simply cannot stop the population growth quickly.

It's almost as if they're saying, "Let them use the condoms or starve. We'll stop the population growth one way or the other."

In reality, we don't have the "luxury" of starving the people and keeping the wildlife. If we don't feed the larger population by raising crop yields, they will feed themselves by plowing down every inch of ground possible.

But many people instinctively react like Senator Bumpers, in his Appropriations Committee hearing. He didn't *want* to hear that more people could be fed.

There seem to be millions of others like him:

- Why else would authors such as Lester Brown, Paul Ehrlich, and the Paddock brothers (*Famine 1975!*) be able to recycle their same failed famine predictions over and over to renewed success in the bookstores?

- Why else would the *Amicus Journal* of the Natural Resources Defense Fund name Brown its "humanitarian of the year" for loudly being antihuman?

- Why else would the National Education Association and the teachers of America make green publications some of the most widely used "outside source materials" in U.S. education?

MYTHS SUPERIMPOSED ON HISTORY:

"Our species once did live in stable harmony with the natural environment. . . . That was not because men were incapable of changing their environment but for some more enveloping and deeper reasons still. The change began between five and ten thousand years ago and become more destructive and less accountable with the progress of civilization. The economic and material needs of growing villages and towns are, I believe, not causes but results of this change. . . . [I]t wrenched the ancient social machinery that had limited human births . . . a kind of failure in some fundamental dimension of human existence . . . a kind of madness."

Paul Shepard, *Nature and Madness*, Sierra Club Books, 1982[9]

Reality Comment: Shepard, perhaps without realizing it, is actually describing the impact of society's shift from hunting to agriculture. That shift began about 10,000 years ago and spread gradually around the world.

It did not occur on account of rising birth rates, as Shepard implies. It was rather the direct result of lower *death* rates—as higher food output from farming staved off more famine and supported additional people. Shepard's willingness to misread birth rate causes and effects emphasizes how deeply antagonistic the environmental movement truly has been to high-yield farming.

HISTORICAL REALITY:

"In 1845, the twenty-eight-year-old Thoreau . . . built himself a little cabin near Walden Pond in Concord, Mass. The land was owned by Emerson and was about as far out of town as the average modern driving range. . . . Thoreau frequently went to dinners and parties in Concord, and according to his list of household expenses in *Walden*, he sent his laundry out to be done. Thoreau lived in his shack for two years, devoting his time to being full of sanctimonious beatnik. . . . And he is the source of the loathsome self-righteousness that turns every kid who's ever thought 'a tree is better looking than a parking lot' into Saint Paul of the Recycling Bin."

P. J. O'Rourke, *All the Troubles in the World: The Lighter Side of Overpopulation, Famine, Ecological Disaster, Ethnic Hatred, Plague and Poverty*, New York, Atlantic Monthly Press, 1994, pp. 129–30

"Urbanization (in ancient times) meant that a number of new problems had to be solved. . . . Urbanization was accompanied by rapid progress in the technology of large-scale construction, transport and agriculture. . . . The inhabitants of large, sparsely populated continents were doomed to be illiterate subsistence producers. Their rich natural resources were of little use to them."

E. Boserup, *Population and Technological Change*, University of Chicago Press, 1981

Even though past population growth has stimulated powerful improvements in human technology and public administration, there is no question that things would be simpler and easier for everybody if it stopped at 6 billion.

Unfortunately, that is wishful thinking. Neither Lester Brown nor Senator Bumpers nor you nor I are making life's intimate decisions for mothers and fathers in the Third World.

Sheer crowding doesn't seem to be frightening Calcutta or Hong Kong. Neither does it seem to frighten the affluent but densely residential Dutch. Even the residents of Manhattan Island, who enthusiastically support Planned Parenthood and the Worldwatch Institute, stack themselves into a population density that makes an anthill seem spacious.

WHO WILL COMPILE THE SUBTRACTION LIST?

"It is imperative that we reduce U.S. population to no more than 150 million and stabilize it there. . . . But the path we are on now is propelling us headlong toward a catastrophic size of 400 million and more."

Mailing from Negative Population Growth, Inc., Teaneck, New Jersey, Spring 1994

Reality Comment: If NPG plans to "reduce" the U.S. population by over 100 million, they will outrank Hitler and Stalin as history's foremost mass murderers.

The Worriers

There are *lots* of handwringers. Nature and evolution have seen to that.

Careful people tend to live longer than daredevils. They are more likely to reproduce. They try to teach their children caution. They surround themselves with fences, zoning ordinances, vaccinations and make sound diet and life style choices.

In the days of the wilderness, it was wise to keep looking over your shoulder—and to both sides. Lethal dangers were waiting just a paw-swipe away.

Today, however, both evolution and technology have left us with far fewer dangers. About all we have left of statistical importance are mistakes like doing drugs, smoking cigarettes, and not wearing seatbelts.

Most of the people reading this book can count on living into their eighties.

If we eat lots of fruits and vegetables and get regular exercise, we can even count on *enjoying* those "golden years."

That leaves a lot of anxiety with no valid place to go.

No saber-toothed tiger. No typhoid or cholera in the chlorinated water. No undulant fever or campylobacter in the pasteurized milk. No fear of personal famine; the supermarket is a few blocks away and probably open at least 15 hours per day even when it snows.

Inevitably, at least some of the anxiety that saved our forefathers from saber-toothed tigers winds up targeted inappropriately at such irrational worries as these:

- All of the good jobs in America will disappear.

- Our children, the best-educated and best-equipped generation in the world's most successful economy, will never live as well as their parents (many of whom remember the threat of polio and the death of playmates from ailments now easily cured by antibiotics).

- Electronic emissions from computer terminals are ruining our health.

There are *lots* of handwringers. Their attention span, however, tends to be short.

The Guilt-Ridden

Many people in the world believe that there's only a limited supply of wealth. Thus, if one person or country has more, they must have gotten it at the expense of another person or country.

More and more people fear that they got their affluence by exploiting more than their share of natural resources. They don't understand that most of the credit goes to their being members of a successful society with constructive institutions, productive values, and lots of knowledge.

There is no rational basis for the "limited-wealth" belief. If the world ever was constrained by "natural resources," that time passed at least 200 years ago. Today, knowledge is creating new wealth out of very unlikely resources—such as sand, which we turn into silicone chips and glass fiber telecommunications cables. Hard work, investment, and commerce are making billions of people affluent all at once.

Two hundred years ago, people understood where wealth came from, because it came directly. The farmer who had the best land and worked the hardest usually got the best crop. The shoemaker who made the best shoes got the business, and so on.

Today, millions of people do such specialized things that those things may not make total sense even to those who do them successfully (such as floor traders on commodity exchanges). Thus, there's lots of uncertainty about wealth and poverty and the reasons for them.

That may be why the environmental movement has had huge success in tapping guilt feelings about wealth, in America and the other affluent countries. When a supposedly selfless, confident-sounding "expert" says that people should feel guilty for living too well and "endangering the spotted owls," lots of them do.

AS AN EXAMPLE:

"Rising expectations for the poor is a cruel joke foisted upon them by the Establishment. As our new economy of use-it-once-and-throw-it-away produces more and more products for the affluent, the share of our resources available for the poor declines. Blessed be the starving blacks of Mississippi with their outdoor privies, for they are ecologically sound, and they shall inherit a nation."

Wayne H. Davis, "Overpopulated America,"
The New Republic, January 1970

True Believers Lacking Something to Believe In

The declining depth of America's religious beliefs has also played an important role in the rise of environmentalism. In fact, environmentalism can be viewed as a return to the pagan nature-worship of pre-Biblical times.

There are many reasons put forward for the declining importance of religion in America and the Western World, all representing arguments that are beyond the scope of this book.

However, if we look at the prescriptions of most religions, we see that they are aimed at telling us how to live our lives productively and harmoniously—and to build our societies rather than tearing them apart. Such messages, which represent thousands of years of learning about human behavior, have now been pitched out in the secular revolution.

Ironically, the very "freedom" delivered by the secular revolt has left many people looking for some larger ideal to justify their lives. "Nature" has become the religious substitute for many of them.

That is unfortunate, mainly because nature-worship is unlikely to be truly satisfying or to deliver a successful society that protects its children and resources. It has already been abandoned time and again in history for visions that offered a more elevated understanding of human beings and their place in the world.

MYTHMAKERS ON HIGH:

"In his compelling work, *Broken Trust, Broken Land*, University of Washington natural resource sociologist Robert G. Lee reminds readers of a point made by many scholars—that environmentalism is a religion. But Lee goes farther, suggesting it to be a reincarnation of Calvinism, the faith of early Puritans. A central tenet of Calvinism . . . is that while 'the elect' are 'predestined to eternal salvation,' everyone else is 'predestined for eternal damnation.' This . . . makes it easier for true believers to inflict pain and ignore suffering. They feel justified punishing, and withholding sympathy from, those deemed to break God's law."

Alston Chase, environmental columnist, "The Election of 1994 Was a Religious War," Creators Syndicate, Inc., November 1994

The Environmental Strategies

How, in the face of its many intellectual and moral weaknesses, has the environmental movement achieved so much? The evidence suggests three answers:

- The movement makes heavy use of fear as a strategy, constantly declaring new crises even though each may be based on shaky science. Environmental zealots attempt to create a constant crisis mentality, with new threats linked to the "dangers" of modern technology. If a "crisis" resonates with the public, they pursue it. If science effectively refutes a danger, they simply shift to a new one.

- The environmental movement has successfully exploited the media and the typical reporter's breathless hunger for scary front-page headlines. Often, this media impact has translated into political power even without the environmentalists having to win elections. Polling results have been more than enough to swing politicians and bureaucrats into line.

- The movement is made up of people who enjoy networking and organizing. Many of them have lots of spare time and energy. When I attended a big national environmental meeting in Louisville, Kentucky, in 1993, the entire meeting seemed to be affluent housewives, retired professionals, and salaried activists.

ON-LINE MYTHMAKING:

"Ecoline, a toll-free, global information system developed by the University of Vermont's Environmental Studies Program and the Together Foundation, connects callers to live operators who can provide the names, addresses and phone numbers of more than 60,000 organizations worldwide working on sustainable development, environmental projects and 'environmentally friendly products.'"

"Good News," *Earth Island Journal*, Winter 1993–94[10]

Beyond Political Correctness

At the Hudson Institute's 1994 agricultural conference in Indianapolis, one of our panelists was Dr. Adam Finkel of Resources for the Future. Dr. Finkel was defending the use of the high-dose rat tests to find the least little scrap of potential "cancer risk" in pesticide residues—but he volunteered that if the use of low-risk pesticides would save wildlife, we should use them. Finkel is a graduate of the Harvard School of Public Health—and a former staffer for Vice President Al Gore!

Small Steps in Reality

Fortunately, some important environmentalists are beginning to recognize the importance of high-yield farming.

In 1997, in a debate held, fittingly, at the zoo in Sydney, Australia, the leader of Australia's biggest environmental group agreed with me that the world urgently needs higher-yield agriculture. In fact, Jim Downey, executive director of the Australian Conservation Foundation, said that getting higher yields from the existing farmland "may be the most urgent part of the conservation equation."

Downey's group had been part of an Australian conservation alliance, along with the national government and Australian farm groups. The coalition had pursued such practical conservation improvements as conservation tillage, and grass filter strips and attempting to rein in the foxes (an alien species there), which threaten to cause the extinction of many of Australia's unique small mammals. Downey had obviously been impressed by his collaboration with farmers.

More recently, Carl Pope, the executive director of the Sierra Club, wrote a letter to *Philanthropy Magazine*, "strongly endorsing" the Center for Global Food Issue's position that the world needs more high-yield plant breeding research, including biotechnology, done by public and philanthropically supported institutions.

Pope says he still dislikes farm chemicals, and fundamentally distrusts big companies that are producing pesticides and commercial biotech products. However, his letter says that a "massive increase in [public] research is absolutely critical; only then can the promise of high-tech breeding be combined with the social and environmental needs of the world."

Mr. Pope, in fact, quoted me on the urgency for research: "If the world's future food production was left to slash-and-burn farming, we could expect to lose half of the remaining tropical forests in the next several decades. To save the wildlands with human dietary change, we might need 50 percent of the world's population to become vegan . . . there is no likelihood of such a global dietary shift occurring voluntarily. Quite the contrary, Chinese meat demand has more than doubled over the last seven years."

Pope concludes: "We have, unfortunately, passed the point in human history where we can adopt any single fix for our problems. We will need to combine social changes such as women's education with family planning to bring down fertility; publicly accountable and oriented research into better plant varieties, with a reduction of excessive reliance on chemical inputs, to increase food production in an environmentally sustainable way; and creative strategies to change farming practices in ways that will accommodate biological diversity alongside food and fibre production."

Pope's letter contrasts sharply with the negative positions of Greenpeace and the World Wildlife Fund, which have been campaigning to have national governments ban the use of biotechnology in food production—even though biotech is swiftly proving to be mankind's biggest new tool in raising crop yields.

Maurice Strong, the Canadian who chaired the Rio Earth Summit in 1992, and who currently chairs the Earth Council, recently told a press conference, "This is the most exciting time in agricultural history. The world is at the cusp of a new scientific revolution in the biological sciences. . . . A compelling moral and ethical imperative underpins our call for a global effort to harness the best of science to meet the needs of the poor and the environment."

Strong noted that biotechnology could (1) permit us to produce more food on the same area of land, thereby reducing pressure to expand farming into the wilderness, rain forests, and marginal lands that are the world's richest storehouses of biodiversity; (2) Reduce post-harvest food losses and thus boost the "realized nutritional yield" of food per acre; (3) Displace costly and resource-intensive inputs such as fuel, fertilizers, or pesticides; (4) encourage the adoption of more sustainable farming practices, such as conservation tillage, precision agriculture, and integrated crop management.

In 1998, Strong had chaired the review commission for the Third World's farm research network, The Consultative Group on International Agricultural Research. He noted that the private sector is spending billions on biotech research, but all under a proprietary regime. He says the key question is how the new science can be harnessed for the needs of the poor and the environment—and this is the challenge to the CGIAR.

Lester Brown, our Malthusian of previous chapters, wrote a chapter on "Struggling to Raise Cropland Productivity," in the *State of the World (1998)*. "One of the obvious conclusions of this analysis is that there is a pressing need for a much greater investment in the agricultural sector," writes Brown . . . "in agricultural research, agricultural extension, soil conservation, and irrigation efficiency."

This is a welcome, though modest, concession from a man who has spent the last 25 years telling us that there was no point in investing in agricultural research because it produced only "the illusion of progress." During that period, Brown maintained that the only solution to the world's problems was population management.

These are concerned conservationists who have wanted to better the world and save its wildlife. Until now, the population ogre has transfixed them. They have too often seen high-yield farming as a pernicious "enabler" of population growth.

To the extent that the environmental movement can move to embrace high-yield conservation, it can offer an enormous boost to the credibility of modern agriculture.

We must help the new converts with gentle urgency into a full appreciation of high-yield farming's massive conservation benefits. As the Pope and Strong letters indicate, we must start with plant breeding and biotechnology. Only gradually will they begin to see virtue in such other important aspects as herbicides, confinement meat production, and frozen foods. Only late in the process will they become eager for free farm trade. But realizing the need for high-yield conservation is the first and most critical step for the environmental movement.

Notes

1. William Tucker, *Progress and Privilege: America in the Age of Environmentalism*, Garden City, New York, Anchor Press/Doubleday, 1982, pp. 151–152.

2. Philip Shabecoff, *A Fierce Green Fire: The American Environmental Movement*, New York, Hill and Wang, 1993, p. xiii.

3. Shabecoff, op. cit., pp. 88–90.

4. Lynn White, "The Historical Roots of Our Ecologic Crisis," *Science*, March 10, 1967.

5. W. McKibben, *The End of Nature*, New York, Random House, 1989.

6. Clifford D. May, "Preservation for Profit," *New York Times Magazine*, September 12, 1982, p. 146.

7. David Brower, quoted by Virginia Postrel in "The Green Road to Serfdom," *Reason*, April 1990, pp. 23–4.

8. Thomas Berry from *The Dream of the Earth*, excerpted in *Learning to Listen to the Land*, Washington, D.C., Island Press, 1992, p. 257.

9. Paul Shepard, excerpt from *Nature and Madness*, reprinted in *Learning to Listen to the Land*, op. cit., pp. 136–149.

10. *Earth Island Journal*, vol. 9, no. 1, San Francisco, Earth Island Institute, San Francisco, p. 5.

24

What Can Farmers Do?

MYTHMAKERS SAY:

"Americans are changing the way they view the land. Increasingly, they regard land less as a commodity for trade and more as a resource to preserve. They no longer consider land solely as a site for homes and buildings or as a place to grow food and fiber. . . . The future of America's Great Plains lies in tapping into this enlarged appreciation of land to create new, restorative uses for the vast stretches that are now agriculturally marginal. We call these new uses the Buffalo Commons."

Frank and Deborah Popper, *Forum for Applied Research and Public Policy*, Winter 1994

"This is considered almost holy work by farmers and ranchers. Kill off everything you can't eat. Kill off anything that eats what you eat. Kill off anything that doesn't feed what you eat . . . the more competitors you destroy, the more humans you can bring into the world, and that makes it just about the holiest work there is."

Daniel Quinn, *Ishmael,* Bantam/Turner books, 1992, p.132

"Given an expanding food supply, any population will expand. This is true of any species, including the human. The Takers [farmers] have been proving this here for ten thousand years. For 10,000 years they've been steadily increasing food production to feed an increased population, and every time they've done this the population increased still more."

Quinn, *Ishmael,* p. 136

"One American burdens the earth much more than twenty Bangladeshis. . . . This is a terrible thing to say. In order to stabilize world populations, we must eliminate 350,000 people per day. It is a horrible thing to say, but it's just as bad not to say it."

Jacques Cousteau, TV show host and advocate for oceans and the environment, *UNESCO Courier*, November 1991

"We have overburdened Earth's carrying capacity, and there is little left to offer the newcomers—especially those who live in poverty-ridden areas of the world. . . . Dwindling resources, compounded with high fertility rates, yield an ever widening deficit of food, water, shelter and basic health care."

Werner Fornos, president, The Population Institute, fund-raising letter, March 1999

REALITY SAYS:

"I strongly endorse [Dennis Avery's] call for a renewed commitment to government and philanthropic funding of agricultural research, including research into conventionally bred or bio-engineered new varieties of crops. A massive increase in such research is, as Avery argues, absolutely critical. Only then can the promise of high-tech breeding be combined with the social and environmental needs of the world."

Carl Pope, Executive Director, Sierra Club, letter to the editor of *Philanthropy Magazine*, December 21, 1998

"This is the most exciting time in agricultural history. The world is at the cusp of a new scientific revolution in the biological sciences. . . . A compelling moral and ethical imperative underpins our call for a global effort to harness the best of science to meet the needs of the poor and the environment."

Maurice Strong, chair of the Earth Council and former secretary-general of the Rio Earth Summit, 1992

"Environmentalists often disdain America's factory farms in favor of the agriculture practiced by traditionalists in the Adirondacks and peasants in the tropics, but those farmers need much more land to grow food. Avery calculates that the increased efficiency of modern farming has saved an area of land equal to the combined size of Europe, the United States and South America. 'Old MacDonald's farm couldn't feed the planet's population without destroying virtually every square mile of wildlife habitat,'

Avery said. 'New York farmers should be exporting the milk so India doesn't dispossess the Bengal tiger.'"

John Tierney, *New York Times*,
March 18, 1999

Farming has never before faced urban hostility. Throughout history most of the population has been farmers and the few who weren't visited their farmer relatives often. Farmers have been praised and honored as society's pioneers and food providers. The small farmer was Thomas Jefferson's ideal citizen, the poster-child of democracy.

After World War II, the Industrial Revolution lured huge numbers of workers from the farms to the cities, and people began to lose firsthand knowledge of where and how food is produced. This is dangerous for agriculture, because modern farming methods have changed as radically as nonfarm technology.

A farmer from biblical times, walking onto an American farm in 1854 would have felt pretty much at home. Even in 1924, that biblical farmer would have understood the farming system though he would have marveled at the speed and power of the horse-drawn equipment.

Our biblical farmer would be lost on today's farm. He would be frightened of the tractors that speed crop planting to get the early-season yield gains. He would marvel at the yield of hybrid seeds, let alone the conservation tillage that looks so much uglier than a well-plowed field. He'd be surprised at artificial insemination, and, perhaps, be happy that confinement feeding kept the pigs and chickens out of the farmhouse.

Today's urban residents understand high-yield farming even less well than would our biblical farmer. And, frankly, we haven't bothered to tell them much about it. Farmers stopped talking to nonfarmers when the nonfarmers moved away to the cities. Farmers started talking to the agriculture committees of their legislatures instead. That worked until the eco-activists trumped our strategy by going directly to urban audiences with their tales of agricultural evil.

Today, about the only messages that urban residents hear on farming come from its enemies.

City folks can't hear messages about hunger. They're not hungry, and can't imagine they will be. City folks can't hear warnings that the supermarket might someday run out of food. They just *know* it won't. City folks can't hear messages about higher-cost food. They know it will cost only a small percentage of the money they earn.

Fortunately, we in agriculture still have one message that reaches a live nerve ending in the city: saving wildlands. They also urgently want to save them. They just don't know that they need modern farming to do it, because we haven't told them.

Their ignorance is dangerous, mainly because in a modern democracy, public opinion sets the regulatory parameters. If the public decides it won't tolerate confinement hogs, it can drive them out of business. If the public truly decides that it doesn't need pesticides, it can ban whole categories of them on politically driven criteria. City folks hold the future of farming in their hands.

Today's urbanites rarely think about farming. When they do, it's usually because the media or one of their neighbors has brought up something unpleasant: the cancer risk of pesticides, the additives in food, or the ugly idea of cramming lots of hogs and chickens into "confinement."

School districts are now mandating that their schools be pesticide free, because activists tell parents that farm chemicals are risking their kids' health. The Audubon Society has a huge membership that suspects they'd have more birds to enjoy if they could ban farm chemicals. Many urbanites think farmers are careless about soil erosion and need to be pushed into farming more safely.

At the same time that city folks are driving cars with fifteen computers inside, and watching fly-by-wire stealth fighter planes win their wars, they want farming to look like Old McDonald's farm of 1875. They don't understand that Old McDonald was killing cabbage worms with lead arsenate, the only pesticide ever found to be a real cancer risk to humans. They don't remember that Old McDonald's cows were often spreading tuberculosis and undulant fever. They've never realized that Old McDonald's low-yield farming had literally worn out the farms of the eastern seaboard, and was making a good start on wearing out the world's biggest chunk of prime farmland, the American Midwest.

What Can Farmers Do?

Agriculture cannot disprove new scare theories as rapidly as activists can construct them. That's been proven over and over again, from the original "cranberry scare" of 1959 through DDT and non-Hodgkin's lymphoma to the current activist campaign in California to "prevent any use of pesticides in or near our children's schools."

Nor can agriculture depend on the urban media—oriented as they are now toward front-page consumer scares—to give in-depth, unbiased coverage of modern farming.

For the first time in history, humanity is in danger of rejecting a more productive farming system. In fact, society seems about to reject the system that has given most of the world's people their first real food security.

The mechanism of destruction will be regulatory strangulation.

Reprinted from BridgeNews Forum, November 20, 1998

Farmers Need to Repair Their Public Image

Switzerland nearly outlawed biotechnology in food production over the summer. Was this an isolated problem, or does it mean more trouble for modern agriculture as it tries to triple world food production for the 21st century?

Switzerland's "Gene Protection Initiative" would have banned the release of genetically modified organisms into the environment. In effect, the initiative would have outlawed biotechnology for food production in Switzerland. And that was the intent.

The referendum resulted from a six-year campaign by a consortium of 70 Swiss organizations that collected 100,000 voters' signatures.

Key organizations in the anti-biotech coalition included Greenpeace, the World Wildlife Fund, Swiss Organic Farmers, and several animal rights groups. Ominously, the big Swiss Lutheran Women's League and Swiss Catholic Women's League were also members.

The well-funded publicity campaign featured full-page newspaper ads that equated biotechnology with nuclear technology. Campaigners also harshly criticized science for experiments involving not only lab rats, guinea pigs, and monkeys, but also worms, flies, and fish.

The vote was expected to be close until just a few weeks before the referendum, when a major campaign was launched by Switzerland's scientists and research-based companies.

Voters were reminded that a ban would drive away many of Switzerland's most notable companies, including the large pharmaceutical industry. The jobs they provide would go elsewhere too.

The biotech proponents also played up the favorable image of biotechnology in human medicine, with posters showing sick children, and Swiss biologists working on well-known diseases.

Agriculture cannot take much comfort from the results in Switzerland, however. The real lesson from the Swiss referendum is that there will be more anti-farming ballot initiatives, and pressure for still-tighter regulation of farming and farm research.

The fact that both the Lutheran and Catholic women's organizations joined the coalition is a huge red flag for the future. The women's church groups have lots of members, and they're not far-out activists like many of the people in Greenpeace.

Obviously, modern agriculture has failed to convince First World mothers that using modern science in food production is a positive for their children.

Why is agriculture losing its reputation when it's doing a better job of feeding the world than ever before? Just a generation ago, one billion Asians were expected to die of famine; now they're eating pork and ice cream. Those numbers may be part of the problem.

Eco-activists have pushed the erroneous idea that more food will mean a world with more people. Meanwhile, Switzerland's seven million people are crowded into the narrow valleys of the Swiss Alps.

Few Swiss understand that we're living in the first period in world history when increased food security means smaller families, not larger ones.

Animal rights have also become a much more emotional issue for city folks. Urbanites no longer want to think about "harvesting" animals as a crop.

Even the Lutheran and Catholic women's groups are apparently uncertain now about the Judeo-Christian concept that man was given "dominion" over the other species on the planet.

The Swiss would rather think of cows being milked than slaughtered. They'd rather not have animals used in research at all.

All of this adds up to trouble for modern agriculture. As the world gets richer and more urbanized, the traditional appreciation of farmers will fade even further.

The image of the hardworking farmer with a hoe is being replaced by an image of a rural tycoon on a big tractor. The fear of famine is being replaced by a more vivid fear of overpopulation.

Today's consumer is distressed that farmers don't treat battery hens and confinement sows like family pets.

Given all this, plus the powerful potential of biotechnology to alter natural organisms and eco-systems, the Swiss electorate nearly rebelled.

Farmers aren't making the environmental argument to the city folks. The only information most city people get about modern farming today comes from activists condemning it.

What to Do:

1. Don't Be Diverted by the "Family Farm" Issue

Farm subsidies were offered for 60 years as a way to "save the family farm." Yet farmer numbers have continued to decline in all the subsidizing countries.

Now organic farming is being offered as a way to "create more farming jobs." Unfortunately, organic farming doesn't create the sort of jobs most young people want. Organic farming jobs involve lots of hand labor, in all kinds of harsh weather, and they don't pay very well.

Recently, Pennsylvania's Rodale Institute reported on a comparison between one of its organic production sites and the mainstream farms surrounding it. After nine years of "preparing the land" the organic production still yielded 21 percent less crop and required 42 percent more labor. Ask the local high school kids if requiring more hand weeding in the hot sun will keep more of them on the farm. And, if and when all farms are required to be organic, the price premium for organically grown commodities will disappear.

Let's face it, most of the kids who leave the farm go because they want higher pay, less drudgery, and more bright lights. If they hadn't left, farmers' incomes would be even lower.

The whole world is moving down the path toward fewer farmers. America has already traveled down that road, from 95 percent farmers to less than 5 percent farmers. The future vibrancy of America's rural communities will depend on higher incomes from improved export opportunities for its farmers; and the new trend toward retirees and decentralized urban workers moving to the country to get more attractive rural lifestyles.

What to Do:

2. Stop Talking About "Farm Surpluses"

Farmers have been talking about food surpluses all of my life. Is it any wonder that city folks think we should put more land back into bird sanctuaries, bison grazing, and wildlife parks? Is it any wonder urbanites think we can mandate that our farming look picturesque and heartwarming? (For a guide to what they want, check the TV ads at Thanksgiving and Christmas. Red barns and Budweiser draft horses thrill them, not confinement hog barns, steel grain bins, and emotionally unsettling pesticide sprayers.)

Farmers are the only people in history who have tried to make a virtue out of being unnecessary. Every time farmers complain that the crops are too big, they make the argument against their own high-yield farming. Every time a farmer says the word "surplus" in public, organically inclined food editors and environmental regulators post the clipping on bulletin boards all over the country. It then becomes their best argument for eliminating the high-yield farm inputs and going back to "natural" farming.

Farm prices are low in today's world because of trade barriers and food self-sufficiency policies in Asia, not because the world is producing more food than consumers want to buy.

In a world where Chinese families have $12,000 in annual purchasing power and want more meat and ice cream than their own farmers can cost-effectively produce, there is no farm surplus. When

India is getting two-thirds of the fodder for 400 million cattle, buffalo, sheep, and goats by stealing branches from its forests and crop residue from its soils, there is no farm surplus. When Indonesia is clearing tropical forest to grow low-yield grain for chicken feed on highly erodible soils, there is no surplus.

There are too many farm trade barriers. There are too many families in Asia paying more than they can afford to give their children high-quality diets. There are too many farmers extending low-yield farming into wildlife habitat.

But there is no surplus of farming resources.

No matter how bad prices get, American farmers should never again mention the word *surplus*. Crying surplus is a call for modern farming's death warrant as it will only hasten the regulatory murder of high-yield agriculture's imputs.

What to Do:

3. Retake the High Moral Ground with Hunger and Conservation Messages

The environmental movement has become the most powerful political force in the First World by taking the high moral ground away from farmers. While farmers still thought they wore white hats because they were "saving the world from hunger," (which, by the way, they *were* doing) activists indicted them for "creating overpopulation" by producing "too much food."

Farmers have always been more comfortable with being hunger-fighters than with the less tangible idea of saving tropical forests somewhere far away, but Americans respond with strong positives to saving wildlife habitat.

In the Third World, the validity of the messages is reversed: Third World publics respond very favorably to more and better food. But saving wildlands is not yet on their priority lists. That's why farmers must present both the hunger and conservation messages.

The First World will hear the conservation message and be slightly uncomfortable with meeting the hunger challenge. The Third World will be grateful for the opportunity to overcome hunger and malnutrition, and accept the conservation that comes with high yields.

Together, conquering hunger and conserving wildlands give high-yield farmers a strong claim to the high moral ground in every part of the world.

What to do:

4. Speak Out

I recently spoke to an audience of high-powered executives from big cities. One drove the latest Porsche and had never lived more than five miles from downtown New York City. Another was heavily involved with maintaining her company's collection of expensive 19th-century oil paintings. Virtually every one of these urban sophisticates came up after the presentation and thanked me for opening a window on the world that they had never looked through before. They suddenly saw farming in a totally new light.

Even though I was once hit with a chocolate pie by a protester, I have more often received standing ovations at liberal arts colleges. Even organic farming groups have to lower their hostility shields and think a bit when you ask them how they plan to produce three times as much food for the world of 2040—without clearing lots more tropical forest to make up for their lower yields.

Urban people need to understand that high-yield farming leaves lots more room for nature. They also need to understand some of the other major points of urban interest outlined in this book:

- The world population is restabilizing quickly, in part due to the food security, which allows parents to feel sure their first two or three children will live to adulthood.

- Very few of the world's people are willing to be vegetarian, let alone vegan.

- The same urban, affluent couples who are willing to stop at 1.7 children demand the highest-quality diets for the children they do have, with lots of meat, milk, and fruit, and extensive cotton wardrobes.

- The pet challenge will include at least 1 billion more cats and dogs, whose owners will demand high-quality diets for their beloved companions.

- There is no proven way to feed the world's children and still save room for wildlife, without still more and still higher-yield farming.

American farmers are intelligent and well educated. If each one of us spoke out on behalf of their profession to their neighbors, newspapers, and community groups it would, indeed, make a difference in how nonfarmers perceive modern agriculture.

What to Do:

5. Recognize the Media Bias (or Know Your Opponents)

The activists are not entirely to blame for the dilemma of modern agriculture.

Another real problem is that the journalists of today have abandoned the old idea of seeking truth. Instead, they seek a sensational "balance" that mainly features the most outrageous activists they can find. Always keep in mind the first priority of a journalist: "if it bleeds, it leads."

As an example, I was recently involved in a "debate" on BBC's World Service. I represented the millions of high-yield farmers in the world who provide perhaps 80 percent of the world's food supply while saving more than 15 million square miles of wildlands from being plowed for crops.

Vandana Shiva, an Indian woman who authored a book called *The Violence of the Green Revolution,* which contends that the 15,000 people who have been killed in the struggle for Sikh independence in India's Punjab were actually fighting over hybrid seeds, was my opponent! It would be a laughable—except that's what passes in modern journalism for "balance."

I've been on a first-name basis with Bill Stephens, the science editor of the *New York Times*, for many years. He quotes me on world hunger issues. He has never dared to touch the concept of high-yield conservation, because it is at odds with the Green views not only of his editorial board, but of his colleagues throughout Western journalism.

When our little Center for Global Food Issues released information showing that the Black River of North Carolina was still pristine despite having 9 million confinement hogs in its watershed, a reporter called and asked if the information came from an industry-financed study. We told her the data came from the State Department of Environment and Natural Resources. "Oh dear," she said, "that's not what my readers want to hear."

No less a personage than Tony Blair, the prime minister of England, has expressed his outrage that the media are ignoring prestigious, science-based views on biotechnology in food production, while eagerly chasing and highlighting anything that will keep the public's biotech hysteria boiling.

The British generally suspect that Prince Charles, the heir to their throne, can afford to be an organic farmer mainly because his family is one of richest in the world. Nevertheless, the British media love to quote Charles' organically based opposition to high-yield farming and biotechnology in food.

The same British media virtually ignore Charles' sister, Princess Anne, who says publicly that it would be immoral not to pursue biotechnology as a way to help overcome hunger and malnutrition in the world. No mainstream British paper covered Princess Anne's speech to the International Federation of Agricultural Journalists in 1997; she pointed out that organic farming takes more land away from nature!

Farmers cannot expect to get a fair presentation of their views in today's TV news shows and print columns. However well written their press releases, however brilliant their spokespersons, the best they can expect is occasional sympathy about bad weather and low prices. They can also expect lots of nostalgia about yesterday's small farms.

What to do:

6. As an Industry, Buy Advertising

This may be the most controversial statement in this controversial book.

Unfortunately, agriculture to date has been unwilling to spend either the money or the organizational effort to protect their image. It's almost as though farmers and agribusiness are still in denial; as though they cannot yet believe that society regards modern farming with distaste. They seem to believe that any minute, city folks will discover what a mistake they've made and come rushing out to Main Street, Nebraska, to apologize.

So far, the only agricultural entity telling urban America anything positive about high-yield farming is ADM, the big corn and soybean processor headquartered in Decatur, Illinois.

ADM hired David Brinkley, long one of the most respected journalists in the Western world, to explain in its TV ads that (1) farmers are still the most indispensable people because we have to have enough food; (2) that we really don't have any choice about providing the milk that the little Asian girl's mother wants her to have; and (3) that high-yield farming is saving millions of square miles of wildlands from being cleared for low-yield crops.

The Brinkley ads have been aired primarily on the Sunday morning "political" shows such as ABC's *Sunday Morning* and *One on One with John McLaughlin*. They are also aired on CNN's *Moneyline* and various other cable networks. Early on, there were complaints from other journalists that Brinkley had "sold out" by doing commercials—but no one has contested the accuracy of the messages Brinkley delivers.

Farmers may say, "It isn't fair that we should have to spend money to justify actions we know are good." It isn't fair. But in the age of environmental concern, the farmers and ranchers who have custody of vast stretches of America's land and most of its water had better be prepared to justify their stewardship to the city folks—or lose control of those resources.

An example of an advertisement for high-yield farming
(Courtesy, Archer Daniels Midland Company)

Who else will speak up for agriculture, now that it's been indicted before the court of public opinion? The land-grant colleges are ducking the public's disapproval, and competing for grants to research low-input farming systems. The U.S. Department of Agriculture has become a junior partner of the Environmental Protection Agency. State governors know there are more votes in the cities than in the countryside, and the governors appoint the heads of the State Departments of Agriculture. And, our schoolteachers owe their allegiance to the National Education Association, which is helping to lead the assault on modern farming.

Farmers used to be able to protect themselves by working with the agriculture committees of their legislatures. But now the urban perception of farming's sins is so ingrained that even farming's legislative friends often don't dare vote with the farmers.

Farmers and agribusiness are all alone. If they want more people to believe they are a major constructive force for people and the environment, they'll have to explain why—and they're already so low in the public's estimation that they'll have to spend money to deliver the explanation.

That's harsh. That's unfair—and it's real.

What others have done:

The Model Ad Campaign of the American Plastics Council

Meanwhile, the American Plastics Council has been conducting a highly effective campaign that has virtually taken plastics out of the environmental debate. The Plastics Council's ads show:

- little kids riding bicycles—wearing plastic helmets to protect their skulls from accidents

- a heart attack victim being saved, with the aid of plastic syringes, a plastic oxygen tent, and many other plastic medical devices

- a mother and son striding through a medieval food market that is full of rotting produce and sweating butchers hacking at fly-blown meat—and then striding out of the frame into a clean modern supermarket with the food protected by plastic

It reportedly costs the American Plastics Council about $20 million per year to present its views to urban America, in a positive and nonconfrontational way. It is costing American agriculture and agribusiness billions of dollars per year to meet the regulatory mandates being imposed because of today's urban misconceptions of farming.

Which seems the better bargain?

Epilogue

Think with me about a future time on the planet Earth. It might be about 2050.

Think about 8 to 9 billion humans who have the wealth and technology to coexist cooperatively, constructively, and enjoyably with their environment. Population growth has now stopped. Birth rates all over the world have now come down to match the low death rates produced by modern medicine. Population growth has leveled out, even with the remarkable death-delaying progress in biotechnology and gene therapy.

Contemplate the following events that brought this about:

- long-term birth control technologies;

- reform of outdated welfare systems, which used to encourage births among the poor of the First World;

- the rapid spread of affluence to all of the Third World except remote parts of sub-Saharan Africa and the high valleys of the Andes, thanks in large part to the open trade and financial flows mandated by the General Agreement on Tariffs and Trade; and

- a broad and fundamental change in the way societies view women and women view themselves, giving full credit for childbearing and child-rearing—but also giving full credit for their entire economic and social potential.

Virtually all of the world's expanded human population now has access to rewarding careers. There is less economic pressure to "work," but new knowledge and information systems have made "work" one of the most interesting human activities. None of these careers depend on cutting tropical rain forests or old-growth fir trees in America's Northwest. Some do involve helping to protect and manage those areas as wilderness/wildlife habitats. Few jobs are on the old "assembly lines," because these have been automated. Most of the new jobs utilize information technologies and provide eagerly sought and well-rewarded services to the billions of other people and creatures on the planet.

Humans now treat their sewage so thoroughly that it no longer overfertilizes surface waters. Landfills are no longer hermetically sealed to preserve their trash, but managed for rapid degradation back into "compost." Nonpolluting energy systems no longer depend heavily on burning fossil fuels. Though the new energy systems have been enormously expensive, they have been phased in slowly enough to avoid stopping economic growth; otherwise there would have been severe suffering and even violent opposition in the Third World.

This large number of people lives on less than 4 percent of the earth's land area—from choice. People will always be basically gregarious. In addition, new transport and housing technologies and the end of the "welfare ghettoes" have made cities more pleasant than ever.

These people produce ample food—in wondrous variety—from less land than they used in 2000. They still have to contend with pests, but they are able to do it more successfully than ever, thanks to the higher productivity, stress tolerance, and pest resistance which have been genetically engineered into most crop plants and domestic livestock and poultry. Any pesticide sprays are used in grams per acre, biodegrade quickly, and have extremely narrow toxicity aimed directly at pest species only. Domestic crop species have been engineered so they are unharmed by the safest pesticides, ensuring that we can use the safest compounds the most broadly.

This world in the future produces ample supplies of renewable lumber, paper, and other forest products from a small amount of land—equal to only 5 percent of the closed forests that existed in the world in 2000. High-yielding hybrid trees grow faster, straighter, and disease-free on tree plantations. The plantations are themselves fine wildlife habitat for much of their growth cycle. Their real purpose, however, it is to totally eliminate human pressures on the other 95 percent of the wild forests. Thus the wild forests do not even have to be logged, let alone clear-cut.

Is this picture too good to be true? Probably not.

I am not an expert in nonfossil energies, nor on automation. I will not attempt to explain how the big improvements can be made in those areas. (People who *are* experts in those fields assure me that the potential is there.) In agriculture and forestry, I know that this "impossible future" is attainable.

It is attainable because of humanity's expanding knowledge.

It is attainable because better knowledge continues to give us higher yields per acre—safely.

Index

About the Author

Dennis T. Avery is a Senior Fellow at Hudson Institute and is director of Hudson's Center for Global Food Issues. He is a recognized expert on international agriculture, specializing in the study of how technology and national farm policies interact to affect farm output.

Mr. Avery grew up on a Michigan dairy farm and studied agricultural economics at Michigan State University and the University of Wisconsin. He holds awards for outstanding performance from three different government agencies and was awarded the National Intelligence Medal of Achievement in 1983.

Mr. Avery served for nearly a decade (1980–88) as senior agricultural analyst for the U.S. Department of State, where he was responsible for assessing the foreign-policy implications of food and farming developments worldwide.

He is the author of *Biodiversity: Saving Species with Biotechnology,* a Hudson Institute Executive Briefing that challenges the conventional wisdom on loss of species, arguing that destruction of habitat, not industrialization, is the primary threat, and that biotechnology and economic growth are the keys to the solution. He also authored *Global Food Progress 1991,* an overview of the state of the world's ability to produce food and a critique of myths about impending global starvation, and is editor of Hudson's *Global Food Quarterly* newsletter.

Mr. Avery's articles have appeared in *The Wall Street Journal, The Readers Digest, Washington Times, Detroit News, Christian Science Monitor,* and many other publications. He is widely quoted in publications such as the *New York Times, USA Today, Time, News-*

week, U.S. News and World Report, Insight, and *Successful Farming,* contributes a weekly column to BridgeNews Forum, and makes frequent TV and radio appearances.

About Hudson Institute

Hudson Institute is a private, not-for-profit research organization founded in 1961 by the late Herman Kahn. Hudson analyzes and makes recommendations about public policy for business and government executives, as well as for the public at large. The institute does not advocate an express ideology or political position. However, more than thirty years of work on the most important issues of the day has forged a viewpoint that embodies skepticism about the conventional wisdom, optimism about solving problems, a commitment to free institutions and individual responsibility, an appreciation of the crucial role of technology in achieving progress, and an abiding respect for the importance of values, culture, and religion in human affairs.

Since 1984, Hudson has been headquartered in Indianapolis, Indiana. It also maintains offices in Washington, D.C.; Madison, Wisconsin; and Brussels, Belgium.